Researching animal research

Manchester University Press

INSCRIPTIONS

Since the very earliest studies of scientific communities, we have known that texts and worlds are bound together. One of the most important ways to stabilise, organise and grow a laboratory, a group of scholars, even an entire intellectual community, is to write things down. As for science, so for the social studies of science: Inscriptions is a space for writing, recording and inscribing the most exciting current work in sociological and anthropological – and any related – studies of science.

The series foregrounds theoretically innovative and empirically rich interdisciplinary work that is emerging in the UK and internationally. It is self-consciously hospitable in terms of its approach to discipline (all areas of social sciences are considered), topic (we are interested in all scientific objects, including biomedical objects) and scale (books will include both fine-grained case studies and broad accounts of scientific cultures).

For readers, the series signals a new generation of scholarship captured in monograph form – tracking and analysing how science moves through our societies, cultures and lives. Employing innovative methodologies for investigating changing worlds, it is home to compelling new accounts of how science, technology, biomedicine and the environment translate and transform our social lives.

Previously published titles

Embodiment and everyday cyborgs: Technologies that alter subjectivity
Gill Haddow

Trust in the system: Research Ethics Committees and the regulation of biomedical research Adam Hedgecoe

Personalised cancer medicine: Future crafting in the genomic era Anne Kerr *et al.*

Sugar rush: Science, politics and the demonisation of fatness Karen Throsby

The elephant and the dragon in contemporary life sciences: A call for decolonising global governance Joy Y. Zhang & Saheli Datta Burton

Researching animal research

What the humanities and social sciences can contribute to laboratory animal science and welfare

Edited by
Gail Davies, Beth Greenhough,
Pru Hobson-West, Robert G. W. Kirk,
Alexandra Palmer and Emma Roe

Manchester University Press

Published by Manchester University Press
Oxford Road, Manchester M13 9PL

www.manchesteruniversitypress.co.uk

British Library Cataloguing-in-Publication Data
A catalogue record for this book is available from the British Library

ISBN 978 1 5261 6575 6 hardback

First published 2024

Typeset
by Cheshire Typesetting Ltd, Cuddington, Cheshire

Contents

Figures and tables

Figures

Table

Contributors

Alistair Anderson is a Research Fellow in the School of Sociology and Social Policy at the University of Nottingham. Alistair's interests include veterinary humanities, antibiotic resistance, vaccine hesitancy, and animal research. His work as part of the Animal Research Nexus Programme investigated the role of Named Veterinary Surgeons in animal research laboratories.

Larry Carbone has worked as a laboratory animal veterinarian/ Named Veterinary Surgeon/animal ethics committee coordinator in US academic research facilities. He is a historian of ethics in laboratory animal welfare policy. Since leaving university employ, he continues serving on a local ethics committee and is a freelance writer in animal ethics.

Bentley Crudgington is an artist, producer, and creative practitioner working at the interface of human, animal, and environmental health. They design processes that aim to make users a little less comfortable with the familiar, known, and connected, and a little more comfortable with that which feels unfamiliar, different, and other. Bentley is interested in participation, agency, complicity, and the ethics of performance.

Gail Davies is Professor in Human Geography at the University of Exeter. Her research charts the changing geographies of biomedical research and seeks to facilitate engagement between differently situated knowledges in science, technology, and health. She was co-principal investigator on the Animal Research Nexus

Programme and chaired the UK review of harm–benefit analysis in her role on the UK Animals in Science Committee (2013–2019).

Ngaire Dennison is a veterinary surgeon and has worked in the field of laboratory animal science for more than 25 years in both Named Veterinary Surgeon and Animals (Scientific Procedures) Inspector roles. Her current interests include education and training of those working with research animals, severity and welfare assessment, and engagement with the vets and the wider public about the use of animals in research.

Amy Fleming is an actor and writer specialising in interactive and immersive theatre. In addition to being a Creative Associate of The Lab Collective, Amy is an associate artist of 6Foot Stories and a member of Degenerate Fox, a theatre company specialising in neo-futurism.

Carrie Friese is Associate Professor in the Sociology Department at the London School of Economics and Political Science. Her research and teaching are rooted in exploring (social) reproduction in science, technology, medicine, and society. She is the author of *Cloning Wild Life: Zoos, Captivity and the Future of Endangered Animals* and is finalising a second book tentatively entitled *More-than-Human Humanitarianism: Bioscience, Care and the Sacrificial Logic of Britain.*

Eva Haifa Giraud is Senior Lecturer in Digital Media and Society at the University of Sheffield, working in media studies, feminist science studies, and non-anthropocentric theory. Her monograph, *What Comes After Entanglement? Activism, Anthropocentrism, and an Ethics of Exclusion* (2019), explores relations between critical theory and political practice, with particular reference to the intersection of animal studies and critical animal studies.

Richard Gorman is a Research Fellow in Bioethics at Brighton and Sussex Medical School, following work on the Animal Research Nexus Programme at the University of Exeter. Richard's research is interested in the social and ethical implications of different healthcare practices, and aims to be responsive to stakeholder interests, engage different publics, and contribute to addressing complex policy issues.

Beth Greenhough is Professor in the School of Geography and the Environment at the University of Oxford. Beth's research examines science–society relations in the areas of health, biomedicine, and the environment. She was co-principal investigator on the Animal Research Nexus Programme and has published widely, including on the culture of care and everyday ethics in animal research.

Amy Hinterberger is Associate Professor of Sociology in the Department of Global Health and Social Medicine at King's College London. She currently leads a Wellcome Trust Investigator Award in the Social Sciences and Humanities called Biomedical Research and the Politics of the Human (www.politicsofthehuman.org).

Pru Hobson-West is Professor of Science, Medicine and Society in the School of Sociology and Social Policy, and an Honorary Professor in the School of Veterinary Medicine and Science at the University of Nottingham. Pru was co-principal investigator on the Animal Research Nexus Programme and has expertise in science and technology studies, medical sociology, and animals and society.

Angela Kerton is the Managing Director of The Learning Curve (Development) Ltd, a biomedical training provider. She particularly enjoys the creative aspects of teaching and sourcing new materials. Angela is a vet with experience of first opinion small animal charity practice and as a Named Veterinary Surgeon at a large London academic establishment. She is passionate about animal welfare and is a fan of the TV programme *All Creatures Great and Small.*

Gabrielle King finished her PhD exploring the everyday experiences of engaging with biomedical research for people affected by Motor Neuron Disease at the University of Edinburgh Medical School in 2021. She has also held research, policy, and care roles in universities, the civil service, and the third sector.

Robert G. W. Kirk is Reader in Medical History and Humanities at the University of Manchester. He was co-principal investigator on the Animal Research Nexus Programme and is currently working on a monograph charting the history of laboratory animal experiment and the science of animal welfare, titled *Reliable Animals, Responsible Scientists: Animal Experiment, Animal Welfare and the Development of the Biomedical Sciences.*

Bella Lear (formerly Williams) is a science communicator and Chief Executive at Understanding Animal Research Oceania (UAR Oceania). She initiated the Concordat on Openness on Animal Research and has since supported people across the research community to put openness at the heart of their communications on animal research.

Louise Mackenzie, PhD, is an Edinburgh-born interdisciplinary artist, curator, and writer. Her practice explores human relationships with the non-human world using experimental and experiential practices, sound, and new/found media. She is co-founder of Alive Together, an interdisciplinary community for research in human–animal relationships. Her work has been exhibited nationally and internationally, including at the ZKM (Germany), BALTIC CCA (UK), and National Library of Madrid (Spain).

Renelle McGlacken worked on the AnNex programme as both a PhD student and a Research Fellow. Completing her PhD in the School of Sociology and Social Policy, University of Nottingham, Renelle used the Mass Observation Project to explore everyday understandings of animal research. As a Research Fellow, she analysed public–professional relations around animal research, focusing on the Named Veterinary Surgeon.

Reuben Message is a Research Fellow in the Science, Technology and Innovation studies subject area at the University of Edinburgh. On the AnNex programme, he worked with Beth Greenhough and Ally Palmer at the University of Oxford to better understand the variety and emergence of new species and spaces in animal research. His research interests lie at the intersections of politics, law, animals, and technoscience.

Dmitriy Myelnikov is a historian of science and medicine with a broad interest in twentieth-century biomedicine, human–animal relations, and science communication, and is based at the University of Cambridge. On the Animal Research Nexus team, he worked with Rob Kirk at the University of Manchester to elucidate the history of the Animals (Scientific Procedures) Act 1986 and its effect on the practices, perceptions, and infrastructures of animal research.

Alexandra Palmer is a social anthropologist and primatologist by training, whose research explores social and ethical questions in human–animal relationships, with a focus on wildlife. Her AnNex work at the University of Oxford examined non-laboratory animal research in the UK (in wildlife field sites, veterinary clinics, zoos, farms, and fisheries), while her other work has focused on invasive species and domestic cat management, orangutan rehabilitation, and zoos. She moved to the University of Auckland to become a Research Fellow in 2021.

Sara Peres worked on AnNex as a Research Fellow at the University of Southampton. Her work explored the organisation and practices related to the breeding, supply, and biobanking of laboratory animals. Her previous research has focused on historical and contemporary approaches to seed banking. She is currently a Data Manager at the Medical Research Council Clinical Trials Unit, at University College London.

Emma Roe is Professor in More-Than-Human Geography at the University of Southampton. Emma was co-principal investigator on the Animal Research Nexus Programme. Her work looks at relations between humans and the non-human world, to inform the way we tackle global challenges around animals, meat, microbes, and most recently viruses, through deepening understanding of cultures, economies, biotechnologies, and societies.

Natalie Scott is a member of The Lab Collective. Natalie works regularly within the outdoor arts sector, has previously been General Manager for Outdoor Arts UK, and has delivered training with companies such as Punchdrunk (*The Third Day*), Specifiq (Zurich UOA MA programme), Dank Parish, and Emergency Exit Arts.

Tess Skidmore completed her PhD in the School of Geography and Environmental Science at the University of Southampton on laboratory animal rehoming. The PhD research, funded through the Animal Research Nexus Programme, studied care, science, and both human and animal welfare in laboratory spaces and domestic homes.

Joe Thorpe is a founding member of Lab Collective. As well as directing credits for The Lab Collective, he has worked with immersive companies such as Punchdrunk (*The Third Day*), Specfiq, Hartshorn Hook (the *Great Gatsby South Korea*), and is a Director of Dank Parish (*Boomtown Fair* and *Peaky Blinders*).

Jordi L. Tremoleda is a Reader in Animal Science and Welfare and the Named Veterinary Surgeon at Queen Mary University London. Jordi's work encompasses all broader aspects associated with the best care and welfare of animals used in translational research. He is actively involved in academic education and public engagement on animal ethics and the 3RS. He is an Editor in Chief for Laboratory Animals and has published widely on refinement for animal research.

Liz Tyson completed her PhD in Animal Law at the University of Essex. Her thesis explored the ways in which animal captivity is regulated and how welfare is impacted as a result. She works as Programs Director for Born Free USA and is responsible for the management of one of the largest primate sanctuaries in the world. Her work largely focuses around ending the exploitation of wild animals in captivity.

Acknowledgements

The Animal Research Nexus Programme has been a collaborative effort from inception to this collected volume. We have worked together in ways that have been immeasurably enriched by the wide range of collaborators, creative professionals, and conversations with both colleagues and critical friends represented in this book. We are indebted to the Wellcome Trust [grant no: 205393] for supporting our work together through a Collaborative Award in Humanities and Social Science from 2017 to 2023. This unique funding scheme enabled Principal Investigators from five institutions, with different experiences of working on the social aspects of animal research, to come together collectively and forge a new collaborative approach. The realisation of this work would not have been possible without the considerable expertise and energy of all the postdoctoral research fellows and PhD students who worked on the programme. We are very grateful for the personal, practical, and technical help provided by our research administrator Fiona French, who has made so much of this activity possible day-to-day. Thank you to John Jackson at The Argument by Design for design and web support and Morgane Colleau for setting up many of our collaborative systems.

Our history of working together extends back to the Wellcome Trust Small Grant in Society and Ethics [104339/Z/14/Z] on 'Developing a collaborative agenda for humanities and social scientific research on laboratory animal welfare', which ran from 2014 to 2015. This was supplemented by funding from the University of Exeter's Humanities Arts and Social Sciences Strategic Development Fund. The Wellcome Small Grant [100899/B/13/Z] awarded to

Greenhough and Roe provided initial support for studying animal research that fed into Chapter 12. Davies would like to acknowledge support from the British Pharmacological Society, especially Anna Zacharia, and the University of Exeter ESRC Impact Acceleration award for formative work on translational research and the evaluation of benefits, which fed into this work on patient involvement.

Our subsequent collaborations have been expanded in key places and moments of possibility by additional support from a variety of sources. The labels engagement activity, discussed by McGlacken and Hobson-West in Chapter 15, benefited from a University of Nottingham ESRC Impact Acceleration Account award [grant no: ES/T501992/1]. This activity also benefited from feedback from the University of Nottingham's Centre for Applied Bioethics, the help of Fiona French in setting up the 'labanimal-labels' website, and assistance from Alistair Anderson and Kathleen Salter from the University of Nottingham in running the final two workshops. Hobson-West's work on ethics benefited from a Collaborative Studentship, with Penny Hawkins/RSPCA, funded by Economic and Social Research Council [Grant number: ES/P000711/1]. Gorman's work on horseshoe crabs was funded by a Wellcome Trust Secondment Fellowship in Humanities and Social Science [grant no: 218323/Z/19/Z] with support from the RSPCA. Keble College [grant no: KSRG074] and the Economic and Social Research Council [1906-FOSS-470 Palmer] supported stakeholder workshops on non-laboratory research and citizen science. The University of Oxford Knowledge Exchange Seed Fund [KCD00043] and ESRC IAA [2105-KICK-666] supported the development of the Care-Full Stories training resource.

We want to say thank you to the following who have provided valuable support as members of our advisory panel for the Animal Research Nexus Programme: Christina Dodkin, Martin Fray, Carrie Friese, Kimberley Jayne, Bella Lear (formerly Williams), Sabina Leonelli, Dave Lewis, Eliot Lilley, Kate Millar, Paolo Palladino, Jane Smith, Sara Wells, and Greg Whelan. Your support has been invaluable, and we are grateful for the many ways in which you have supported and grounded our work across many different strands.

We have been further guided in individual projects by kind assistance from many others working in and around the animal research community. The team at the University of Exeter would additionally like to thank Henry Buller, Emma Dorris, Olwen Goodall, Bec Hanley, Penny Hawkins, Elliot Lilley, Richard Milne, Natasha Ratcliffe, Cheryl Scudamore, Kristina Staley, Kaushik Sunder Rajan, and Jo Welsman for their insights into the changing contexts and practices of research involvement, for helping with workshops, reading reports, and providing feedback on outputs, and other warmly welcomed forms of professional and personal support. In addition to our shared collaborators mentioned by other teams, the team at the University of Manchester would like to add a special thank you to Jenny Bangham, Jonathan Bushell, Tone Druglitrø, Carrie Friese, John Gluck, Adam Johnson, Patrick Kerwin, Claas Kirchhelle, Douglas Macbeth, Chris Magee, Hugh Marriage, Graham Morrissey, Neil Pemberton, Terry Priest, Jonathan Pledge, Edmund Ramsden, Andrew Rowan, Mike Simonsen Jackson, Mette Nordahl Svendsen (and the amazing 'Life Worth Living' team), Sarah Waitz, and Duncan Wilson. Those named, with many others who prefer to remain anonymous, went above and beyond with their generosity of time and effort in supporting our work and improving our understanding of animal research, past and present, whilst daring to venture answers to our impossible questions about possible futures. Thank you.

The team at the University of Nottingham would particularly like to thank Sarah Wolfensohn, Lucy Whitfield, and Anja Petrie for their advice on project planning. They would also like to thank Nickie Charles and the University of Warwick for hosting Hobson-West's Visiting Academic role in 2019, and the workshop on the Mass Observation Archive. The team at the University of Oxford would like to thank those who participated in the POLEs workshop and panel on citizen science regulation, particularly Julie Lane, Jim Reynolds, Roger Dickey, and Jamie Lorimer who were involved in both events, and Lucy Hawkes and Matt Witt for their efforts to advance thinking around animal welfare in wildlife research. They are also deeply grateful for the advice and assistance of all those who worked on the Care-Full Stories training resource, especially James Bussell, Manuel Berdoy, Ida Berglöw Kenneway, Mark Gardiner, Jackie Harrison, Penny Hawkins, Angela Kerton, Sally Robinson, Jordi L. Tremoleda, Sara Wells, and Lucy Whitfield, as well as Hibba Mazhary (research assistant extraordinaire).

The team at the University of Southampton would like to show their sincere appreciation to Keith Davies, Sara Wells, and Martin Fray for their support and engagement. They would also like to thank the Institute of Animal Technology for hosting the Moving Animals workshop in 2018 at their annual conference, which engaged stakeholders in shaping the direction of research and the making of The Mouse Exchange (MX). They would like to recognise the inspiring artistic contribution of Paul Hurley to MX and the many conversations with him over the couple of years it took to devise it. They would like to thank the support and encouragement from the University of Southampton's Public Engagement and Research Unit, particularly Steven Dorney and the team of people who organise the Southampton Science and Engineering Fair and Hands-on Humanities Festival. They thank those who have joined the team of MX facilitators, along the way – Charlotte Veal, Tess Skidmore, and Paul Hurley. Finally, they thank Understanding Animal Research for its recognition of MX with a 2020 Openness in Animal Research Award, and its interest in promoting it to the animal research community. They also appreciated feedback from industry audiences that included the Laboratory Animal Veterinary Association meeting in Manchester 2019 and a UK University Animal Welfare and Ethics Review Board in 2020.

As a whole team, we have learnt so much from collaborative workshops with cognate projects and want to thank Sabina Leonelli and Rachel Ankeny for bringing our work into dialogue with research on the 'Organisms and Us' project, and Carrie Friese and wonderful colleagues working on the 'Care as Science' project. We have benefitted from ongoing conversations with Tone Druglitrø and others working on the 'Resisting bodies: The practices and politics of the immune system' project. We would like to thank Andy Stirling for talking us through all things 'nexus' in our opening event in 2018, and Anna Olsson and Louise Mackenzie for co-organising the 'Interdisciplinary dialogues' session at the Federation of European Laboratory Animal Science Associations (FELASA) in 2022. All academic research is a collective effort, and we are indebted to the many conversations with colleagues, at conferences, in committees, and other contexts, which have made this work possible. We are also grateful for the extended networks that have supported the contributions of the commentators here. Carbone would particularly like to thank David Takacs for helpful comments on his chapter, and the Brooks McCormick Jr. Animal Law & Policy Program at Harvard University for financial support. We would like to thank all the commentators for responding so positively to the invitation to add to this volume, and for bringing such rich and diverse perspectives.

Above all, we would like to thank those individuals who gave up their time to share the experiences and perspectives of working in and around animal research that we draw on in the various chapters in this book. We have learnt so much from all of you about how to bring curiosity and care to challenging conversations. Some of you have now been working with us as individual researchers, as well as collectively, for a while. We cannot name you because of the confidentiality offered, but we hope you recognise how much we value your continued support. Responsibility for the interpretations here resides with chapter authors, but none of this would have been possible without all your extraordinary generosity and openness to engaging with us.

Editing this book has been a genuinely collective effort and we have all benefited from the opportunity to present our work to each other and receive comments that helped clarify what we were trying to say and reach beyond our disciplinary positions. Collecting the research contributions and conversations together into this book has only been possible thanks to a further set of individuals and organisations. We would like to express our gratitude to the editors at Manchester University Press, including Thomas Dark, Laura Swift, and Shannon Kneis, and series editors Des Fitzgerald and Amy Hinterberger. We want to offer our warmest thanks to the different reviewers who engaged so generously and constructively with the book proposal and the draft manuscript. We hope you will see your comments reflected in the final text as an additional formative voice in creating the animal research nexus we map out here. We are so grateful to Madeleine Hatfield and Jenny Case at Yellowback for their careful reading and suggestions for making our work more accessible and for their invaluable help in preparing the final manuscript. Finally, we would like to thank the Wellcome Trust for providing the funds to make this book open access online, and to Huw Golledge who arranged for any royalties from the book to be donated to the Universities Federation for Animal Welfare (UFAW) in recognition of the foundational interdisciplinary work of William Moys Russell and Rex Burch.

Introduction

Gail Davies, Beth Greenhough, Pru Hobson-West, Robert G. W. Kirk, Alexandra Palmer, and Emma Roe

Why research animal research?

The collaborative Animal Research Nexus Programme (AnNex), funded by the Wellcome Trust from 2017 to 2023, sought to deliver new research and engagement to increase understanding of the social relations around animal research in the UK. The starting point for this work is that animal research is part of a complex research infrastructure made up of humans and animals, practices inside and outside of the laboratory, formal laws, and professional norms, as well as social imaginaries of the past and futures of medicine and multispecies health. The aim of this book is to demonstrate an interdisciplinary approach for investigating this wider infrastructure, and to examine the changing social relations through which UK animal research is constituted. In doing so, it shows how the humanities and social sciences can contribute to understanding laboratory animal science and welfare and foster conversations about animal research across the sector. Creating opportunities for conversations that enable meaningful exchange between diverse perspectives, in ways that acknowledge the power relations between different positions and entities, was a key aim of the collaborative agenda-setting paper[1] that preceded the current collaboration. This goal continues to inform the way the authors work together. This introduction explores the disciplinary and conceptual trajectories that have underpinned the research, introduces the range of organisations and actors that animal research involves, and outlines the methods used to engage these.

AnNex took place across five UK universities (University of Exeter, University of Manchester, University of Nottingham,

University of Oxford, and University of Southampton), and involved six distinct, but interrelated, thematic strands of work. These were: 1) *History and Cultures*, focusing on the recent history of animal research in the UK, with an emphasis on the development and passing of the Animals (Scientific Procedures) Act, or ASPA; 2) *Species and Spaces*, examining the challenges that new or different species (zebrafish) and sites outside the laboratory (known as 'Places Other Than Licensed Establishments') introduce into animal research; 3) *Markets and Materials*, exploring the changing networks relating to research animals before (breeding and supply) and after (rehoming) their time in the laboratory; 4) *People and Professions*, investigating challenges faced by veterinarians working in the laboratory, and how publics and laboratory workers imagine each other; 5) *Engagement and Involvement*, charting the changing interfaces between patient representatives and those who use animals in research; and 6) *Collaboration and Communication*, using nexus thinking and findings from all other strands to engage stakeholders, publics, and researchers from different disciplines in productive dialogue. This book presents some key results from these projects, while also bringing AnNex work into conversation with perspectives from beyond the programme team. Our work is focused on the UK, which has a long history of both regulatory and interdisciplinary discussions about the acceptability and operation of animal research, to which we hope this volume makes a positive contribution.

This book is written as a collaboration between historians, geographers, sociologists, anthropologists, science and technology scholars, and creative professionals, with invited commentaries from the arts, social sciences, and animal research sector. Each author brings their own disciplinary interests to the topic, which means that different issues and actors are foregrounded. For example, sociologists are attuned to look at professional identities; science and technology scholars often focus on knowledge controversies; while geographers may draw attention to the flows of materials or expertise between places and people. The book draws on the specialist literatures from these disciplines, and the interdisciplinary field of human–animal studies, in a way that facilitates exchange between these fields and is accessible to those who are not from them.

Despite different disciplinary backgrounds, the authors also hold much in common, including a commitment to in-depth case studies and to the importance of contextualisation. Animal research has featured as a case study for many of the book's chapter authors and commentators in their previous research,[2] and for other scholars working in these disciplines. A few prominent examples include work focusing on animal research in the US by Lesley Sharp and Arnold Arluke;[3] in Europe by Tara Holmberg and Tone Druglitrø;[4] and in the UK by Lynda Birke and Mike Michael.[5] Some key concepts used throughout the book derive from these literatures. These include: *coproduction*, an approach to looking at how ideas in science and of society are generated together through regulatory forms and institutional norms (see Kirk in Chapter 4); *care*, a concept derived from feminist scholarship (see Greenhough in Chapter 8); *expertise*, which has been discussed and critiqued extensively in science and technology studies (see Hobson-West in Chapter 13); and *public engagement*, which is often construed as the transfer of knowledge from experts to uninformed publics, but is increasingly (including in our own work) constructed to be more multidimensional (see Roe in Chapter 17).

Another central concept used in this book is that of *nexus*, which figured prominently in the title of the research programme. The animal research 'nexus' refers less to an object of study than a way of approaching the complex web of connections that make up animal research.[6] As a way of thinking, a nexus approach has been used within the social sciences (and particularly environmental studies) to better understand how complex, multi-dimensional arrangements develop over time through the interaction of interdependent and sometimes conflicting interests, processes, objects, and values.[7] By focusing attention upon connections and interdependencies, nexus approaches can reveal transformative ways of thinking about a topic by connecting constituent parts in novel or unexpected ways, while exposing neglected drivers, influences, and obstructers of change. Implicit within this approach is the assertion that changes in animal research also involve the reworking of social relations: natural and social order are co-produced as opposed to *a priori* assumption that they are different in kind. As a historical process, animal research can be mapped in these terms by identifying how particular constituting parts – for instance, legal instruments and practices of care – shape and change each other over time. This happens as different parts of the nexus of human health, animal welfare, and research governance shift around each other. For example, wider changes in societal expectations as to what animals are (for example, whether they are property or how far they are sentient), or what science and medicine should be doing in the pursuit of collective health and wellbeing, will change policies and practices around animal research but are also informed by them. A nexus is always in motion, and as such, the chapters that follow are intended to serve as an invitation for others to enter and remake the animal research nexus, or to explore how far these ideas developed in the UK can be applied to different contexts and situations.

Of course, discussion of animal research cannot take place without consideration of ethics. Ethical debates around animal research have taken place since the nineteenth century in the UK,[8] and these ethical debates have recently been enlivened by scholars in the humanities and social sciences and the field of critical animal studies.[9] Many of the commentators, collaborators, and research participants undertake advocacy from a particular ethical standpoint. The book seeks to include a diversity of voices in response to the work, with invited commentaries from a critical perspective by Giraud and Tyson, alongside voices from within the animal research community, with commentaries by Dennison, Carbone, Tremoleda and Kerton, and Lear. Chapter authors also bring a diverse set of ethical commitments and things they care for through their current and previous work, including care for animal welfare policy,[10] research governance,[11] scientific training,[12] patients and publics,[13] social and historical methods,[14] and those who do the work of care.[15] Where relevant the book seeks to reflect on how these differences in expertise, ethics, and past experience inform the work (see for example Hobson-West, Chapter 13).

Yet the goal of AnNex as a whole is not to champion a particular 'side' in ethical debates, nor to resolve difficult outstanding questions about animal research. Rather, methods from the humanities, arts, and social sciences are used to understand how ethics are enacted in animal research – for example, how institutions seek to define and enact a 'culture of care' for both animals and humans (Chapter 6). In doing so, the book aims to situate ethical debates in their wider regulatory, institutional, and constitutional contexts to better understand their consequences. This suggestion is not an attempt to undermine the importance of ongoing and energetic ethical debates around animal research. However, the book proposes that there is value in focusing on the *constitution* of animal research, where constitution refers to social, economic, and political processes that shape and allocate identities, authority, and rights and obligations to certain actors.[16] Arguments around the ethics of animal research have travelled internationally in often important ways, but they are also frequently illustrated through case studies of earlier time periods or high-profile examples that do not translate meaningfully from place to place. This can affect a barrier to conversations with those involved in the day-to-day oversight and activities of animal research, or strategic policy development. The book suggests that these dialogues are more likely to take place through a focus on the constitution of animal research, and how ethics – or other processes like public engagement – work in practice in specific contexts. As such, the approach seeks to prompt reflection and nuanced discussion about animal research.

Who is involved in animal research?

Given our interest in the constitution of animal research, a key starting point for our empirical explorations is to understand the diversity of actors and interests that animal research involves, the variety of ways in which they are held together, and what and who is included and excluded. The range of different actors and agencies that UK animal research involved during the period of research can be illustrated through the UK government's 2014 delivery plan on 'Working to reduce the use of animals in scientific research'.[17] This report was produced during the period of the Liberal Democrat and Conservative coalition government in the UK, which ran from 2010 to 2015. Its acknowledgements range widely, referring to input from a long list of departments, agencies, and advisory bodies, from the Home Office (where the UK regulation for animal research resides) through to those involved in overseeing health, business, science and funding, food, human and veterinary medicines, and more, including the National Centre for the Replacement, Reduction and Refinement of Animals in Research (NC3Rs). The list encompasses a broad set of government responsibilities, professional roles, economic sectors, national and international interests, and human, environmental, and animal stakeholders. As the report says on page 18, 'Everyone has a part to play' when it comes to reducing the use of animals in scientific research.[18]

However, while everyone has a part to play, these roles are not equal and are often carried out in prescribed ways. While relations between the actors and agencies involved in animal research are not solely shaped by law and licensing – informal and changing social and political norms around expertise and authority are also crucial[19] – the current legal framework of the Animals (Scientific Procedures) Act 1986 (ASPA) is used to ground our work in UK animal research. As Myelnikov (Chapter 1) demonstrates, while certain voices were excluded from the law's development, ASPA also signalled a reduction in the authority of scientists compared with its predecessor. ASPA placed greater emphasis on the knowledge and expertise of laboratory animal veterinarians, animal technicians and technologists, animal welfare scientists and advocates, and sometimes selected lay members in its operation. This law, and its revision following the European Directive 2010/63/EU in 2013, continues to regulate the use of animals used for research in the UK.[20] It sets out the three-fold licensing system of UK animal research, with its requirements for people, places, and projects to be licensed before research can be carried out. It incorporates the need for a harm–benefit analysis, whereby the potential harms to animals have to be weighed against the likely benefits of the research for each project. It also mandates that researchers have to apply the 3Rs to their research, considering ways to replace, reduce, and refine the use of animals. Supporting this complex work of regulation are a raft of documents and named roles that have to be in place and consulted throughout the research, from the original idea to final reporting.[21]

Chapter 1 sets the scene for many of the chapters that follow; it explores how relationships between different actors involved in animal research are framed, rather than simply determined, by law. Many chapters explore how boundary cases reveal both the central importance of law and its openness to interpretation at the margins, such as the flexibility required to apply ASPA to wildlife science in the work of Palmer (Chapter 10), the proposed arguments about extending the 3Rs to horseshoe crab welfare in work by Gorman (Chapter 2), or the challenges of accounting for animals bred but not used in scientific procedures in the work by Peres and Roe (Chapter 12). Looking at these margins can help understand where new kinds of coalitions around future legislative change for animal research may occur. Other chapters look specifically at the changing expression of professional voices, such as the Named Veterinarian Surgeons (NVS) required under ASPA in work by Anderson and Hobson West (Chapter 9), or the skilled role of the animal technologist in facilitating both science and care in chapters by Kirk (Chapter 5), Greenhough and Roe (Chapter 6), and Message (Chapter 7). Work by Davies, Gorman, and King (Chapter 11) looks at growing interest in the role of patient voices in animal research. Patient representatives are increasingly empowered as stakeholders in health research, but still occupy an ambiguous position in conversations around animal research – the hesitancies here perhaps mirroring the uncertain engagement between the regulatory body for UK animal research and the Department of Health and Social Care.

As well as looking at regulatory control, many chapters are also concerned with the changing flows of knowledge and distribution of expertise throughout animal research. Animal research involves many diverse practices. The formal work of scientific research includes generating ideas, designing research projects, securing funding, conducting experiments, writing up, and disseminating research. Then there are a whole series of wider material and organisational practices that support this endeavour, including the provision of physical and administrative infrastructures, processes of peer and ethical review, the breeding and care of laboratory animals, the training of staff, the provision of funding, the creation and maintenance of spaces for disseminating work through scientific publications, and stakeholder and public engagement. Many of the voices now challenging animal research are emerging from within these practices. Joining ongoing societal debates over animal research are scientists and others raising questions around how far frameworks like the 3Rs or harm–benefit analysis are able to encapsulate the range of issues around contemporary animal research, from reporting bias to translational validity.[22] Where once scientists were presumed to hold authority over both the scientific value of the research and the process of assessing animal suffering, now these judgements are increasingly subject to scrutiny from without and within, adding ever-changing demands to research governance.

Our focus has been on the social dimensions of these practices: the relations between humans, and between humans and the animals they work with. These relations involve particular kinds of skills and competencies, which our work and methods have been developed to pay close attention to, including the embodied skills animal technologists draw on to sense and respond to the care needs of their animals,[23] and the communication skills that different stakeholders in animal research use to engage both with each other and with wider publics.[24] This focus has allowed key conceptual questions around care to be explored (see Part II), as well as emerging concerns around the need to develop and sustain a strong culture of care in animal research[25] – a task which includes addressing issues such as the impact of workplace cultures,[26] emotional labour,[27] and compassion fatigue.[28] As Hilgartner suggests, the constitutional basis of knowledge production allocates entitlements, but also burdens.[29] Inspired by the work of feminist care scholars, we hope our work has been attuned to how these entitlements and burdens are distributed, in often unequal ways. In doing so, we examine who benefits from the knowledge produced and who carries out the work of care across the complex nexus of animal research.

Of course, animal research also involves animals. The social sciences and humanities are often said to have undergone an 'animal turn' since the late 1990s:[30] a move that has meant studying not just how animals act as 'mirrors and windows' into human social life,[31] but also how animals are themselves agents who influence the world and our interactions with them. This interest in animal agency has inspired proposals for new methods for studying human–animal relationships,[32] for example by observing animals,[33] and for historians (not without difficulties)[34] to look for animal agency in archives.[35] It has also encouraged the rethinking of human–animal relationships in the research laboratory, for example via attention to how research animals shape both science[36] and animal care,[37] and how they act as 'workers' rather than merely 'lab tools'.[38] We take inspiration from this thinking throughout the book. While moving out from the empirical sites of regulation means our work here rarely involves dedicated observations of research animals, their agency is apparent in how their aesthetic and corporeal charisma,[39] or lack thereof, shape care practices. Skidmore (Chapter 3) writes of how dogs are disproportionately rehomed compared with other species, while Message (Chapter 7) and Gorman (Chapter 2) both articulate some of the challenges of extending care to the cold-blooded, or blue-blooded in the case of horseshoe crabs. The power of individualisation and bonds with specific animals for shaping science, relationships, and care are also explored, though Message reminds us that individualisation is not the only route to care.

Finally, while most animal research takes place in a world of certified experts and named individuals (those with specific responsibilities under ASPA legislation), the policy and practice of animal research also creates particular ideas or imaginaries of lay publics. These imaginaries range from publics as beneficiaries of the outcomes of animal research, as patients or consumers of medical knowledge and products, to images of publics as citizens with a stake in animal research who ought to be consulted, for example through the use of national opinion polling.[40] Less commonly, some individuals considered part of the contested category of 'publics' are also involved in animal research, for example as lay members of ethical review processes,[41] as citizen scientists (Chapter 10),[42] or as participants in research involvement (Chapter 11). However, a nexus approach also encourages us to look at these imaginaries together: it is important to consider how the very category of 'publics' is created or performed through discourses around animal research,[43] and how scientific imaginaries of what the public think or want impact directly on the policy and practice of animal research.[44] These public voices are usually opaque within formal requirements of regulation; it is often not evident when different members of the public have shaped the workings of animal research in a meaningful way, nor the routes for others to make their contribution. How we might act carefully to create these wider connections motivates our work on patient involvement (Chapter 11) and public engagement (Part IV).

How did we research animal research?

The individual contributors to this volume come from a variety of disciplines, ethical orientations, and positions in relation to animal research. Each chapter thus reflects a slightly different focus, methodology, and goal. However, there are common approaches and threads that hold together many of the book's contributions, such as a commitment to, wherever possible, collaboration with stakeholders to co-produce questions. This included working with researchers, regulators, animal technologists, veterinarians, and others whose day-to-day practices revolve around the questions raised by doing animal research. The book includes a selection of voices from these individuals in the commentaries that conclude each part. We also engaged with those affected by animal research in direct or indirect ways, but who are currently held outside or at the margins of these conversations, like citizen scientists in wildlife research, patient representatives in research involvement, or diverse participants in public engagement events. Representing this multiplicity in the book is intended to reflect the value of bringing different perspectives and disciplines into conversation with one another.

Given the history of animal rights activity in the UK, there has been a tendency to both construct and imagine the worlds of animal research as closed and secretive. However, our work has also coincided with a growing openness agenda around animal research, through which institutions have sought to enable people who work in the field to talk more openly about their work,[45] as well as increasing public transparency and open science. This has made an important contribution to making our work possible, but its focus on facilitating communication across the sector still leaves some questions and voices beyond its scope.[46] The boundaries between inside and outside in this field are still significant, but they are shifting. Our approach was necessarily shaped by who would engage with us, which has implications for the kinds of accounts that are presented here. The convention of offering anonymity in the social sciences may have facilitated people's engagement as research participants, but it also brings challenges for historians and others who are used to putting names on record. All methodological choices, whether in the social sciences or sciences, have consequences for the way a field is constructed and what evidence is produced. Particularly enriching were conversations with those who were similarly speaking in a number of different directions in their work, such as the veterinarians who aim to support researchers and protect animals, or animal welfare scientists who generate knowledge but also new controversies in their work around what matters to animals in care or training practices. We have also learned much from those looking backwards or forwards, from those involved in past policy changes, and those who are developing new research directions.

The work in this book draws primarily on qualitative rather than quantitative research methodologies. Qualitative research is most suitable for exploring the interrelationships and networks that co-produce research policy and practice across the animal research nexus. Qualitative research is widely recognised as the most appropriate approach to 'answer questions about experience, meaning and perspective'.[47] Within this qualitative and constructivist paradigm, the chapters in this book use a range of methods. First, some of the chapters utilise documentary analysis of historical and contemporary documents, such as political debates, reports, and newspaper and magazine articles. Second, many of the chapters draw on in-depth qualitative interviews with those engaged in animal research, such as researchers, regulators, animal technologists, and laboratory veterinarians. Third, many of the projects also utilise ethnographic observation of key sites and spaces such as research facilities and industry events. Finally, the book's authors also report on more creative attempts to think within the topic of animal research. Part IV in particular reports from a series of more experimental methodologies, including the use of drawing activities, crafting, and immersive theatre to explore the different ways in which these might facilitate participants to engage with animal research.

The qualitative data from this work are analysed in a variety of ways; for example, social science chapters typically made use of inductive thematic analysis, whereby text from interviews and fieldnotes is assigned codes based on themes emerging from within the data (rather than being pre-defined ahead of time).[48] Where relevant, individual chapter authors have described their analysis methods in more detail. Working together, albeit on different individual projects, also meant agreeing some methodological strategies in advance, such as anonymising transcripts and allocating letters for the pseudonyms given to research participants across the AnNex programme to prevent different individuals being given the same name. Despite some differences in analytical methods, we have adopted common strategies for identifying central themes in each chapter, and across the book as a whole. Firstly, we explored *connections* – for example, the displacement of certain forms of expertise cuts across contributions focusing on patients (Chapter 11), citizen scientists (Chapter 10), animal care staff (Chapters 4, 6, and 12), publics (Part IV), veterinarians (Chapter 9), and activists (Chapters 4 and 8). Themes were also identified through exploring *comparisons* – between the treatment of certain species (Chapters 2, 3, and 7) and between different research sites (Chapter 10). *Absences* offered a third set of central concerns, for example the lack of attention to certain species or locations in animal research statistics and regulations (Chapters 2 and 12), and the absence of certain voices from policy conversations (Chapters 1 and 11). Finally, we examined instances of *change* over time, such as the expectations around care for laboratory animals and staff in the past (Chapter 5) and looking forward into the future (Chapter 6).

Reflecting on our collaboration for the purposes of this introduction, several recurring questions come to the fore, including what we understood by 'interdisciplinarity', and how to shape this elusive concept into a pragmatic collaborative tool that unified rather than fractured our research.[49] Working with stakeholders from the animal research community for many of us was interdisciplinary; what else could working across the humanities, arts, social sciences, and natural sciences be? Nevertheless, our work was often haunted by unasked and unanswered questions, like what distinguished the humanities from the social sciences. Other frictions emerged at the boundaries between fields. One consequence of living with rather than resolving these uncertainties was that the further along the collaborative process we travelled, our trust in other ways of working grew as our certainty as to how best to define our own expertise and disciplinary identity diminished. Navigating collaboration became a matter of pragmatic decision-making and compromise, driven by an emphasis on interactive research events as opposed to interdisciplinary research.[50] Whether this reflected a trust that ambiguity could be productive, or something less seemly, was less important than the fact that our collective act of faith moved our research forward.[51] We knew that we could collaborate with each other and with the animal research community. We knew the latter saw value in what the humanities and social sciences could contribute to animal research. We knew the key to success was attention to language and, if not quite developing shared understanding, at least being sensitive to the ways in which the same words could be understood differently across different audiences.[52] And we knew that our premise was neglected but not original: no less a figure than W. M. S. Russell had drawn on the humanities and social sciences in the original formulation of the 3Rs.[53]

If you reach for any dictionary, the word 'annex' is usually given at least two different definitions. In the first, an annex is defined as an extension, or a supplement, suggesting something added later. All too often the humanities, arts, and social sciences have been construed as this sort of addition to the primary work of science and technology,[54] either playing a supporting role for the operational business of science, or imagined as the means through which to enrol wider publics more effectively into supporting scientific research or demonstrating scientific literacy.[55] A second meaning of annex also refers to movements over space and of meaning, referring to a territory that is annexed or appropriated by another. Here too there are some resonances with past engagements between science and social science, and in particular the perceived threat that the emerging field of science and technology studies was seen to present to scientific authority.[56] As the AnNex programme comes to a close, we suggest that while our work could be seen to have elements of each of these meanings, it is in fact something rather different.

Humanities scholars who read Derrida will know that the extension or supplement comes from outside and is necessarily different, while also having the capacity to change understandings of the original.[57]

Understanding the meanings of animal research for different people, and enabling communication across these meanings, have been prominent themes in our work. One key aim was to generate new cultures of communication between stakeholders across animal research, but this was not limited to communicating the results of animal research to a wider public. We have sought to create new opportunities for dialogue through which knowledges, understandings, meanings, and experiences are exchanged in many different directions, which we hope is reflected in the contributions and commentaries in this volume. Our work has included drawing curious members of the public to make a mouse and learn more about the breeding and supply of lab animals (Chapter 14), to experience the complex decision-making processes of ethical review (Chapter 16), and to design mock labels for medicines (Chapter 15). It has involved working with scientists and patient groups to develop new guidelines for patient and public involvement in animal research (Chapter 11)[58] and helping researchers, named people, animal technologists, and others develop cultures of care and mutual respect.[59] We have sought to bring insights from the worlds of animal research into conversation with wider debates about how science is transformed, whether through the histories of scientific innovation and regulation,[60] the practices of translational research,[61] public engagements with science,[62] changing professional roles,[63] or supporting care infrastructures. Similarly, while we did seek to extend our work with scientists by co-producing an agenda for social science and humanities engagement with animal research,[64] we sought to do so in a way that was collaborative, curious, and respectful, as well as critical. For us, above all the Animal Research Nexus is a meeting point, where we arrive with a commitment to 'working with' (as opposed to either annexing or being annexed by) the wide range of stakeholders and publics invested in animal research. In doing so, we hope that the Animal Research Nexus delivers new thinking, research, and engagement to increase understanding of the social relations around animal research, and generates new connections, conversations, and cultures of communication between them.

Introducing the book

While these motivations held our work together, each of the chapters makes a distinctive contribution. The chapters are organised into four sections. Part I focuses on 'Changing and implementing regulation', Part II on 'Culturing and sustaining care', Part III on 'Distributing expertise and accountability', and Part IV on 'Experimenting with openness and engagement'. Each part includes three or four chapters showcasing research completed by members of the AnNex programme. Each part also closes with a chapter featuring three different commentaries, two from invited experts who are not members of the AnNex team, and one from the AnNex team member who brought these different voices together to discuss the contents of that section. In incorporating these commentaries we have sought to replicate in print some of the dialogic methods that characterise our work, and challenge ourselves and others to consider how nexus thinking constructs what is inside or outside of any conversation. These commentaries feature the voices of practitioners active in shaping the practices and policies around animal research (whether as veterinarians, activists, artists, or engagement experts) and academics who have worked on animal research (from both cognate and critical perspectives). Some commentaries are written by those with whom we have worked directly, others by those who have been instrumental in opening up the contexts in which our work takes place, and yet others by those who have provided a conceptual and political challenge that we have valued throughout. We asked each person to introduce themselves within their commentaries, which interestingly came more readily to stakeholders than academics. The commentaries have enriched our understanding of what it means to position our work in relation to and speak across the animal research nexus, even as they raise many questions that are still to be answered.

Part I centres on the theme of regulation and policy practices. It begins with Myelnikov's historical account of how ASPA came to be (Chapter 1). Myelnikov demonstrates that while the voices of animal advocates perceived as 'extreme' were excluded from the policy-making process, ASPA did respond to the concerns of such groups and change practice on the ground. Gorman (Chapter 2) then provides an account of the exclusion of horseshoe crabs from animal research regulation and social imaginations, despite their blood being widely used for checking that medical products are free from contamination. In doing so, Gorman challenges us to expand the scope of oversight and care to include these forgotten creatures. Finally, Skidmore (Chapter 3) extends the theme of inclusion and exclusion in care and regulation by considering why some species, such as dogs, are disproportionately considered candidates for rehoming after life in the laboratory. In the commentaries in Chapter 4, Tyson offers a critique from an animal rights perspective, picking up on the themes of invisibility and exclusion of certain animals, harms, and voices from animal research discourses and policy. Hinterberger examines what Part I's chapters can tell us about the institutional life of animals: how laws, regulations, and other forms of governance shape the definition and care of research animals, in the past, present, and future. Finally, Kirk brings these threads together by considering how regulation both shapes, and is shaped by, popular perceptions of research animals, including their charisma.

Part II focuses on the theme of culturing and sustaining care, beginning with Kirk's historical account of how the role of the professional laboratory animal technician developed alongside that of the purpose-bred research animal (Chapter 5). Kirk explores the difficulty of standardising and formalising the care provided by technicians, a theme picked up by Greenhough and Roe (Chapter 6) in their discussion of the emergence and meaning of 'cultures of care' in animal research. A lesson from this piece is that different forms of care – such as 'cold' institutionalised care versus 'warm' everyday care – may not sit easily alongside each other. Finally, Message (Chapter 7) explores how aquarists seek other ways of enacting 'good care' when faced with the difficulties of empathising and developing emotional connections with individual zebrafish. In the commentaries (Chapter 8), Tremoleda and Kerton draw on this work and their own professional experience to reflect on the challenges of defining, developing, harmonising, and assessing cultures of care across institutions. Giraud then poses a challenging question: are the notions of care put forward by those within the research laboratory fundamentally incommensurable with the forms of care proposed by activists, who view care as necessitating the end of animal research? Concluding Part II, Greenhough reflects on how care is constituted in both academic scholarship and animal research practice.

Part III turns to the question of how expertise and accountability are distributed across the animal research nexus. Anderson and Hobson-West (Chapter 9) explore the complex regulations and expectations placed on Named Veterinary Surgeons (NVS), with a focus on how NVSs view themselves as at the margins of the veterinary profession, and how they negotiate the laboratory–clinic boundary. Developing these themes of marginalisation and lab-field borders, Palmer (Chapter 10) examines the role of 'citizen scientists' in wildlife research. Palmer demonstrates that while citizen scientists feel excluded from securing ASPA licences themselves, the law's valuable flexibility allows expert non-professionals to significantly contribute to research. Davies, Gorman, and King (Chapter 11) then explore the involvement of another group often incorrectly perceived as non-expert: patients. Taking aim at four common assumptions about patient involvement, the authors highlight the value of including patient voices in biomedical research, and further opportunities for doing this. Chapter 12, by Peres and Roe, demonstrates that when it comes to the problem of 'avoidable surplus', those who bear the emotional burden of culling surplus animals are both the most aware of the problem, and often experts in surplus avoidance. The commentaries in Chapter 13 from Carbone and Dennison both draw on personal experience as laboratory animal veterinarians. Carbone uses this experience to reflect on the value, and challenges, brought about by outsiders – citizen scientists, patients, and social scientists – watching, participating in, and critiquing animal research. Dennison explores lessons from the chapters about the importance of open, honest, and meaningful conversations around animal research. Hobson-West closes the part by highlighting the cross-cutting themes of active navigation of regulation; expertise and space, and the creation of insider/outsider boundaries; and the narration of expertise, including how we do this as academics.

Part IV focuses on our efforts to experiment with openness and engagement. It begins with an account by Roe et al. of The Mouse

Exchange (Chapter 14) – a participatory, curiosity-driven activity in which publics are encouraged to explore the breeding and care of laboratory animals through making research mice from felt. McGlacken and Hobson-West (Chapter 15) describe their efforts to engage in open discussion through asking publics: should medicines developed via animal testing be labelled, and if so what should the label look like? Finally, Crudgington and co-authors from The Lab Collective (Chapter 16) discuss the interactive theatre performance *Vector*, in which publics are placed in the shoes of an Animal Welfare and Ethical Review Board and asked to decide how to use and care for research animals when developing treatment for a fast-spreading zoonotic disease. Together, these activities demonstrate the value of grounding discussions about animal research in people's experiences, moving conversations beyond the experiment itself, and developing two-way engagements in which publics are participants in conversations rather than merely recipients of information. In Chapter 17, the two external commentaries contextualise these experiments through contributions involving personal biography and professional practice. Lear writes from her perspective as a driver of developments in openness within the UK life sciences sector, providing insights into the trajectories through which openness has emerged as both an opportunity and a challenge. Mackenzie explores the provocations animal research provides to an artist: for honesty, for openness, for experimentation, and perhaps for work that 'playfully jumps off the fence and dives deep into ... imaginative possibility'. Roe's closing commentary connects questions of openness to those of care and ethical responsibility, asking what kinds of openness would benefit the animals in animal research.

Finally, we are grateful to Carrie Friese for the Afterword, which puts her reading of AnNex into a wider context. We are indebted to Friese in many ways. She has collaborated with team members in organising events and journal special issues, but she is also able to offer an overview of our work that is hard to perceive and articulate when you are in the middle of staging sometimes fraught conversations. We hope readers of this text are able to follow her in moving out of the polarisation cycle, to hold sometimes different possibilities side by side, working together to exchange experiences across them. We will continue to work on these possibilities beyond this book. We close with a collated bibliography of outputs from the AnNex programme to date, introducing the academic articles, stakeholder reports, and public websites through which we hope to continue these conversations.

Notes

1 Gail Davies et al., 'Developing a Collaborative Agenda for Humanities and Social Scientific Research on Laboratory Animal Science and Welfare', *PLOS ONE*, 11.7 (2016), 1–12, DOI: 10.1371/journal.pone.0158791.

2 Many of the authors discuss their prior work in later chapters. Prior work from commentators which has informed the way we have put this volume together include: Larry Carbone, 'Open Transparent Communication about Animals in Laboratories: Dialog for Multiple Voices and Multiple Audiences', *Animals*, 11.2 (2021), 368, DOI: 10.3390/ani11020368; Annabella Williams, 'Caring for Those Who Care: Towards a More Expansive Understanding of "Cultures of Care" in Laboratory Animal Facilities', *Social & Cultural Geography*, 24.1 (2021), 31–48; Amy Hinterberger, 'Marked "h" for Human: Chimeric Life and the Politics of the Human', *BioSocieties*, 2017, DOI: 10.1057/s41292-017-0079-7; Eva Giraud and Gregory Hollin, 'Care, Laboratory Beagles and Affective Utopia', *Theory, Culture & Society*, 33.4 (2016), 27–49, DOI: 10.1177/0263276415619685.

3 Lesley A. Sharp, *Animal Ethos: The Morality of Human–Animal Encounters in Experimental Lab Science* (Berkley, CA: University of California Press, 2018). Arnold Arluke and Clinton Sanders, *Regarding Animals* (Philadelphia, PA: Temple University Press, 1996).

4 Tora Holmberg, 'Mortal Love: Care Practices in Animal Experimentation', *Feminist Theory*, 12.2 (2011), 147–163, DOI: 10.1177/1464700111404206. Tone Druglitrø, '"Skilled Care" and the Making of Good Science', *Science, Technology, & Human Values*, 43.3 (2018), 649–670, DOI: 10.1177/0162243916688093.

5 Lynda Birke et al., *The Sacrifice: How Scientific Experiments Transform Animals and People* (Lafayette, IN: Purdue University Press, 2007). Mike Michael and Nik Brown, 'Scientific Citizenships: Self-Representations of Xenotransplantation's Publics', *Science as Culture*, 14.1 (2005), 39–57, DOI: 10.1080/09505430500041769.

6 Gail Davies et al., 'Animal Research Nexus: A New Approach to the Connections between Science, Health, and Animal Welfare', *Medical Humanities*, 46.4 (2020), 499–511, DOI: 10.1136/medhum-2019-011778.
7 Davies et al., 'Animal Research Nexus'.
8 A. W. H. Bates, *Anti-Vivisection and the Profession of Medicine in Britain: A Social History* (London: Palgrave Macmillan, 2017), DOI: 10.1057/978-1-137-55697-4; Davies et al., 'Animal Research Nexus'.
9 Eva Haifa Giraud, *What Comes after Entanglement? Activism, Anthropocentrism, and an Ethics of Exclusion* (Durham, NC: Duke University Press, 2019).
10 Henry Buller and Emma Roe, *Food and Animal Welfare*, Ebook (Bloomsbury Publishing, 2018).
11 Gail Davies, 'Harm–Benefit Analysis: Opportunities for Enhancing Ethical Review in Animal Research', *Laboratory Animals*, 47.3 (2018), 57–58, DOI: 10.1038/s41684-018-0002-2; Penny Hawkins and Pru Hobson-West, *Delivering Effective Ethical Review: The AWERB as a 'Forum for Discussion'* (Horsham: RSPCA, 2017), www.science.rspca.org.uk/documents/1494935/9042554/AWERB+forum+for+discussion+booklet.pdf/36fdb4db-8819-cbd3-89ec-c9a7e67bf07c?t=1583938525299 [accessed 1 February 2023].
12 James W. E. Lowe et al., 'Training to Translate: Understanding and Informing Translational Animal Research in Pre-Clinical Pharmacology', *TECNOSCIENZA: Italian Journal of Science & Technology Studies*, 10.2 (2020), 5–30; Pru Hobson-West and Kate Millar, 'Telling Their Own Stories: Encouraging Veterinary Students to Ethically Reflect', *Veterinary Record*, 188.10 (2021), e17, DOI: 10.1002/vetr.17; Beth Greenhough and Hibba Mazhary, *Care-Full Stories: Innovating a New Resource for Teaching a Culture of Care in Animal Research Facilities*, An Animal Research Nexus Report, July 2021, www.nimalresearchnexus.org/publications/care-full-stories-innovating-new-resource-teaching-culture-care-animal-research [accessed 1 February 2023].
13 Pru Hobson-West, 'The Role of "Public Opinion" in the UK Animal Research Debate', *Journal of Medical Ethics*, 36.1 (2010), 46–49, DOI: 10.1136/jme.2009.030817; Gail Davies et al., 'The Social Aspects of Genome Editing: Publics as Stakeholders, Populations and Participants in Animal Research', *Laboratory Animals*, 56.1 (2021), 88–96, DOI: 10.1177/0023677221993157.
14 Gail Davies et al. (eds), 'Science, Culture, and Care in Laboratory Animal Research', *Science, Technology & Human Values*, 43.4 (2018), 603–621, DOI: 10.1177/016224391875703.

15 Emma Roe and Beth Greenhough, 'A Good Life? A Good Death? Reconciling Care and Harm in Animal Research', *Social & Cultural Geography*, 24.1 (2021), 49–66, DOI: 10.1080/14649365.2021.1901977; Beth Greenhough and Emma Roe, 'Attuning to Laboratory Animals and Telling Stories: Learning Animal Geography Research Skills from Animal Technologists', *Environment and Planning D: Society and Space*, 37.2 (2019), 367–384, DOI: 10.1177/0263775818807.
16 *Science and Democracy: Making Knowledge and Making Power in the Biosciences and Beyond*, ed. by Stephen Hilgartner et al. (New York: Routledge, 2015), p. 8.
17 *Working to Reduce the Use of Animals in Scientific Research*, 2014, www.nls.ldls.org.uk/welcome.html?ark:/81055/vdc_100041489742.0x000001 [accessed 14 August 2018].
18 It is also notable as the first, and so far only, report on UK animal research to have the Department of Health on the cover, and to incorporate the perspectives of those who are the promised and potential beneficiaries of animal research.
19 Gail Davies, 'Locating the "Culture Wars" in Laboratory Animal Research: National Constitutions and Global Competition', *Studies in History and Philosophy of Science Part A*, 89 (2021), 177–187, DOI: 10.1016/j.shpsa.2021.08.010; Kristin Asdal, 'Subjected to Parliament: The Laboratory of Experimental Medicine and the Animal Body', *Social Studies of Science*, 38 (2008), 899–917; Tone Druglitrø, 'Procedural Care: Licensing Practices in Animal Research', *Science as Culture* (2022), 1–21, DOI: 10.1080/09505431.2021.2025215.
20 Animals (Scientific Procedures) Act, 1986, www.web.archive.org/web/20190906163847/, www.legislation.gov.uk/ukpga/1986/14/contents [accessed 1 February 2023].
21 See Davies 'Locating the "Culture Wars"'.
22 Malcolm Macleod and Swapna Mohan, 'Reproducibility and Rigor in Animal-Based Research', *ILAR Journal*, 60.1 (2019), 17–23, DOI: 10.1093/ilar/ilz015; Hanno Würbel, 'More than 3Rs: The Importance of Scientific Validity for Harm–Benefit Analysis of Animal Research', *Laboratory Animals*, 46.4 (2017), 164–166, DOI: 10.1038/laban.1220.
23 Carrie Friese, 'Intimate Entanglements in the Animal House: Caring for and about Mice', *The Sociological Review*, 67.2 (2019), 287–298, DOI: 10.1177/0038026119829753; Druglitrø, '"Skilled Care" and the Making of Good Science'; Beth Greenhough and Emma Roe, 'Exploring the Role of Animal Technologists in Implementing the 3Rs: An Ethnographic Investigation of the UK University Sector', *Science, Technology, & Human Values*, 43.4 (2017), DOI: 10.1177/

0162243917718066; Gail Davies, 'Caring for the Multiple and the Multitude: Assembling Animal Welfare and Enabling Ethical Critique', *Environment and Planning D: Society and Space*, 30.4 (2012), 623–638.

24 Richard Gorman and Gail Davies, 'When "Cultures of Care" Meet: Entanglements and Accountabilities at the Intersection of Animal Research and Patient Involvement in the UK', *Social & Cultural Geography* (2020), 1–19, DOI: 10.1080/14649365.2020.1814850; Davies et al., 'Animal Research Nexus'.

25 Sally Robinson and Angela Kerton, 'Contributing to Your Culture of Care', *Animal Technology and Welfare*, 20.3 (2021): 211; Sally Robinson et al., 'The European Federation of the Pharmaceutical Industry and Associations' Research and Animal Welfare Group: Assessing and Benchmarking "Culture of Care" in the Context of Using Animals for Scientific Purpose', *Laboratory Animals*, 54.5 (2019), 421–432, DOI: 10.1177/0023677219887998.

26 Wellcome Trust, *What Researchers Think about the Culture They Work In*, 2020, https://wellcome.org/reports/what-researchers-think-about-research-culture [accessed 1 February 2023].

27 Arnold Arluke, 'Uneasiness among Laboratory Technicians', *Occupational Medicine*, 14.2 (1999), 305–316; Keith Davies and Duncan Lewis, 'Can Caring for Laboratory Animals Be Classified as Emotional Labour?' *Animal Technology and Welfare*, 9.1 (2010), 1.

28 Megan R. LaFollette et al., 'Laboratory Animal Welfare Meets Human Welfare: A Cross-Sectional Study of Professional Quality of Life, Including Compassion Fatigue in Laboratory Animal Personnel', *Frontiers in Veterinary Science*, 7.114 (2020), DOI: 10.3389/fvets.2020.00114.

29 Stephen Hilgartner, *Knowledge and Control in the Genomics Revolution* (Cambridge, MA: MIT Press, 2017), p. 9.

30 Donna Jeanne Haraway, *When Species Meet* (Minneapolis, MI: University of Minnesota Press, 2008).

31 Molly H. Mullin, 'Mirrors and Windows: Sociocultural Studies of Human–Animal Relationships', *Annual Review of Anthropology*, 28 (1999), 201–224.

32 Timothy Hodgetts and Jamie Lorimer, 'Methodologies for Animals' Geographies: Cultures, Communication and Genomics', *Cultural Geographies*, 22.2 (2015), 285–295, DOI: 10.1177/1474474014525114; Henry Buller, 'Animal Geographies II: Methods', *Progress in Human Geography*, 39.3 (2015), 374–384, DOI: 10.1177/0309132514527401.

33 Alexandra Palmer et al., 'Accessing Orangutans' Perspectives: Interdisciplinary Methods at the Human/Animal Interface', *Current Anthropology*, 56.4 (2015), 571–578.

34 Philip Howell, 'Animals, Agency, and History', in *The Routledge Companion to Animal–Human History*, ed. by Hilda Kean (London: Routledge, 2018), pp. 197–221, DOI: 10.4324/9780429468933-9.

35 Chris Pearson, 'Dogs, History, and Agency', *History and Theory*, 52.4 (2013), 128–145, DOI: 10.1111/hith.10683.

36 Vinciane Despret, 'The Body We Care for: Figures of Anthropo-Zoo-Genesis', *Body & Society*, 10.2–3 (2004), 111–134, DOI: 10.1177/1357034X04042938.

37 Beth Greenhough and Emma Roe, 'Ethics, Space, and Somatic Sensibilities: Comparing Relationships between Scientific Researchers and Their Human and Animal Experimental Subjects', *Environment and Planning D: Society and Space*, 29.1 (2011), 47–66, DOI: 10.1068/d17109.

38 Jonathan L. Clark, 'Labourers or Lab Tools? Rethinking the Role of Lab Animals', in *The Rise of Critical Animal Studies: From the Margins to the Centre*, ed. by Nik Taylor and Richard Twine (Abingdon: Routledge, 2014), pp. 139–164; Haraway, *When Species Meet*.

39 Jamie Lorimer, *Wildlife in the Anthropocene* (Minneapolis, MN: University of Minnesota Press, 2015).

40 Ipsos MORI, *Public Attitudes to Animal Research in 2018* (Report for the Department of Business, Energy & Industrial Strategy, 2018).

41 Maggie Jennings and Jane Smith, *A Resource Book for Lay Members of Ethical Review and Similar Bodies Worldwide* (Horsham: RSPCA, 2015).

42 Alexandra Palmer et al., 'Animal Research beyond the Laboratory: Report from a Workshop on Places Other than Licensed Establishments (POLEs) in the UK', *Animals*, 10.10 (2020), 1868, DOI: 10.3390/ani10101868.

43 Renelle McGlacken and Pru Hobson-West, 'Critiquing Imaginaries of "the Public" in UK Dialogue around Animal Research: Insights from the Mass Observation Project', *Studies in History and Philosophy of Science*, 91 (2022), 280–287, DOI: 10.1016/j.shpsa.2021.12.009.

44 Pru Hobson-West and Ashley Davies, 'Societal Sentience: Constructions of the Public in Animal Research Policy and Practice', *Science, Technology, & Human Values*, 43.4 (2018), 671–693, DOI: 10.1177/0162243917736138.

45 Wendy Jarrett, 'The Concordat on Openness and its Benefits to Animal Research', *Laboratory Animals*, 45.6 (2016), 201–202, DOI: 10.1038/laban.1026.

46 Elisabeth H. Ormandy et al., 'Animal Research, Accountability, Openness and Public Engagement: Report from an International Expert Forum', *Animals: An Open Access Journal from MDPI*, 9.9 (2019), 622, DOI: 10.3390/ani9090622; Carbone, 'Open Transparent Communication about Animals in Laboratories'.
47 K. Hammarberg et al., 'Qualitative Research Methods: When to Use Them and How to Judge Them', *Human Reproduction*, 31.3 (2016), 498–501, DOI: 10.1093/humrep/dev334.
48 Margaret D. LeCompte and Jean J. Schensul, *Analysis and Interpretation of Ethnographic Data: A Mixed Methods Approach* (Lanham, MD: AltaMira Press, 2012).
49 In this effort we were guided by Felicity Callard and Des Fitzgerald, *Rethinking Interdisciplinarity across the Social Sciences and Neurosciences* (Houndmills, Basingstoke: Palgrave Macmillan, 2015).
50 Mike Michael, *The Research Event: Towards Prospective Methodologies in Sociology* (London: Routledge, 2021), DOI: 10.4324/9781351133555.
51 See Callard and Fitzgerald *Rethinking Interdisciplinarity*, pp. 124–126.
52 Davies et al. (2016) 'Developing a Collaborative Agenda for Humanities and Social Scientific Research on Laboratory Animal Science and Welfare'.
53 Robert G. W. Kirk, 'Recovering the Principles of Humane Experimental Technique: The 3Rs and the Human Essence of Animal Research', *Science, Technology, & Human Values*, 43.4 (2018), 622–648, DOI: 10.1177/0162243917726579.
54 Andrew Barry et al., 'Logics of Interdisciplinarity', *Economy and Society*, 37.1 (2008), 20–49, DOI: 10.1080/03085140701760841.
55 Hans Peter Peters, 'Looking Back and Looking Ahead', *Public Understanding of Science*, 31.3 (2022), 256–265, DOI: 10.1177/09636625221094165; Sarah R. Davies, 'STS and Science Communication: Reflecting on a Relationship', *Public Understanding of Science*, 31.3 (2022), 305–313, DOI: 10.1177/09636625221075953.
56 Thomas F. Gieryn, *Cultural Boundaries of Science: Credibility on the Line* (Chicago, IL: University of Chicago Press, 1999).
57 Jacques Derrida, *Of Grammatology* (Baltimore, London: Johns Hopkins University Press, 1976).
58 See also Gorman and Davies, 'When "Cultures of Care" Meet'; Gail Davies et al., *Informing Involvement around Animal Research: Report and Resources from the Animal Research Nexus Project*, 2022, www.animalresearchnexus.org/publications/informing-involvement-around-animal-research [accessed 1 February 2023].

59 Greenhough and Mazhary, *Care-Full Stories*.
60 Dmitriy Myelnikov, 'Cuts and the Cutting Edge: British Science Funding and the Making of Animal Biotechnology in 1980s Edinburgh', *The British Journal for the History of Science*, 50.4 (2017), 701–728, DOI: 10.1017/S0007087417000826; Robert G. W. Kirk, 'Recovering The Principles of Humane Experimental Technique: The 3Rs and the Human Essence of Animal Research', *Science, Technology, & Human Values*, 43.4 (2018), 622–648, DOI: 10.1177/0162243917726579.
61 James W. E Lowe et al., 'Training to Translate: Understanding and Informing Translational Animal Research in Pre-Clinical Pharmacology', *TECNOSCIENZA: Italian Journal of Science & Technology Studies*, 10.2 (2020), 5–30; Gail Davies, 'What is a Humanized Mouse? Remaking the Species and Spaces of Translational Medicine', *Body & Society*, 18.3–4 (2012), 126–155, DOI: 10.1177/1357034X12446378.
62 McGlacken and Hobson-West, 'Critiquing Imaginaries of "the Public"'; Davies et al., 'Animal Research Nexus'.
63 Vanessa Ashall and Pru Hobson-West, 'The Vet in the Lab: Exploring the Position of Animal Professionals in Non-Therapeutic Roles', in *Professionals in Food Chains*, ed. by Svenja Springer and Herwig Grimm (Wageningen: Wageningen Academic Publishers, 2018), pp. 291–295, DOI: 10.3920/978-90-8686-869-8_45; Hobson-West and Davies, 'Societal Sentience'.
64 Gail Davies et al., 'Developing a Collaborative Agenda for Humanities and Social Scientific Research on Laboratory Animal Science and Welfare', *PLOS ONE*, 11.7 (2016), 1–12, DOI: 10.1371/journal.pone.0158791.

Part I

Changing and implementing regulation

1

A 'fragile consensus'? The origins of the Animals (Scientific Procedures) Act 1986

Dmitriy Myelnikov

Introduction

The Animals (Scientific Procedures) Act 1986 (ASPA) regulates the conduct of most scientific work that involves non-human vertebrates in the UK, and is overseen by the Home Office.[1] ASPA had replaced the 1876 Cruelty to Animals Act, a law that had governed scientific research for 110 years, through a period of profound transformations in science and medicine. Since its passage in 1986, ASPA has undergone some updates and changes, and has been harmonised with EU legislation via European Union Directive 2010/63/EU. Its core principles, however, have remained largely consistent with its original spirit. ASPA involved difficult negotiations, balances, and exclusions across numerous interest groups invested in its outcome. It ended up reshaping moral obligations, legal commitments, and approaches to regulating research. The history of ASPA offers a case study for investigating the kinds of multidirectional entanglements that we call the animal research nexus.[2] This chapter explores the compromises and construction of consensus involved in the making of ASPA, and outlines how its implementation put pressure on that consensus.

To date, histories of ASPA have been written by figures involved in the process, and from animal rights perspectives.[3] Robert Garner's study, drawing on published sources and communication and interviews with some of the key players, offers an excellent account on the passage of ASPA and the role of moderate welfare voices in its design.[4] Using recently opened government and personal archives and supplemented by interviews with historical actors as evidence, the historians in the Animal Research Nexus Programme have been able to revisit this crucial bit of legislation, and answer several questions.[5] Why did ASPA originate when it did, under Margaret Thatcher's governments that had, at best, a passing interest in the issue? How did the civil servants and politicians involved in its design navigate the broad range of positions and interests over the highly divisive issues; which voices did they prioritise, and which did they exclude? How did ASPA operate on the ground, and what impact did it have on science, animal care, and activism? Finally, how do these past decisions, specific to their time and context, resonate today?

After introducing the political context of animal campaigns in the 1970s, the chapter traces strategies involved in consensus-building across three key contexts: the alliance of the moderate animal welfare groups that ended up representing the position of animal welfare in ASPA design; the Home Office E Division that drafted two white papers (in 1983 and 1985),[6] trying to balance diverse stakeholder interests while ensuring continuity in implementation and excluding the more radical voices; and the Cabinet, where animal research was a marginal and inconveniently controversial concern, yet pressure existed to enact reform. Paradoxically, given the extent of polarisation over animal research that intensified in the 1970s, consensus was a key theme in deliberations over ASPA. Home Office civil servants sought to establish it across the stakeholders; ministers and MPs sought to create workable law and assuage societal concerns; various activist groups found different ways of arriving at a meaningful set of reforms for them. Despite Margaret Thatcher's rhetorical disavowals of consensus in favour of conviction politics, her government actively sought to construct and claim a common position, while excluding voices that could threaten their agenda. The fragility of this consensus, moreover, proved not so much a weakness, but a political tool that ensured the smooth passage of ASPA through Parliament.

Putting animals into politics

Why did reform take place in the 1980s? Animal experimentation has been controversial throughout its history, especially in

Britain. While the intensity of public debate waxed and waned, the opposition to animal experimentation never disappeared, and the legal protections afforded to laboratory animals were routinely criticised.[7] Throughout most of the twentieth century, amid dramatic changes in medicine and the life sciences, experiments remained governed by the Cruelty to Animals Act (1876). Despite multiple calls for reform, subsequent governments delayed any substantial action, reluctant to engage with controversy and happy with ad hoc arrangements within the Home Office. It took the new wave of opposition to animal research in the 1970s to push for reform, as it combined activism and media campaigns with targeted lobbying of subsequent governments.

The Cruelty to Animals Act (1876) was a product of Victorian debates about morality, animal sentience, politics, and society.[8] Yet it created a series of institutions that persisted for over a century in their oversight. It gave the responsibility over animal experimentation to the Secretary of State for the Home Department ('the Home Secretary') and instituted an Inspectorate within the Home Office. Qualified applicants were granted licences to perform experiments, based on the assessment of their overall competence, and inspectors monitored compliance. As the biological and medical sciences went through spectacular transformations in the twentieth century, the system kept up through the executive branch; any issues were dealt with internally through the licensing system, stripping scientists of their licence in the most egregious cases.[9] The justifications under the 1876 Act – extending knowledge of physiology, prolonging life, or alleviating suffering – were interpreted broadly. Numbers of animal experiments, which had to be reported annually under the Act, grew steadily through the twentieth century, reaching the peak of 5.3 million in 1972.[10] At the same time, the 1876 Act required reduction of pain, and the majority of work was done under anaesthesia; dogs, cats, horses, and donkeys received extra protections.[11]

The fact that old legislation was regulating a field where much was changing drew growing criticism, especially in the post-war years. In 1963, Harold Macmillan's Conservative government intervened, initiating a committee to review legislation, chaired by the lawyer Sir Sydney Littlewood. The Littlewood report found the 1876 Act 'generally effective' but outdated and in need of reform, especially when it came to the scope of procedures, licensing, the powers of the Inspectorate, and the lack of emphasis on veterinary expertise within laboratory animal care.[12] These recommendations were well received, but subsequent governments preferred to ignore them. Many changes to care and husbandry happened without regulatory oversight; the University Federation for Animal Welfare (UFAW, founded in 1938), which promoted better care, and published crucial handbooks, was influential, as were new associations of animal technicians (Institute of Animal Technology, f. 1950) and of groups of vets specialising in laboratory animal case (British Laboratory Animal Veterinary Association, f. 1963).[13] In addition, in the 1970s a small number of scientists were making the case for greater use of alternatives to animals in testing toxins or carcinogens, addressing fellow researchers.[14] Still, while many scientists may have agreed it was time to update the 1876 Act, the pressure for reform came largely from without.

In the 1970s, campaigns for animal welfare grew in both numbers and ambition. The civil rights movement of the 1960s and second wave feminism were inspiring some to extend similar arguments to non-human animals, and new provocative texts were attracting broad readership. In Britain, a new generation of activists, now known as the Oxford Group, published a collection of essays in 1971 making a case against the wide-ranging mistreatment of non-human animals.[15] One of their key members, the psychologist Richard Ryder, coined the term speciesism, by analogy with racism and sexism. In 1975, Peter Singer's *Animal Liberation* made a philosophical case for animal rights, and was a subject of wide-ranging and intense discussions, and Ryder's *Victims of Science* appeared in the same year, focusing specifically on the scientific uses of animals.[16] The new wave of campaigners targeted established institutions. The Royal Society for the Prevention of Cruelty to Animals (RSPCA) had more active and radical members, Ryder among them, pushing the body to campaign more vigorously despite the tension this caused with its Conservative rural patrons, for instance in campaigns against hunting. The year 1976 was the Animal Welfare Year, with numerous events and interactions set up by welfare groups.

Public controversies in the 1970s helped turn animal experimentation into one of the key issues for the new animal advocacy.

The smoking beagles exposé was a landmark media event. When working undercover at the ICI at Alderley Park near Macclesfield, Mary Beith obtained striking image of dogs restrained and made to inhale smoke in research aiming to design a safer cigarette. Published in the *Sunday People* in 1975, the images caused an uproar – the rather trivial purpose of the experiments, the use of charismatic dogs, and the fact that the photographs did not depict gory procedures and could thus circulate widely all contributed to their effect.[17] Beith's investigation happened with Ryder's input, who suggested ICI as a site to investigate, and his *Victims of Science*, appearing shortly after, benefited from ongoing controversy around the beagle case.[18] Cases like the smoking beagles helped galvanise public opinion against animal *testing* – the preferred emphasis of anti-vivisection campaigns – and made political inaction more of a liability.

In parallel to public campaigns, activists also embraced closer engagement with politicians, strategically pursuing the issue across party lines. Certain key figures proved essential go-betweens, notably Douglas Houghton, Baron Houghton of Sowerby, a veteran Labour politician by then in the House of Lords. As Ryder put it in retrospect, 'Houghton's main achievement at this time was to deflect our attentions away from backbench MPs and onto the Government itself.'[19] One of the early interactions, stemming from the Animal Welfare Year in 1976, was a joint paper between Lord Houghton and Lord Platt, an eminent surgeon, which outlined a series of changes that would be acceptable to both the scientific establishment and welfare groups. As a Labour peer, Houghton organised a meeting between the Labour Home Secretary Merlyn Rees and some of the key figures and MPs invested in animal welfare, including Clive Hollands, Ryder, and Lord Platt. In February 1977, the group presented the Houghton/Platt memorandum and called for executive changes in the administration of the 1876 Act, namely expanding the Advisory Committee, tightening conditions around pain on personal licences held by scientists, and promoting alternatives to animal testing.[20] The Houghton group soon became the Committee for Reform of Animal Experimentation (CRAE), led by Hollands with Houghton's patronage. While animal activism overall was getting more radical, CRAE pursued pragmatic and moderate goals. The Houghton/Platt memorandum paved the way for an active campaign with tangible proposals, which flourished in the late 1970s and involved multiple stakeholders, including anti-vivisection societies. 'Put Animals into Politics' was one of the slogans, and by the 1979 General Election, Labour, Conservative, and Liberal manifestos had all committed to reform. Labour's proposals went furthest, with a dedicated policy paper on various issues in animal welfare, 'Living without Cruelty'.[21] The charter was a significant move for the party, whose socialist wing had been suspicious of animal welfare as an essentially middle-class concern and a distraction from human social issues.[22] Avoiding change was no longer the easy option.

The Conservatives' 1979 manifesto may have committed to reforming animal experimentation, but it was not a priority for Thatcher's first government. Animal welfare groups continued the pressure for reform, increasingly focused on legislation rather than administrative changes of practice. The House of Lords became an initial site of contention, as shortly after the General Election, a Bill was introduced there by the Earl of Halsbury, who had had significant involvement in industrial research and development. The Halsbury Bill effectively promoted the position of pro-research lobby groups, including the Research Defence Society, and sought to pre-empt too radical a change, while proposing significant administrative reform. In response, Peter Fry MP presented an alternative Bill to the Commons in November 1979, which had input from the RSPCA and positioned itself towards animal protection.[23]

Despite the antagonism, the two Bills had much in common, showing both sides willing to make concessions in the face of looming reform. The Bills proposed new systems of licensing with stronger scrutiny and a need for two sponsors; an empowered statutory Advisory Committee on animal procedures; greater powers for the Inspectorate; requirement to euthanise animals in severe pain. The key difference was that the Fry Bill sought to limit purposes under which animal testing could be done to medicine only, and to create five grades of licences based on the levels of pain to the animal. Neither Bill passed the committee stage, but the government appeased both sides, promising to design its own legislation, initially in the form of the White Paper that appeared in 1983.[24] In addition to Parliament, pressure came from the European Communities, as the European Commission was in the process of setting up its own directive on animal experiments that would establish minimal standards across the member states.[25] Between 1983 and 1985, tense discussions across multiple stakeholders shaped the government Bill, with a Supplementary White Paper published in 1985.[26] ASPA then finally made it onto the parliamentary calendar in 1986, and after a relatively smooth passage through the Commons and Lords, received royal assent on 20 May 1986.

The triple alliance

The voices that the government was willing to engage were what it designated as the 'moderate' animal welfare groups, on the one hand, and the scientific bodies on the other. Three organisations, acting in a coalition and willing to push for incremental, moderate change, ended up representing and articulating the animal welfare case. This 'triple alliance' comprised CRAE alongside the British Veterinary Association (BVA), and the Fund for the Replacement of Animals in Medical Experiments (FRAME). How did these disparate groups come together, and how did they claim the political centre ground of reform?

FRAME, founded in 1969, pursued promotion of alternatives to animals within the biomedical research community. It was itself involved in lobbying MPs to promote the use of alternatives, especially in the light of the Advisory Committee on the Cruelty to Animals' Act investigation into the controversial LD50 test.[27] Houghton and Hollands interacted with members of FRAME Toxicity Committee in parliamentary groups, and were eager to build an alliance with a group that was dedicated to one of the main issues in the CRAE proposals, and one that could lend CRAE scientific credibility.[28] Michael Balls, a toxicologist who became Chairman of FRAME's trustees in 1981, was the key figure in the process.

The third member of the alliance, the BVA, joined in 1982. The veterinary profession had been divided on the animal research debate in Britain. One the one hand, veterinary medicine relied on animal experiments and some veterinarians had built expertise in working with laboratory animals. The Royal College of Veterinary Surgeons, the regulatory body in charge of registration and professional gatekeeping, made representations defending scientific interests.[29] The British Veterinary Association – the professional association – took a more welfare-focused approach, driven in part by its constituent, the British Laboratory Animal Veterinary Association (BLAVA). In 1982, BLAVA Secretary John Seamer articulated the BVA's policy position on animal research, which recognised the need for animal experiments but called for tighter regulation and interpretation of pain, extending protections beyond the moment of experiment, and replacing animals with alternatives where practical. Given the many affinities in policy with CRAE and FRAME, and CRAE's interest in introducing veterinary care more directly into animal experimentation, Hollands and Balls approached the BVA to join on the draft response to the White Paper.[30]

The figure of the veterinarian – what became known as a Named Veterinary Surgeon (NVS) – was a compromise that appealed to many parties, including the various representatives of the profession.[31] To CRAE, they were potential 'animal friends', there on behalf of the animals.[32] The BVA members also allowed the triple alliance to boost its scientific authority over a key point – the interpretation of pain in experiments. In negotiating the new constitution of animal research, knowledge of pain was contested. The CRAE/FRAME/BVA coalition recruited alternative sources of evidence and authority, notably the neuroscientist and RSPCA member Patrick Wall, Professor of Anatomy at University College London and editor of the journal *Pain*. Wall argued that pain could interfere with most experiments in the first place, and that a workable assessment could be made using the principle his journal had adopted: 'pain in animals is manifested by abnormal behaviour which can be alleviated by analgesic procedures which relieve pain in humans.'[33] Laboratory animal veterinarians positioned themselves further across the diverse interests by claiming expertise and practical approaches to interpreting pain and distress – especially for rodents, central in research but often unfamiliar to ordinary veterinarians. In assessing pain, the Home Office *Code of Practice* relied on David Morton, Peter Griffiths, and Paul Flecknell's research on recognising distress and promoting consistent pain relief (analgesia) alongside anaesthesia (Morton represented the BVA and at the late stages was on the RSPCA animal experiments committee).[34] Voices from the laboratory animal veterinary community therefore offered a palatable position in codifying the new governance of research that straddled divisions.

In choosing to pursue compromise, the triple alliance was placed in a precarious position with respect to other animal welfare and rights groups. Major anti-vivisection groups – the British Union for the Abolition of Vivisection, National Anti-Vivisection Society, Animal Aid and Scottish Anti-Vivisection Society – launched a campaign of 'Mobilisation Against the Government White Paper', seeking to ban cosmetic testing, the controversial LD50 and Draize Eye Irritation toxicity tests, military animal research, and invasive psychological research. Increasingly through the consultations, Mobilisation criticised CRAE's position, but had little engagement from the Home Office.[35] Despite protests from others in the animal protection community as to the illegitimacy of such narrow representation, CRAE took a self-consciously moderate position. To the RSPCA, Lord Houghton presented CRAE not as a competitor, but as 'the channel of communication to Ministers and Government'.[36] The alliance's press-release insisted, 'the operative words here are "animal welfare" and not "animal rights" or "animal liberation"'.[37]

The triple alliance's compromise position was largely behind its influence on reform. It was not, however, always consistent within. The Home Office civil servants suggested that FRAME and the BVA were easier to negotiate with than CRAE; some issues caused considerable differences, notably the re-use of animals that had been anaesthetised. The 1876 Act forbade any re-use under those conditions – death was always a preferable outcome to pain and suffering in its logic. In designing ASPA, civil servants sought to adjust the clause to allow carefully controlled re-use of animals that had fully recovered from anaesthesia in further experiments, and in some cases, to allow rehoming or release into the wild of animals after experiments.[38] All sides agreed that re-use after anaesthesia would have limited effect as the cases where this could happen were few. FRAME and BVA members favoured re-use in some cases, as it would lower the number of cats, dogs, and primates used in research – species that would receive special protections under ASPA and engendered most public concern.[39] CRAE argued against such concessions, arguing that the system might be open to abuse, and that the move could be interpreted as lowering protections compared to the 1876 Act.[40] During the Lords debate on the government's Bill in January 1986, Lord Houghton successfully argued against the amendment that would allow re-use:

> This proposal is going into highly contentious politics very soon indeed – as soon as this Bill arrives in another place [House of Commons]. In these circumstances it is absolutely essential from the point of view of continued goodwill and collaboration that this be dropped here and now. Let us have discussions about this for the future. Let us see whether we can rationalise the situation; let us see whether we can get proposals for the re-use of animals, if that becomes a feasible political proposition, and consider under what conditions it may be done.
>
> [...]
>
> However irrational it may be and however difficult it may be to justify, even in ethical terms, let us at least refrain from trying to reverse the understanding that was reached in collaboration on the consensus Bill.[41]

The removal of re-use, except in very few circumstances under explicit permission from the Home Secretary, also marginalised some attempts to introduce re-homing of animals post-recovery, where animals would be 'returned to the farm or the wild or be found suitable homes as pets, depending on the species'.[42] In his speech, Houghton argued the issue must not be raised to avoid dividing the opinion in Commons, since re-use and rehoming to reduce numbers was in fact a special case for cats and dogs, which were already receiving extra protection continuing from the 1876 Act, and that rehoming would imply scientist had ownership over the animals in their charge, which he believed was misguided.[43] The final text of ASPA did not mention rehoming, while the guidance published in 1990 only euphemistically referred to animals being 'dispatched into private care' or 'released for non-scientific purposes' and mentioned rehoming as pets only in Appendix IV, requiring that a project licence must explicitly allow for releasing animals as 'pets' in advance.[44] As Skidmore shows in Chapter 3, while rehoming is technically enabled by the law, it remains a rare practice. The key focus on pain as the primary welfare concern in ASPA has meant that humane killing in accordance with the Act is still the most common outcome.

In emphasising collaboration and consensus, Houghton rallied against allowing re-use but also expressed a hope for future flexibility in interpreting and amending the law. Such a compromise approach was consistent with the aims and strategies of the triple alliance. By accepting the licensing reform and the new Animal Procedures Committee (APC) as useful moves from the Home Office, and focusing on core issues – pain, alternatives, and reuse – the alliance was a major voice in shaping ASPA. Its anticipation for flexible interpretation of the law proved justified, as ASPA implementation changed through the actions of the Animals Procedure Committee (with many key players of the triple alliance as members) and, in 1993, through an amendment to include *Octopus vulgaris* alongside protected vertebrate species.[45] By being willing to relinquish certain issues to the executive rather than legislative sphere, these could be further negotiated with less public pressure. But how were these concerns and strategies addressed and managed by the Home Office, whose civil servants also had to balance multiple perspectives from scientific bodies, industry, other government departments and its own Inspectorate?

Curating the consensus: inclusions and exclusions at the Home Office

Civil servants in the Home Office 'E' Division oversaw the design of the new legislation, under the political auspices of David Mellor, the junior Home Office minister assigned to the mission. Geoffrey de Deney was the senior civil servant, while Nigel Johnson coordinated most of the drafting, with the aid of Martha Woolridge. While most stakeholders worked on the assumption that the system was woefully outdated, this view was not necessarily shared by the E Division. While the 1876 Act may have been archaic and arcane, the system of supervising animal research that had developed ad hoc and with little parliamentary oversight appealed to civil servants, working side-by-side with the Inspectorate.

As Nigel Johnson put it, 'the present system works. It works well. It would be deplorable to replace it by a new system which did not work.'[46] This commitment did make the Home Office much more receptive to the scientific and industrial lobbies, but not blindly so. The civil servants were eager to expand the degree of oversight and to spell out concrete expectations and parameters. Mainstream scientific groups, moreover, had publicly accepted the need for reform, and while they campaigned against tying the levels of pain to the perceived importance of research, they largely accepted a tougher licensing system. In many ways, the resulting regulation challenged existing arrangements greatly and at high cost. The old settlement, in which scientists held the authority over both scientific value of the research and assessing animal suffering, while the Home Office represented society's interest as regulator, no longer held. Veterinarians, animal technicians, and moderate welfare advocates acquired a degree of expertise, control, and responsibility.[47]

Several issues took centre stage in the discussions of the White Paper and subsequent responses to it in drafting legislation. Some of the key features were largely well-received, notably the three-part licensing system, which required licences for an individual as a competent experimenter, for the institution and facilities where the research took place, and, crucially, for the actual project as being justified on scientific grounds and in its experimental design. The proposals broadened the law's scope from just the experiment to a much broader 'procedure', reflected in the Bill's name, and extended protections to housing and breeding of animals, which now had to be included in statistics. While this meant a greater regulatory burden for institutions and animal suppliers, it was one of the key commitments of the Home Office, which responded to pressures to ascertain that only purpose-bred animals could be used in experiments. The scope of protected species caused some debate – the legislation remained constrained to vertebrates, largely for convenience, even though octopuses had their advocates for inclusion and did end up being covered by the law's extension in 1993. Primates received special protections, alongside cats, dogs, horses, and donkeys. For the first time, foetal and larval forms were protected, too, after certain developmental stages. Finally, new figures became significant, as animal technologists received more responsibilities and institutional leverage, and veterinarians were given a formalised role.

In setting out the scope of the White Paper, the Home Office had a disparate field of stakeholders to deal with. From the scientific bodies, the Royal Society was a major participant, as well as multiple representatives of universities and the pharmaceutical industry. The combative pro-animal experimentation Research Defence Society (formerly Physiological Society) took a back seat in liaising with the Home Office, except to emphasise laboratory break-ins and the intimidation that researchers increasingly faced from more radical activists. At the other end were anti-vivisection societies, who mobilised public campaigns of letter writing; some opposed all forms of animal experimentation, while others were willing to negotiate. The RSPCA, as a more moderate body combining multiple interests, was struggling with internal conflict – often public – between the more radical new members and the moderate old guard.[48] Mellor met with RSPCA representatives regularly, and pain proved a major source of contention, but the Society was not closely involved in the legislative process until the late stages. There were other professions and groups that sat uneasily across the two extremes. Animal technicians and scientific experts in care and husbandry had some representation, through the Institute of Animal Technologists and UFAW, especially when it came to the details of care and possible guidelines. Finally, as discussed above, the veterinary profession provided a diversity of inputs.

From the outset, Home Office civil servants decided to exclude anti-vivisection groups and anyone opposed to animal experiments in principle. As de Deney put it:

> Nothing we can responsibly say will satisfy people who are convinced that it is morally wrong to use animals for scientific purposes, but we need to get it across as strongly as possible to people with an open mind that we are committed to preserving and strengthening the protection given to animals used in scientific procedures, and will extend control to the breeding and supply of animals.[49]

Through selective engagement and ready exclusion of disruptive voices, the Home Office civil servants narrowed the extensive range of opinions to manageable options. One point, however, caused major contention, and required most effort from the Home Office in aligning the scientific interests and welfare arguments. It was the question of animal pain – how it could be defined in law, and whether levels of pain induced in an experiment could be made conditional on the potential importance of research to human and animal health.

The issue of pain was a major point of contention, and one that illuminates the mechanisms underlying the design of new legislation. By 1984, the E Division civil servants were exhausted by the issue. Venting his frustration to de Deney, Nigel Johnson commented in a brief note, 'Personally I should welcome a period of say six months' silence in public discussion of the pain condition, to give people interested in the topic a chance to consider ascertaining what the facts are and whether there is anything new to say about it.'[50] As a key subject for negotiation, discussion of pain demonstrates the strategies that the Home Office used to construct consensus opinion, and how it balanced formal legal language in an Act of Parliament with other kinds of legal instruments such as guidelines and the Code of Practice, and less formal arrangements with the Inspectorate.

Animal welfare groups objected to severe pain being permissible, and sought to tie procedures to goals, so that high levels of pain could only be allowed for important medical research with concrete outcomes. Both the RSPCA and the CRAE/BVA/FRAME coalition were also starting from a point of preventing any pain to animals through the use of anaesthesia, including in lethal amounts if severe pain was likely on recovery, in line with the 1876 Act system. A number of scientists and institutions advocated that some experiments that caused severe pain had to be allowed, at least in principle, and argued that the Home Office could not judge potential importance of research before it had taken place. The Royal Society, charged with creating the Code of Practice to accompany the law, questioned the basis for gradations of pain. In negotiating with CRAE/BVA/FRAME, the Home Office civil servants quoted the words of the Royal Society president, Sir Andrew Huxley, that 'criteria [of pain] cannot usefully be incorporated in an Act of Parliament because a phase like "substantial pain or distress" cannot be given a precise definition; no one has yet discovered a distress indicator in the blood to which a legal limit of concentration can be ascribed'.[51]

In navigating the pain condition, the Home Office was eager to keep CRAE and especially Lord Houghton as allies, who could convey legitimacy among moderate welfare groups and welfare-minded MPs and Lords. The E Division civil servants argued that many of the demands were already being carried out in practice by the Inspectorate – that very painful experiments would not be allowed without a good reason, and the Chief Inspector conceded that informal 'pain–benefit analyses' had been happening routinely.[52] Reluctant to legislate levels of pain without clear precedent, and eager to maintain interpretive flexibility, the proposed solution avoided including explicit discussions of pain levels in ASPA, but committed to administrative procedure that would correlate pain to potential benefit at the level of assessing project licence applications. The decision was similar to the Police and Criminal Evidence Act 1984, which was being drafted by the Home Office in parallel, and also relied on a code of practice in major reform of policing in Britain.[53] The guidance to the implementation of ASPA, published in 1990, defined three levels of pain severity – mild, moderate, or severe – and urged that harm to the animal be weighed against potential scientific benefits.[54]

In negotiating the 'pain condition', the Home Office pushed the parties to maintain what it represented as consensus between scientific and moderate welfare interests. It relied on resources beyond legislation, by seeking to consolidate new agreements at the level of regulatory practice and committing to extend the Inspectorate's powers to make decisions. This quest for consensus position, carefully balanced by civil servants, not only ensured potential legitimacy; it also served as a political instrument to push legislation through Parliament.

The uses of consensus in the Cabinet

The connotations of 'consensus' were fraught in the 1980s. In the broadest sense, Thatcher's governments self-consciously put an end to the 'post-war consensus' in political economy, ripping apart the tacit agreement between Labour and Conservative politicians on the importance of the welfare state, Keynesian economic measures, high levels of nationalisation, and strong trade unions.[55] Thatcher loathed the word in its political sense, and frequently emphasised that she was not a consensus politician, but one driven by conviction.[56] Yet in practice, as Brian Harrison put it, 'in some areas Thatcher's combative rhetoric and style did not preclude even the direct pursuit of consensus, let alone the indirect one'.[57] Animal research, loaded with potential controversy, was one such area, and the pleading from the fragile consensus standpoint paid off.

David Mellor and the Home Office civil servants spent considerable energy on maintaining a viable middle ground across the stakeholders they had included in deliberations. But they also capitalised on this delicate balancing act to push the Bill through Parliament. The new Bill was not a priority for Thatcher's government, which was eager to minimise publicity on the sensitive subject of animal experimentation.[58] Despite reform being a manifesto commitment, the government was reluctant to include the Bill in the 1985–86 agenda, as other major and complex legislation took priority,[59] to the frustration of the Home Office. Leon Brittan, the Home Secretary, urged the Cabinet to commit to reform in the 1984–85 Queen's Speech to Parliament that outlined the forthcoming legislative agenda, and to pass the Bill through Parliament the following year. He stressed that the 'considerable consensus we have achieved with moderate animal welfare interests', was bound to 'slip away' with delay, and, moreover, that it was 'perhaps surprising that any sort of consensus was achieved in the first place'.[60] Brittan also stressed the urgency of reform in the face of public pressure, and the political costs of inaction:

> We must defuse the high level of public and political concern. Home Office Ministers now receive over 150 letters every month from MPs about the animals legislation. In the last year we have received 4,000 letters from members of the public. The uncertainty is making backbenchers restive, in response to considerable constituency pressure. The Opposition Parties, too, may be compelled to drift to a more extreme position.[61]

These efforts failed in 1984, but Brittan renewed his commitment to pass the law the following year. In a February 1985 policy summary written for Thatcher, the Home Office insisted reform was widely expected and could no longer be delayed, and summarised the political strategy succinctly:

> Public confidence must be restored in the Government's desire and ability to protect animals from abuse, and to enable scientists to get on with their work in peace ... The proposed legislation would clip the wings of the increasingly violent extreme anti-vivisectionist movement by isolating them from moderate opinion ... to decide not [to] legislate now would leave the Government in the worst possible situation, facing criticism from all those who have taken an active part in the debate.[62]

Writing to Willie Whitelaw, his predecessor as Home Secretary and by then Lord President of the Council, Brittan restated that 'The absence of this measure from the Queen's Speech will ... put a severe strain on the fragile consensus between the moderate animal welfare lobby and the scientific community.'[63] In tandem, Janet Fookes, a Conservative MP closely involved with the RSPCA, wrote to Thatcher to express alarm that 'the unprecedented consensus of opinion voiced by the scientific and veterinary community, together with the responsible element in the animal welfare movements may be lost if there is no Bill before Parliament in the next session'.[64] The multi-directional appeal paid off. The Bill returned to the parliamentary schedule for 1985/86, and a promise to reform animal experimentation concluded the Queen's Speech on 6 November 1985.[65] The efforts the Home Office had gone to in maintaining agreement meant ASPA passed smoothly, receiving royal assent on 20 May 1986.

In pursuing an appearance of consensus, Thatcher's government not only aimed to marginalise the increasingly vocal anti-vivisectionist groups, but also create a new settlement for industry, academia, and society. Once ASPA passed, Sir Robin Ibbs, Director of ICI, wrote to Douglas Hurd, the third Home Secretary in position during the drafting of ASPA, to praise David Mellor's efforts in galvanising agreement and pushing the parliamentary process. Echoing Brittan's views, Ibbs suggested that 'a greater consensus of opinion ... resulted in the isolation of the extremists and the emergence of legislation which, while much more rigorous in its control of animal experiments, is nevertheless still workable by all serious research departments in industry and academia'.[66] Forwarding the letter to Thatcher, Hurd emphasised that while the Home Office 'deliberately [had] not blown trumpets about this', ASPA was 'a major achievement of your Gov[ernmen]t in an area of enormous difficulty'[67] and Thatcher agreed, noting it was 'a great credit to all concerned that [ASPA] had such a smooth passage'.[68]

Implementing ASPA

Did ASPA meet expectations from civil servants, inspectors, and animal welfare groups, and did the Conservative government succeed in retaking the middle ground from animal rights campaigners? The more radical activists distanced themselves from CRAE in the later stages of deliberations. Thus, in late 1985, Richard Ryder resigned from the RSPCA's internal Animal Experimentation Advisory Committee in vocal opposition to the Bill, driven in part by the increasing engagement of veterinarians as representatives of the RSPCA who pushed for compromise.[69] In a letter to Peter Singer, Ryder described the forces behind ASPA as a 'conspiracy of vets, vivisectors and Tories'.[70] A breakaway group from the RSPCA criticised the Act as a free-for-all 'Vivisectionists' charter'.[71] In the immediate aftermath of the passage of ASPA, direct action and attacks on laboratories intensified, and persisted through the 1990s. As ASPA was being implemented through its five-year transition period, however, some campaigners changed their assessment. The more moderate RSPCA became heavily involved in promoting welfare standards. Welfare advocates and scientists sought common ground, notably via the Boyd Group, designed to enhance dialogue between critics and proponents of animal research, which produced a number of influential reports.[72] By 1996, Ryder saw ASPA as achieving some of the goals he had envisioned, an opinion he upheld in a recent interview.[73]

As a piece of legislation, ASPA has been successful in changing some practices of animal experimentation and breeding. With an elaborate system of surveillance represented by the Home Office Inspectorate, the cost of non-compliance could affect careers (but almost never resulted in prosecution). Scientists could lose their licence and therefore their ability to perform experiments. In addition, the Inspectorate offered recourse to Animal Technologists, newly empowered by the legislation, and as a body in charge of licensing as well as surveillance, it curated what counted as best practices across establishments. From the rather informal assessment, much of which was up to the individual inspector, ASPA extended bureaucratic oversight. The Code of Practice favoured specific and measurable parameters, such as pen size, temperature, and humidity. Some parameters came from ethological observations or knowledge of laboratory species in the wild, but many were derived from what their designers saw as existing best practices. Sometimes the decisions were made haphazardly – as a senior animal technician recalled, on request from a Home Office inspector, they 'measured the dog pens and that size became enshrined in the legislation, because [the inspector] says, "well, I know it works"'.[74] As a result, ASPA drove up costs for animal research. The animals themselves became more expensive as regulation extended to breeding establishments, especially species for which dedicated laboratory supply had been less developed, such as farm animals. In addition, many animal units had to be updated, consolidated, or moved into purpose-built facilities.

Infringements and how they were dealt with can shed some light on the extent of regulatory control, and the kind of resistance inspectors and Home Office civil servants faced. Overall, most institutions were eager to comply with regulations, backed by university management, some of whom had personal responsibilities as named individuals under ASPA. In some cases, senior researchers could be recalcitrant, but the Inspectorate had the ability to intervene and redress the balance of power within laboratories.[75] Almost all cases of infringement were handled internally by the Inspectorate and institutions. In at least one case, the Home Office did refer the violation to the police. This had less to do with the nature of the violation – the case involved re-using animals for a relatively minor surgical procedure, explicitly banned by the Act – but with the reticence of the senior experimenter who refused to comply.

The prosecution, however, was not pursued by authorities. The core motivation for this decision, as explained to the E Division, was that prosecution would lift the confidentiality afforded by the Home Office process and involve naming the individual and their institution, exposing both to real risk of attacks. The authorities expressed a strong preference for dealing with the issue through administrative action by the Home Office.[76]

Some of the promises within ASPA did not materialise or were carried out only in part. Dedicated funding for alternatives to animal research – one of the major RSPCA demands and the main campaign goal for FRAME – was one. Initially, the Home Office set up a small funding programme to develop alternatives. After a trial, most scientific experts brought in to assess applications via the Medical Research Council disparaged this alternative stream of public funding with what they argued were inferior referee requirements.[77] The recalcitrance and negative feedback on applications was detrimental to financing alternatives. Once the money was not spent, it became reabsorbed into the Home Office budget and was no longer available in subsequent years. While the Home Office funded some FRAME research – notably early work on INVITTOX, the database of reliable *in vitro* methods in toxicology – most funding for alternatives came from FRAME campaigns and from Europe.

In the forged consensus, some arrangements and commitments were side-lined, while others developed into major regulatory subjects. While the Act was prescriptive and detailed, and shaped by guidelines and the Code of Practice, its interpretation remained flexible. From the outset, ASPA was envisioned as a 'living' piece of legislation, adaptable both through executive powers of the Home Office and the Home Secretary, and through legislative amendments as the research landscape changed. One instance of such flexibility is the 1993 extension of protections to the octopus, the first invertebrate animal to be covered.[78] The reformed and empowered APC was the most significant instrument for such flexibility. From an ad hoc group designed to advise the Home Office minister in charge of animal experiments and mostly filled by scientists, the APC became a statutory body within the Home Office, which combined voices of scientists with veterinarians, animal care experts, several welfare advocates (initially, Michael Balls, Clive Hollands, and the RSPCA's Judith Hampson and Thomas Field-Fisher) and a few lay members, and a lawyer chair. The APC played a key role in some of the crucial executive reform in the 1990s, notably the ban on cosmetic testing (1998), formal prohibition of experiments on the great apes (1998, although no such experiments happened in the UK since at least 1986). Finally, it was the APC, influenced by Balls and Hollands, that put the principle of the 3Rs – reduction (of numbers), refinement (of painless procedures and experimental design), and replacement (with alternatives) – on the agenda. While first articulated as early as 1959, the 3Rs were marginal in the design of ASPA.[79] Through the APC, and building on new interactions between scientists and welfare groups, they became a crucial principle in regulating British animal research.

Conclusion

In settling the new consensus, both the civil service and the government sought to exclude abolitionist voices and reclaim the middle ground of public opinion, increasingly exposed to arguments from animal protection groups. Yet within this consensus, scientists' autonomy over conducting research was challenged, too, as other voices received greater authority – veterinarians, animal technicians, moderate welfare advocates, and selected lay members. Such expansion of oversight and challenge to scientists' professional authority is consistent with other contentious areas of research, especially that on human embryos, where bioethicists became a new voice brought in to balance scientific interests.[80] More generally, the change is consistent with Thatcher's broader policies to curtail professional self-regulation, a less-remembered aspect of Thatcherism. In changing society and liberating markets, Thatcher also targeted the 'vested interests' that professions represented – especially those funded by the state.[81] Much of this broadening of authority and expertise happened in the implementation of ASPA, and in mechanisms and institutions that it enabled but did not predetermine, notably the empowered APC.

The political scientist and anti-vivisection activist Dan Lyons has argued that ASPA was created by 'an élitist policy community environment that is strategically selective in favour of "animal use" groups to the disadvantage of animal protection actors', abandoning any significant change to animal research for minor reform, in what he describes as 'dynamic conservatism'.[82] As far as the exclusion of anti-vivisectionist groups is concerned, archival evidence offers clear support for this narrative. The Home Office civil servants and the government sought to exclude any abolitionist voices from the outset and started from the presumed need to maintain animal experimentation, and indeed sought to marginalise radical groups by building a perceived consensus as the middle ground. While animal protection advocates of the 1970s pushed reform firmly onto the political agenda, through a combination of public campaigns and targeted lobbying, most were excluded from participating in its design. CRAE, seeking a moderate compromise, lost support of many activists in the late stages of deliberations, and its ability to represent the views of the animal protection groups was challenged.[83] Contrary to Lyons's assessment, however, the triple alliance was not a 'peripheral insider'.[84] Negotiations with the Alliance were a crucial component of the Home Office efforts, and the group had extensive access to civil servants and influence on the decision-making process, in order to forge a compromise position that the Home Office could bring to the Cabinet and MPs.

Was ASPA an example of dynamic conservatism? ASPA was not in any major way a Conservative project, but responded to external demands of animal protection groups, without embracing them. It remained permissive in its core, showing many continuities with the 1876 Act. Within the settlement, however, the dynamism mattered, and was built into the law and systems of its implementation. While not meeting demands of the more radical groups, and not ending dramatic public controversy over animal experiments, ASPA changed practice on the ground. It diversified expertise over animal welfare and care, extended the reach of regulation, and formalised internal enforcement at the level of the Inspectorate and licensing. While scientific voices constituted the majority of the APC in numbers, the Committee took a strong role in shaping decisions around welfare issues, defining key concerns for animal research since 1990. ASPA did not make provisions for the 3Rs, moves to increase openness, local institutional ethics review, or to encourage the 'culture of care', but it made these new concerns possible by giving space and authority to voices that championed these issues. Some of the following chapters demonstrate how these dynamics continue to play out (on the 3Rs, Gorman, Chapter 2; on cultures of care, Kirk, Chapter 5, Greenhough and Roe, Chapter 6, Message, Chapter 7; on new voices, Palmer, Chapter 10, and Davies, Gorman, and King, Chapter 11). Constructed as a self-consciously partial consensus position for specific political and pragmatic needs of the 1980s, ASPA created new legal commitments and moral obligations, both explicit and tacit.

Notes

1 Since 1993, ASPA has also protected *Octopus vulgaris*, but no procedures under the Act have been reported with the species.
2 Gail Davies et al., 'Animal Research Nexus: A New Approach to the Connections between Science, Health and Animal Welfare', *Medical Humanities*, 46.4 (2020), 499–511, DOI: 10.1136/medhum-2019-011778.
3 Clive Hollands, 'Achieving the Achievable: A Review of Animals in Politics', *Alternatives to Laboratory Animals*, 23 (1995), 33–38, DOI: 10.1177/026119299502300107; Richard Ryder, *Animal Revolution: Changing Attitudes Towards Speciesism*, 2nd edn (Oxford: Berg, 2000); Michael Balls, 'FRAME, Animal Experimentation and the Three Rs: Past, Present and Future', *Alternatives to Laboratory Animals*, 37 (2009), S1–6, DOI: 10.1177/026119290903702S02; Dan Lyons, *The Politics of Animal Experimentation* (London: Palgrave Macmillan, 2013).
4 Robert Garner, *Political Animals: Animal Protection Policies in Britain and the United States* (Basingstoke: Macmillan, 1998), pp. 176–201.
5 Robert G. W. Kirk and Dmitriy Myelnikov 'Governance, Expertise, and the "Culture of Care": The Changing Constitutions of Laboratory Animal Research in Britain, 1876–2000', *Studies in History and Philosophy of Science*, 93 (2022), 107–122, DOI: 10.1016/j.shpsa.2022.03.004.
6 Home Office, *Scientific Procedures on Living Animals* (Cmnd. 8883) (London: HMSO, 1983); *Scientific Procedures on Living Animals: Supplementary White Paper* (Cmnd. 9521) (London: HMSO, 1985).
7 Ryder, *Animal Revolution.*

8 See Richard French, *Antivivisection and Medical Science in Victorian Society* (Princeton, NJ: Princeton University Press, 1975); Hilda Kean, 'The "Smooth, Cool Men of Science": The Feminist and Socialist Response to Vivisection', *History Workshop Journal*, 40 (1995), 16–38, DOI: 10.1093/hwj/40.1.16; Shira Shmuely, 'Curare: The Poisoned Arrow that Entered the Laboratory and Sparked a Moral Debate', *Social History of Medicine*, 33 (2020), 881–897, DOI: 10.1093/shm/hky124; Tarquin Holmes and Carrie Friese, 'Making the Anaesthetised Animal into a Boundary Object: An Analysis of the 1875 Royal Commission on Vivisection', *History and Philosophy of the Life Sciences*, 42 (2020), 50, DOI: 10.1007/s40656-020-00344-9; Kirk and Myelnikov, 'Governance, Expertise, and the "Culture of Care".

9 Kevin Dolan, *Laboratory Animal Law: Legal Control of the Use of Animals in Research*, 2nd edn (Oxford: Blackwell, 2008).

10 Home Office, *Experiments on Living Animals: Return of the Experiments Performed under the Cruelty to Animals Act 1876, during 1972* (London: HMSO, 1973).

11 Shmuely, 'Curare: The Poisoned Arrow'; Kirk and Myelnikov, 'Governance, Expertise, and the "Culture of Care".

12 Home Office, *Report of the Departmental Committee on Experiments on Animals* (London: HMSO, 1965).

13 Robert G. W. Kirk, 'A Brave New Animal for a Brave New World: The British Laboratory Animals Bureau and the Constitution of International Standards of Laboratory Animal Production and Use, circa 1947–1968', *Isis*, 101 (2010), 62–94, DOI: 10.1086/652689.

14 A landmark collection stemming from a Royal Society symposium held in 1978 is *The Use of Alternatives in Drug Research*, ed. by Andrew N. Rowan and C. J. Stratman (London; Basingstoke: Macmillan, 1980).

15 *Animals, Men and Morals: An Inquiry into the Maltreatment of Non-Humans*, ed. by Stanley Godlovitch et al. (London: Victor Gollancz, 1971); see Robert Garner and Yewande Okuleye, *The Oxford Group and the Emergence of Animal Rights: An Intellectual History* (Oxford: Oxford University Press 2020).

16 Richard D. Ryder, *Victims of Science: The Use of Animals in Research* (London: HarperCollins, 1975). The British edition of Singer's book appeared as Peter Singer, *Animal Liberation: A New Ethics for our Treatment of Animals* (London: Jonathan Cape, 1976).

17 Mary Beith, 'Death for the 30-a-Day Dogs', *Sunday People*, 26 January, 1975; Dmitriy Myelnikov, 'The Smoking Beagles', *The Animal Research Nexus Blog* (2021), www.animalresearchnexus.org/blogs/smoking-beagles [accessed 25 September 2021].

18 Richard D. Ryder, 'Putting Animals into Politics', in *Animal Rights: The Changing Debate*, ed. by Robert Garner (London: Palgrave Macmillan, 1996), pp. 166–193.
19 Ryder, 'Putting Animals into Politics', p. 171.
20 The British Library: Richard Ryder (Ryder papers): C/V/8, Ryder to Houghton, 18 February 1977.
21 The Labour Party, *Living without Cruelty: Labour's Charter for Animal Protection* (London: The Labour Party, 1978).
22 Ryder, 'Putting Animals into Politics', p. 167.
23 Protection of Animals (Scientific Purposes) Bill, H. C. 10 1979–1980 (1979); Laboratory Animals Protection Bill, H. L. 240 1979–1980 (1980).
24 Home Office, *Scientific Procedures*.
25 The Council Directive 86/609/EEC came into effect in November 1986, the same year as ASPA, but the latter was far more extensive in its requirements and protections.
26 Home Office, *Supplementary White Paper*.
27 The LD50 test to estimate the toxicity of a substance, usually for regulatory purposes, by finding a concentration at which half of the animals in a sample died on administration of the substance. See Home Office, *Report on the LD50 Test* (London: Home Office, 1979).
28 People's History Museum, Manchester: Papers of Lord Douglas Houghton (Houghton papers): DHO/1/26 Minutes, CRAE meeting, 18 March 1981.
29 The National Archives of the UK (TNA): HO 285/179/2, 'Summary of comments on the White Paper', November 1983.
30 Chrissie Howting, interviewed by the author in October 2019.
31 Kirk and Myelnikov, 'Governance, Expertise, and the "Culture of Care"'.
32 Ryder papers: C-V-10, CRAE, 'Proposals for the change in the legislation ...' November 1979.
33 TNA: HO 285/182, Wall to Mellor, 22 February 1984.
34 David B. Morton and Peter H. M. Griffiths, 'Guidelines on the Recognition of Pain, Distress and Discomfort in Experimental Animals and an Hypothesis for Assessment', *Veterinary Record*, 116 (1985), pp. 431–436; TNA: HO 285/190, Morton to Richards, 26 July 1984.
35 Lyons, *The Politics of Animal Experimentation*, pp. 228–232.
36 'RSPCA: Minutes of the AEAC meeting held on 16 March 1977', Ryder papers: C/V/8.
37 Houghton papers, DHO/1/14/1, CRAE, 'Commonsense on Animal Experiments', 18 September 1984.

38 Home Office, *Supplementary White Paper*.
39 Houghton papers: DHO 1/14/1, 'Notes on BVA/CRAE/FRAME meeting held on Friday 24 January 1985'.
40 TNA: HO 285/206, Note of a Meeting Held on 11 December 1985.
41 HL Deb 16 January 1986, vol 469, cols. 1244–1245.
42 Home Office, *Supplementary White Paper*, p. 14.
43 HL Deb 16 January 1986, vol 469, cols. 1243–1244.
44 Home Office, *Guidance on the Operation of the Animals (Scientific Procedures) Act 1986* (London: HMSO, 1990), pp. 9, 52, 55.
45 Animals (Scientific Procedures) Act (Amendment) Order 1993 (no. 2103). London: HMSO; See Kirk and Myelnikov, 'Governance, Expertise, and the "Culture of Care". The Animal Procedures Committee, reconstituted in 1987, included Hollands (CRAE), Balls (FRAME), and Judith Hampson (CRAE/RSPCA).
46 TNA: HO 285/182, Johnson to Sutton, April 1984.
47 Kirk and Myelnikov, 'Governance, Expertise, and the "Culture of Care".
48 Richard North, 'The RSPCA. Fur Flies among the Animal Lovers', *Times*, 25 January 1982, p. 6.
49 TNA: HO 285/179/1, De Deney to Rawsthorne, 30 March 1983.
50 TNA: HO 285/182, Johnson to de Deney, 5 March 1984.
51 TNA: HO 285/182, 'Pain and Mr Hollands', April 1984.
52 TNA: HO 285/182, 'Pain and Mr Hollands', April 1984; Johnson to de Deney, 7 March 1984.
53 TNA: HO 285/182, Head to Johnson, 16 April 1984. See Ed Cape and Richard Young, Introduction to *Regulating Policing: The Police and Criminal Evidence Act 1984 Past, Present and Future*, ed. by Ed Cape and Richard Young (Oxford: Hart, 2008).
54 Home Office, *Guidance*.
55 Brian Harrison, 'The Rise, Fall and Rise of Political Consensus in Britain since 1940', *History*, 84 (1999), 301–324, DOI: 10.1111/1468-229X.00110; see also *The Myth of Consensus: New Views on British History, 1945–64*, ed. by Harriet Jones and Michael D. Kandiah (London: Palgrave Macmillan, 1996); David Edgerton, *The Rise and Fall of the British Nation* (London: Allen Lane, 2018), esp. pp. 442–464.
56 Harrison, 'The Rise, Fall and Rise of Political Consensus', pp. 307–308.
57 Harrison, 'The Rise, Fall and Rise of Political Consensus', p. 312.
58 TNA: PREM 19/1794, Hurd to Thatcher, 29 May 1986.
59 These were the Police and Criminal Evidence Act, the Data Protection Act, the Prevention of Terrorism (Temporary Provisions) Act, the Cable and Broadcasting Act, and the Repatriation of Prisoners Act, all of which passed in 1984.

60 TNA: HO 285/189, Brittan, 'Animal Legislation', 12 June 1984.
61 TNA: HO 285/189, Brittan, 'Animal Legislation', 12 June 1984.
62 TNA: PREM 19/1794, Pantling to Flesher, 27 February 1985.
63 TNA: HO 285/194, Brittan to Whitelaw, 10 June 1985.
64 TNA: HO 285/194, Fookes to Thatcher, 21 June 1985.
65 HC Deb 06 November 1985, vol. 86, cols. 3–5.
66 TNA: PREM 19/1794, Ibbs to Hurd, 19 May 1986.
67 TNA: PREM 19/1794, Hurd to Thatcher, 21 May 1986.
68 TNA: PREM 19/1794, Thatcher to Hurd, 29 May 1986.
69 Ryder papers, C-V-12, 'RSPCA Row', 27 November 1985.
70 Ryder papers: A-16, Singer to Ryder, 1 January 1986.
71 RSPCA Action Group (RAG), *RSPCA Tomorrow*, 1 (1985), p. 1.
72 For example, Jane A. Smith amd Kenneth Boyd, *Lives in Balance: The Ethics of Using Animals in Biomedical Research* (Oxford: Oxford University Press, 1991); The Boyd Group, *Ethical Review of Research Involving Animals: A Role for Institutional Ethics Committees?* Archived version, 1995, https://web.archive.org/web/20010630235406/www.boyd-group.demon.co.uk/ethicscomms.htm [accessed on 30 November 2020].
73 Ryder, 'Putting Animals into Politics'; Richard Ryder, interviewed by the author in July 2019.
74 'Adam', a retired senior animal technician, interviewed by the author in 2018.
75 Note of Meeting held at [redacted university] on 19 October 1989. Courtesy of Hugh Marriage.
76 Note of Meeting on 16 April 1991. Courtesy of Hugh Marriage.
77 TNA: HO 285/215, Evans to Alwen, 6 July 1984; Johnson to Richards, 24 June 1987.
78 There were designs to include octopuses and other cephalopod molluscs into the original ASPA, but the Home Office civil servants decided against it, on the grounds that moving away from the simple vertebrate definition might open time-consuming debates over scope, and with the idea that protections could be later extended, as they were. The Animals (Scientific Procedures) Act (Amendment) 1993; Kirk and Myelnikov, 'Governance, Expertise, and the "Culture of Care".
79 William M. S. Russell and Rex L. Burch, *The Principles of Humane Experimental Technique* (London: Methuen, 1959); Robert G. W. Kirk, 'Recovering *The Principles of Humane Experimental Technique*: The 3Rs and the Human Essence of Animal Research', *Science, Technology, & Human Values*, 43 (2018), 622–648, DOI: 10.1177/0162243917726579; Carrie Friese and Nathalie Nuyts,

'From *The Principles* to the Animals (Scientific Procedures) Act: A Commentary on How and Why the 3Rs Became Central to Laboratory Animal Governance in the UK', *Science, Technology, & Human Values*, 43 (2018), 742–747, DOI: 10.1177/0162243917743792.

80 Duncan Wilson, *The Making of British Bioethics* (Manchester: Manchester University Press, 2014), pp. 160–173; Jon Agar, *Science Policy Under Thatcher* (London: UCL Press, 2019), pp. 129–136; Kirk and Myelnikov, 'Governance, Expertise, and the "Culture of Care"'.

81 Brian Harrison, 'Mrs Thatcher and the Intellectuals', *Twentieth Century British History*, 5 (1994), 206–245, DOI: 10.1093/tcbh/5.2.206; Evan J. Evans, *Thatcher and Thatcherism*, 4th edn (London: Routledge, 2018), pp. 80–93.

82 Lyons, *The Politics of Animal Experimentation*, pp. 237, 239.

83 For example, Ryder papers, C-V-12, 'RSPCA Row', 27 November 1985.

84 Lyons, *The Politics of Animal Experimentation*, p. 320.

2

Outside of regulations, outside of imaginations: why is it challenging to care about horseshoe crabs?

Richard Gorman

Introduction

> If you've received a vaccine, used insulin injections, or had an IV in the hospital, you are a consumer of horseshoe crab blood. If you've ever vaccinated a pet, or had a pace maker, stent, or joint replacement implanted, you were also dependent on horseshoe crab blood. In one way or another, we are all consumers of horseshoe crab blood and all of us can play a role in conservation.[1]

The quote above from the Ecological Research and Development Group (a non-profit conservation organisation focused on horseshoe crabs) demonstrates how the biomedical use of these animals benefits huge swathes of the public every day. Moore suggests that 'probably every human since the 1970s has directly or indirectly benefited from horseshoe crab blood'.[2]

The Atlantic horseshoe crab (*Limulus Polyphemus*) is unlikely to be the first animal that people think of when they think about laboratory animals. Yet these enigmatic invertebrates are intricately entangled with the supply chains of modern health and medicine. When the blood of these animals encounters endotoxins, a coagulation process occurs, trapping and containing bacteria. For the animals, this is a defence mechanism against pathogens, a response developed to cope with 350 million years of living in an oceanic stew of bacteria. Endotoxins are bacterial components that can cause systemic toxicity if they enter the bloodstream. Testing for the presence of endotoxins is vital for the safe use of vaccines, injectable medicines, and medical devices in human and veterinary medicine. In North America and Europe, the primary method for endotoxin testing is currently the Limulus Amoebocyte Lysate (LAL) test, a critical component of which is the blood of Atlantic horseshoe crabs. Procuring the raw materials for this testing involves collecting and bleeding over 500,000 crabs from wild populations each year as they breed on the shores of the north-eastern US each spring. Efforts are made by manufacturers to return the crabs to the sea alive following the collection of their blood. However, there are increasing discussions about the impact that capture and bleeding can have on crab health and mortality, along with debates around horseshoe crab sentience and capacity to suffer.[3]

Despite the global reliance on this ancient species, the pharmaceutical utilisation of horseshoe crabs is rarely viewed through the ethical, regulatory, and conceptual frameworks that shape and manage the scientific use of animals more broadly. Frameworks such as the 3Rs – the ambition to reduce, refine, and, where possible, replace the use of animals – are now globally established and widely accepted as the best framework for governing animal-dependent science. The 3Rs have become central to how the use of animals in science is socially understood, politically imagined, and (inter)nationally regulated.[4] In the UK the principles of replacement, reduction, and refinement are embedded within the Animals (Scientific Procedures) Act 1986 (ASPA). This Act requires that research involving the use of animals demonstrates the application and implementation of the 3Rs principles. That is, that animals are replaced with non-animal alternatives wherever possible, that the number of animals is reduced to the minimum needed to achieve the results sought, and that, for those animals which must be used, procedures are refined as much as possible to minimise suffering. The 3Rs exist as a means of weaving together 'good science, good care, and socially acceptable practices'.[5] Yet, the further the 3Rs have become established, the greater the interest in their potential has become. Rather than just principles that should be applied within 'experiments', the 3Rs can increasingly be interpreted as an ethical proposition that might be extended to encompass new species and new spaces (an alternative argument might contest this as worrying conceptual drift). There is a need to be aware that there are different cultural norms to how the 3Rs are performed. For example,

Hobson-West has shown how the 3Rs are conceptualised in different ways by different stakeholders – a route to managing ethical dilemmas, as part of good scientific practice, or as a politically strategic means of achieving consensus in controversial domains.[6]

In the case of horseshoe crabs, the interest in the 3Rs emerges shaped by the professional identities and wider cultures at play within the global pharmaceutical sector. Other ethical frameworks are obviously available (and indeed may have more relevance for a 'wild' animal), but it is the 3Rs centrality in scientific imaginaries that have led it to beginning to be adopted by industry stakeholders searching for a way to initiate conversations about care, welfare, and alternatives. The 3Rs has a level of familiarity, it is deployed in other aspects of corporate work on animals – and indeed, celebrated. It provides a language and framework to grasp for more ethical modalities of relationship with horseshoe crabs. One that is, importantly, accessible in imagining its applicability – at least, on an abstract level. This is important when the specifics of horseshoe crab welfare can be hard to relate to, and when people are greatly removed from day-to-day encounters with the animals. At the same time, an interviewee from the biotechnology sector explained, 'an increased philosophical approach to corporate sustainability has also contributed to more recent questions around the use of horseshoe crabs' (interview, 2020). There have long been calls for a 4th R of 'responsibility'.[7] More recently McLeod and Hartley have suggested that the governance of animal-dependent science could be enhanced by examining the 3Rs through alignment with the Responsible Research and Innovation agenda.[8] However, the adoption (or co-option) of the 3Rs by the logics of 'corporate-social-responsibility', a form of self-regulation amongst international businesses, raises new questions and potential directions of travel for the concept that has been so influential in changing attitudes, behaviours, and cultures of animal research.

Different stakeholders see different value and possibilities in each of the individual 'Rs', to the point of substantial friction between those who advocate focus on 'replacement' above 'reduction'.[9] Thinking about horseshoe crabs through the 3Rs is not easy. This chapter explores why horseshoe crabs frequently fall outside of current regulations and social imaginations, and the challenges that this exclusion presents to those who want to make arguments about more sustainable approaches to endotoxin testing.

How did we end up bleeding horseshoe crabs?

Richard Pfeiffer, the German bacteriologist, is credited with introducing the concept of endotoxin in the late nineteenth century.[10] We now know that endotoxins, synonymously known as lipopolysaccharides, are molecules found as part of the outer membrane of Gram-negative bacteria,[11] and are released upon bacterial cell lysis. Endotoxins are ubiquitous in the natural environment but can become pyrogenic when released into the bloodstream or other tissue where they are not usually found, producing a complex pattern of systemic toxicity in mammals that ranges from fever to life threatening effects such as hypotension and shock.[12] It was such endotoxins that complicated the early use of intravenous therapies, which routinely included unpredictable 'injection fever' side effects. Efforts to remedy these reactions saw initial experiments with rabbits as a predictive fever model in 1912 by Hort and Penfold[13] – a model later built on by Bourn and Seibert in 1925 who demonstrated that measuring the febrile response of rabbits could provide a means of monitoring contaminated intravenous solution.[14] The high demand for intravenous solutions during the Second World War led to increased use of the test and of the adoption of the Rabbit Pyrogen Test (RPT) as a regulatory requirement by the US Pharmacopeia in 1942.[15] The RPT involved clear demands on resources – in the form of animals, facilities, and trained personnel.

Interest in horseshoe crabs began to grow towards the end of the nineteenth century. Horseshoe crabs were large, readily available, and easy to maintain in laboratory aquaria.[16] The establishment of the Marine Biological Laboratory in 1888 at Woods Hole, Massachusetts, close to the breeding grounds of the Atlantic horseshoe crab, inspired a variety of early studies on the species which was 'soon recognised as an animal well suited to morphological and physiological research'.[17]

Between 1950 and 1951, Frederik Bang injected various bacteria into the circulatory system of horseshoe crabs. Bang had speculated that species of ancient origin might reveal primitive immunological functions.[18] He found that Gram-negative bacteria produced intravascular clotting in horseshoe crabs – but Gram-positive bacteria did not.[19] Bang realised that further investigation by a haematologist might provoke a productive collaboration, and Jack Levin joined Bang at Woods Hole. Levin himself reflects that at that point he had 'never previously heard of, much less seen, a horseshoe crab'.[20] Their initial experiments together proved difficult, as samples of horseshoe crab blood would often clot spontaneously. Levin recalls that: 'samples of blood which were liquid when I left the laboratory in the evening were solidly clotted by the next morning'.[21] However, their collaborations would reveal that endotoxin was the key factor in the clotting of horseshoe crab blood, and by 1968, Levin and Bang had recognised that the sensitivity of the system would make a highly applicable method for assaying bacterial endotoxin, with the reaction being able to be adapted to a convenient *in vitro* test.[22]

The 'replacement' of the RPT by LAL as the 'gold-standard' for endotoxin detection involved a lengthy and complex process. Industry acceptance and adoption was slow, with many companies having years of experience with the RPT. The RPT, though expensive, was familiar, and viewed as more straightforward. There was also fear that LAL, through its greater sensitivity, would result in a greater rejection of products, or introduce a greater threshold of regulation.[23] Large studies were needed to document efficacy of LAL. Replacement involved a slow, phased approach, which saw data and confidence in LAL accumulate.[24] Draft guidelines for the utilisation of the LAL test as an end-product testing method for endotoxins in human and veterinary injectable drug products were published in 1980 by the US Food and Drug Administration (FDA), though it was not until 1987 that these were finalised.

Of course, concern about crabs is not new. Attempts to embed a regulatory expectation of care for horseshoe crabs can be witnessed as early as 1978 with the formal introduction of regulations that aimed to provide additional standards governing the manufacture of LAL – in what became known as the 'return to the sea' policy. The additional standards stated that 'to guarantee that the manufacture of LAL will not have an adverse impact on existing crab populations, the horseshoe crabs shall be returned alive to their natural environment after a single collection of their blood'. Additionally, the regulatory note stressed the importance that the horseshoe crabs 'from which blood is collected for production of the lysate, shall be handled in a manner so as to minimise injury to each crab'.[25] This policy was written into the Federal Register from 1978 until 1996, acting as part of the licensing requirements to be a manufacturer of LAL, until it was rescinded as part of the Clinton administration's 'Reinventing Government' reforms – though remains honoured and considered best practice by many biomedical companies today.[26] While framed at ensuring the sustainability of supply chains, the introduction of this policy was a significant move in representing initial expectations amongst policy-makers for a 'culture of care' for horseshoe crabs. For some, however, this was not enough, and in a 1980 issue of the Federal Register the FDA noted they had received a letter requesting 'a restriction be imposed on the collection and bleeding of the crabs during their spawning seasons'[27] – an early indicator of concern for crabs and the impact of this new technology on horseshoe crab populations. The FDA's response was a blunt rejection of this idea, despite acknowledging that 'at present little is known about the effects of bleeding on crabs returned to the wild' (a situation many conservationists might argue remains to this day).[28]

Making connections visible through collaborative social scientific research

Our 'nexus' approach involves using collaborative research to open up new ways of thinking about the relations, regulatory logics, and social categories that constitute and co-produce contemporary policies and practices of animal use.[29] Controversies over animal-use do not only take place within a scientific context; they involve other stakeholders and forms of expertise: policy-makers, industry, media, and members of relevant communities.[30] Language, history, and narratives shape what regulations are, what they do, and their potential for change. Discourses, or socially organised frameworks of meaning, define categories and specify domains of what can be said and done.[31] Discourses govern the way that a topic can be meaningfully talked about, and hence, how ideas are put into practice and used in the production of regulations.[32] Thus, the way in which horseshoe crabs are 'discursively positioned' has important implications for ambitions towards cultivating a level of welfare for these animals, currently invisible and outside of many regulations and imaginations of care. This also requires a historical perspective to see how debates, policy, and practice have been shaped over time, and what is constructed and erased in the telling of history. Many of the discourses and narratives produced to enable and affirm the continued collection of horseshoe crab blood rely on historicised comparisons and justifications.

This research derives from a collaboration with the Royal Society for the Protection of Animals (RSPCA). The RSPCA has an interest in understanding the processes by which humane alternatives are developed, validated, and accepted, so that they can be aware of, and work to resolve, obstacles to implementation that may be due to factors such as perceptions of risk and resistance to change. As a long-term and respected actor within policy debates about the biomedical use of animals, the RSPCA were approached by a number of stakeholders and asked to respond to regulatory consultations about alternatives to the LAL test.

A synthetic substitute to horseshoe crab blood was introduced in 2001 – laboratory-synthesised genetically engineered recombinant Factor C (rFC), becoming commercially available in 2003. Initial uptake of this replacement was extremely limited due to the availability and market-dominance of the LAL test, combined with concerns about a single-source and supply of the synthetic, cautions over the validation of the alternative, and a lack of regulatory requirements to consider alternatives to testing in non-vertebrates. More recently, there has been a renewed attention on replacements to the LAL test, emerging as a result of the aforementioned increasing concerns relating to the impact on horseshoe crab populations, and as recombinant reagents have become commercially available from multiple manufacturers. With some of the manufacturers of rFC new players keen to enter the endotoxin testing market, a cynical perspective might attribute the surge in interest in the 3Rs here to a level of commercialism and marketeering. Several reviews of the performance of rFC as an endotoxin detection method suggest it is equivalent to, or better than, LAL in terms of the ability to detect and quantifiably measure bacterial endotoxin.[33] Others, however, have been less positive about the potential to move to this alternative on a routine or commercial basis, citing concerns about the ability of the alternative to achieve adequate specificity.[34] The topic of alternatives here has generated much discussion, with debates becoming increasingly polarised. Given these conflicting views and interests circulating around this topic, the RSPCA felt that an approach informed by social science could be helpful in understanding the drivers for, and barriers to, adopting alternatives to animal-derived endotoxin assays.

In practice, this involved documentary and policy analysis alongside thirteen empirical interviews across the broad spectrum of international groups with a stake in the biomedical use of horseshoe crabs: manufacturers, biotechnology companies, regulators, pharmaceutical scientists, conservationists, animal-welfare groups, and academic researchers. The research was reviewed and approved by the University of Exeter's Geography Ethics Committee. Participants provided informed consent to participate in this study and share their experiences. Analysis involved a thematic approach, comparing key connections and questions across different perspectives in order to make visible the challenges and opportunities that exist for change regarding the biomedical use of horseshoe crab. Through examining different perspectives about why the use of the horseshoe crab has remained outside of, and resistant to, an engagement with the 3Rs, we can begin to make suggestions as to what frameworks might help to enable better conversations about more sustainable approaches to endotoxin testing. Suggestions for discussing the future use of horseshoe crabs in terms of the 3Rs were bought together at the end of the research in a sector report[35] and summary infographic created in collaboration with the RSPCA (Figure 2.1). However, questions around the history, species, harms, and location of these procedures outlined below demonstrate the ongoing challenges of changing this conversation.

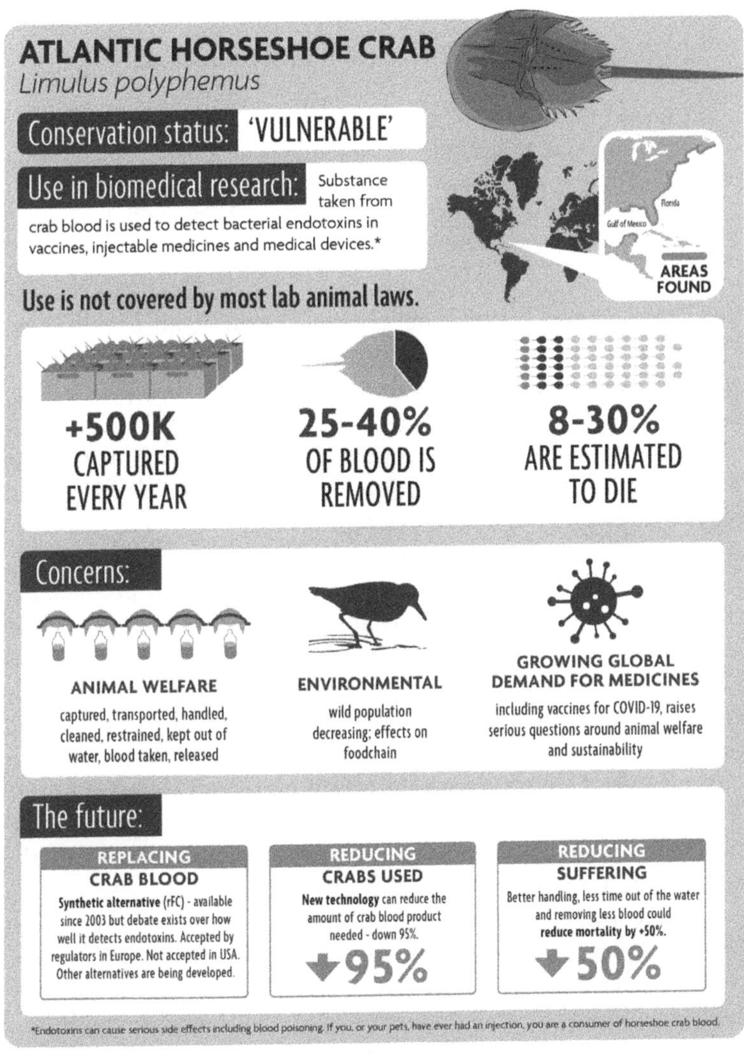

We believe horseshoe crabs can suffer and want to see genuine commitment to replacing their use in endotoxin tests and to conserving them in the wild.

Find out more at: tinyurl.com/HorseshoeCrabReport

Figure 2.1 Horseshoe crabs and the 3Rs, an infographic (Copyright: RSPCA/AnNex. Source: https://animalresearchnexus.org/index.php/publications/horseshoe-crab-use-biomedical-research).

Horseshoe crabs: a successful *in vitro* alternative?

> We knew obviously that the lysate test was a move away from the rabbit pyrogen test, that had been the requirement beforehand, so it was seen as less of an animal test than perhaps the rabbit pyrogen test and other whole animal testing, shall we say. (Pharmaceutical sector employee, interview, 2020)

LAL is commonly understood and positioned as an *in vitro* test, serving as a 'replacement' in itself as an alternative to the *in vivo* RPT, the previous regulatory standard for pyrogen testing. Hartung, for example, describes LAL as 'a historic breakthrough in the replacement of animal experiments'.[36] Similarly, Flint suggests that 'the LAL test, fulfils the objectives of reducing, refining and replacing animals in medical experiments'.[37]

The move to LAL for bacterial endotoxin testing certainly enabled the phasing out of large colonies of rabbits, producing a significant reduction in harm to animals – and to the technicians responsible for injecting the rabbits with samples. However, as Zhang notes, 'the LAL test still required the use of animals, but the grisly process of sticking needles into animals became hidden and outsourced to a different part of the supply chain'.[38] Essentially, an offshoring and outsourcing that causes crabs to disappear – both from laboratory spaces, but also within the types of reporting, statistics, and ultimately, regulatory responsibility, associated with the biomedical usage of animals.

Despite its later celebration as a 'unique example of a successful *in vitro* replacement of an animal toxicity test',[39] the LAL test was 'not discovered as a result of a directed effort to develop an alternative to the rabbit pyrogenicity test'.[40] This contrasts with how LAL is marketed by many companies today. For example, Charles River Laboratories argue: 'The *in vitro* Limulus Amoebocyte Lysate (LAL) test was developed in the 1970s, as part of a 3Rs initiative to replace the need to use hundreds of thousands of rabbits each year in the testing of human and animal products for contamination'.[41] Indeed, an interviewee from the biotechnology sector remarked:

> Of course, one of the social arguments was, 'We're going to replace a rabbit, a warm blooded furry thing with a lab test', which was great and that had a certain impact on the acceptance of this test. But behind the scenes, it was a whole different story, people were not that much concerned with replacing a rabbit. (Biotechnology sector employee, interview, 2020)

Moves to discursively position LAL as a 3Rs solution to the RPT only began to emerge around the time that concerns about the impact of LAL on horseshoe crab populations started to gain more attention. While the LAL test was frequently referred to as an 'Alternative for Rabbit Pyrogen Test' whenever referenced during the slow journey from inception through to regulatory approval, this use of 'alternative' is not embedded in discussions about animal welfare. For example, the 1977 Federal Register notice recognised that LAL was more economical, required a small amount of product, and enabled a larger number of tests to be performed by a single technician.[42] Notably absent in descriptions of advantages are the potential 3Rs benefits that a move to LAL testing could have in terms of reducing the numbers of rabbits used. The LAL test appears to be a victim of a confusing conceptual crossover between two different renderings of 'alternative'. It is referenced initially as an alternative to the rabbit test in the sense of another possibility or choice of methodology, while later reference to LAL as an alternative appears to have segued into the more specific sense of methods that replace, reduce, or refine animal use in specific procedures.

This discursive categorisation positions the LAL test as a 3Rs solution, rather than a problem requiring further attention from the 3Rs. Its very status as having 'replaced' the rabbit test acts to produce a level of confident, comfortable, assurance surrounding it being an ethical choice. The tendency to refer to this transition as a 'replacement' is challenging in that the test still involves the use of cells taken from animals. LAL exists in the category of what Herrmann and Jayne refer to as 'alternatives that still exploit animals'.[43] The discursive framing affirming LAL as an *in vitro* alternative makes the ongoing role that live animals play in its production less visible. Of course, this depends on what is taken to constitute an 'animal' here. At present, horseshoe crabs are outside of the scope of most formal legislation regulating animal use and are not considered a 'protected animal'. Positioning LAL as an, or even *the* alternative, in this space limits capacities for engagement with the 3Rs. Indeed, framing the use of horseshoe crabs as 'replacement' can undermine how the concept of replacement is publicly represented.

Horseshoe crabs: wild, strange, and unregulated invertebrates?

> I was speaking to a colleague, and one of her comments was that she was okay with lab rabbits but she was more uncomfortable with the wild animal aspect. (Pharmaceutical sector employee, interview, 2020)

How people think about the ethics of animal care differs depending on spatial contexts.[44] That is, what publics expect, and think of as acceptable practices, varies according to the imagined spaces in which people 'place' animals. Crabs' positioning as 'wild' within human classifications and imaginations confuses the ethics that are invoked in our utilisation of them. The welfare of wild animals is comparably neglected when contrasted with the focus applied to ensuring the welfare of captive animals, particularly laboratory animals.[45] Concern for wildlife is often expressed through a focus on preventing death rather than attempts at improving welfare directly.[46] The idea that a framework like the 3Rs might apply in the context of wild animals is new, and while there is some discussion, this is often framed under the auspices of 'research' and 'field studies' on wild animals.[47] And again, horseshoe crabs fall through the gaps.

> Another thing that complicates it, or at least has not been in favour of applying the 3Rs, is the species. Horseshoe crabs are invertebrates, so they are not really covered by animal laws. (Biotechnology sector employee, interview, 2020)

Renderings of the 3Rs frequently involve the idea of replacing the use of a 'protected' animal with another species – often an invertebrate – and as such, the LAL test can appear a solid example of the 3Rs' success. Invertebrates are generally exempt from formal animal welfare legislation (with the recent exception of cephalopods), though there is growing interest amongst animal welfare and protection groups and broader publics in campaigning for the humane treatment of crabs, lobsters, and other decapod crustaceans.[48] However, while horseshoe crabs might resemble crustacea, they actually belong to a separate subphylum of arthropods (*Xiphosura*), making their place within law, policy, and regulatory imaginations complex.

The horseshoe crab is an example of how concepts like the 3Rs can be used as a way of reinforcing the status quo of animal use, rather than a means of affecting change. It also demonstrates how what is included and excluded from regulation and protection is intricately linked to social and economic relations of power.

> [Company] has gone head on against the 3Rs, to inform their customer base, which is global and huge, that the utilisation of horseshoe crabs for LAL is not in conflict with the 3Rs because the way the ethical rules are written, it doesn't include arthropods. (Conservation sector employee, interview, 2020)

Corporate secrecy plays a large role in limiting the development and uptake of 3R approaches for horseshoe crabs in endotoxin testing. While openness and transparency about the use of animals within the biomedical industry is increasingly a staple of best practice, this has not permeated into discussions about crab use. An interviewee from the pharmaceutical sector concluded, 'perhaps the improvements or the changes haven't been well publicised, they're perhaps their own worst enemy, I don't know' (interview, 2020). The lack of transparency here exacerbates efforts to cultivate – or even question – crab welfare, with few opportunities to tell the stories about care that are crucial to maintaining the social contracts that underpin the biomedical use of animals.[49]

Yet also, by not talking about crabs, they remain invisible, outside of imaginations, and thus, unlikely to be the focus of public concern. Attitudes towards invertebrates are complex. One interviewee from the biotechnology sector explained the 'alien-ness' of horseshoe crabs: 'It's a scary thing, it's an unknown, it always freaks people out' (interview, 2020). Yet another interviewee, working in the communications sector contested this, 'There's a kind of misnomer that people aren't interested in things that aren't cute and fluffy and yet actually, when you start showing pictures and talking to them about it, it's like "wow this creature's really funky" and you get them interested that way' (interview, 2020). Social expectations of care are not just for those animals we find deeply familiar or appealing, but also for enigmatic invertebrates like horseshoe crabs.

What does it mean to bleed a horseshoe crab?

> They're gently folded and you'll see the hinge on the back that exposes the cardiac membrane, that is sanitised with alcohol and then it's just like bleeding people. (Biotechnology sector employee, interview, 2020)

Intertwined with stances on whether invertebrates can feel pain, there is an oft-cited idea that 'bleeding the crabs does not appear to harm them in any way'.[50] As Moore argues, there is a concerted effort by industry to position the bleeding of crabs as easy, harmless, and quick.[51] The framing of the collection of horseshoe crab blood as harmless, non-invasive, and comparable to human blood donation introduces ambiguity and uncertainty around the priority of this area as a 3Rs issue.

> Sourcing the LAL is not an animal trial so therefore, it has been somewhat excluded from the 3Rs concept or only picked up by very few companies. (Biotechnology sector employee, interview, 2020)

The idea of the 'experiment' is central to the 3Rs concept, and to the wider regulatory frameworks that govern the scientific use of animals.[52] As Palmer et al. have argued, the borders of regulated animal research are, in part, maintained by classificatory decisions about what 'counts as' science.[53] As the quote above explains, horseshoe crabs are left in a rather liminal space in this regard.

IACUC-approved (Institutional Animal Care and Use Committee) protocols are required when animals are used for research, biological testing, or research training. Whether horseshoe crabs are 'used' for biological testing is a matter of interpretation; certainly, they are not used *in vivo*, as discussed earlier, but acquiring their blood is vital in underpinning endotoxin testing. Similarly, the EU's Directive 2010/63/EU asserts that 'wherever possible, a scientifically satisfactory method or testing strategy, not entailing the use of live animals, shall be used instead of a procedure'. There is obvious room for confusion and contestation here, about whether LAL 'entails the use of live animals'. Definitional angst abounds about what counts as 'use', 'science', a 'procedure', and even 'an animal'.

This obscurity is commonplace in the case of animal-derived reagents, where, as Gray et al. note, 'the animal use is buried several layers deep in the production process'.[54] One interviewee from the biotechnology sector described the challenge of trying to get users of LAL 'to at least understand that this is not manna from heaven, this is derived from a living animal' (interview, 2020). Possibilities for relationships with the crabs whose blood forms the basis of the reagent are restricted, with many of those who use the lysate unaware of its animal origins. Similarly, another interviewee from the biotechnology sector confidently stated, 'I would say it's probably not even irrational to say that 95% of the people using LAL don't know its origin and that's probably our fault for not educating them' (interview, 2020). There are few opportunities to drive change when faced with the materiality of a vial of reagent rather than an animal. Despite LAL being used in laboratories and pharmaceutical sites around the world, crabs' presence in the spaces of pharmaceutical science is always at a distance, and invisible.

A fishery, not a laboratory?

> The Atlantic States Marine Fishery Commission, they're the regulatory body around the horseshoe crab. (Pharmaceutical sector employee, interview, 2020)

Despite their pharmaceutical use, as a wild-caught species, horseshoe crabs are most often imagined through the regulatory lens as a fishery, rather than a laboratory species, and managed with a similar set of ethics to other fisheries. Collection is controlled by a patchwork of intra and interstate regulations, but there is great variability in enforcement.[55] Being conceptualised as a fishery is important, as the spatial imaginations invoked can alter both the practice and representation of science, and the ensuing social expectations and public accountabilities.

Fisheries management plans have their own vocabularies, concepts, and priorities. Horseshoe crab use is discussed in terms of 'maximum sustainable yields' and 'mortality rates' rather than the language of 'replacing, reducing, and refining' or 'harm–benefit analysis' that underpins (and indeed, enables) the laboratory use of many other species. This demonstrates the ways in which different regulatory regimes act to configure ways of practising and performing care and welfare. Horseshoe crab care is primarily considered through a productionist lens of farming. It is perhaps a crude measure, but the phrase 'welfare' does not appear once in the Atlantic States Marine Fishery Commission's 271-page '2019 Horseshoe Crab Benchmark Stock Assessment and Peer Review Report'.[56] Similarly, much of the regulation of horseshoe crabs has been driven by 'pressure from conservation groups'.[57] This is in contrast to much of the policy and regulation around the biomedical use of animals, which is more commonly associated with pressure from animal welfare or animal rights groups. As a result, a different ethic is produced, one that is concerned with a species level vulnerability and sustainability, rather than focused on the experiences of the individual crab.

There is also something of an artificial separation of regulation, with horseshoe crabs themselves managed as a fishery, but the product of their blood, LAL, still firmly entrenched within the regulatory systems of various national pharmacopeias and in the US, the FDA. This subtle division – between the animal source and the animal-derived product – creates a level of 'distanciation', encouraging focus not on the animal, but rather on the substances it yields.[58] The disconnect prompts questions as to who should be doing the work of the 3Rs here – the companies who bleed horseshoe crabs or the end-users who utilise the product?

The 'biomedical fishery' is also not the only fishery of horseshoe crabs. There is also a parallel and larger 'bait fishery', which uses horseshoe crabs to catch eel and conch (whelk). The two fisheries are often juxtaposed as being in competition, with occasionally moments of collaboration. The multiple use of horseshoe crabs makes regulation – and establishing ethical positions – complex. The bait fishery is also used for comparison in discussions about the biomedical use of crabs. Arguments focus on the much larger number of crabs utilised as bait (994,491 harvested in 2017), a process that involves a 100% mortality rate. Indeed, many within the biomedical sector feel aggrieved about the attention that their use of crabs attracts in contrast to the practices of the bait industry:

> And the bait industry doesn't get any media attention, that takes a million crabs and chops them up every year. It's like can the bait industry reduce their reliance first? (Pharmaceutical sector employee, interview, 2020)

To confound things further, the biomedical use of horseshoe crabs is frequently positioned as being 'good' for the crabs themselves. The Atlantic States Marine Fishery Commission report argues that 'the biomedical community has had a positive impact on HSC [horseshoe crab] populations through 45 years of consistent conservation practices'.[59] Conservation effort is often positioned in direct contrast and opposition to the 3Rs. For example, Charles River Laboratories, a manufacturer of LAL, suggest on their website that:

> Many companies think that moving away from LAL to reduce the use of the HSC will help them comply with 3R principles. This idea should be carefully evaluated, as moving to rFC could produce counterproductive effects by endangering the conservation efforts … Without the need for LAL in biomedical use, the legal protection of the horseshoe crab is not guaranteed in the future, and they would again fall prey to overfishing and use as bait.[60]

This invokes the concern that a turn to synthetic alternatives might, somewhat ironically, actually result in more harm to horseshoe crab populations. Rather than being a high-value 'catch and release' asset within the biomedical economy, the rise of alternatives may shift the crab's status as a commodity further into the realms of fishing bait.

> There's one company in particular that spends an awful lot of marketing energy trying to convince the world that if it were not for them, there would be no horseshoe crab conservation. (Conservation sector employee, interview, 2020)

The idea that horseshoe crabs are only afforded protection and conservation by an ongoing exploitation of the species is one which will rankle with many stakeholders and publics. Yet it is also a message that has purchase. Interviewees within pharmaceutical settings reported being reassured by the messaging around biomedical companies' investment in conservation. Conservation here becomes a way of practising care, performing stewardship, and offsetting harms to some crabs through providing affordances to the species at large. In some ways, conservation efforts here perform the same role that the 3Rs do in laboratory animal research at large (enabling the continued social acceptability based on assurances around minimisation of harms to animals) and thus, it could be argued, have negated the need for a more specific 3Rs focus.

Conclusion

The case of horseshoe crabs highlights how the ways in which animals are discursively positioned within regulatory logics and social imaginations can limit efforts for cultivating, or even questioning, animal welfare. Questions around the history, species, harms, and geographies of these procedures demonstrate the ongoing challenges of changing this conversation.

While there may be a broad social acceptability for the laboratory usage of animals, this is conditional on steps being taken to minimise any harms to animals.[61] Without these assurances, and openness, ideas about socially acceptable scientific practices involving animals are liable to change. Importantly, the assurances being sought here are around how horseshoe crabs are treated within the laboratory specifically, not the wider species context and conservation work that is enabled by the use of animals, but exactly how welfare is ensured within their use itself.

Horseshoe crabs, and the LAL that is produced from their blood, are a reminder that concepts like the 3Rs can become normative, naturalising the biomedical use of certain animals through historicised discourses and narratives of replacement. For example, Blechová and Pivodová suggest that 'one of the most important aspects' of the LAL test is that it 'is in accordance with the latest demand of the European Pharmacopoeia Commission for the replacement of the animal-based tests in favour of alternative methods'.[62] The crux here is at what point does an alternative stop being an alternative, and simply become the new normal? When are alternatives needed to the alternative? Central to this is whether 3R objectives are viewed as a continual process, and what animals 'count' within an ethical framework like the 3Rs.

Notes

1 Ecological Research & Development Group, 'Consumers of LAL/TAL and Their Role in Conservation', 2013, www.horseshoecrab.org/med/consumers.html [accessed 21 August 2021].

2 Lisa Jean Moore, *Catch and Release* (New York: New York University Press, 2017), p. 104, DOI: 10.18574/nyu/9781479876303.001.0001.

3 Rebecca L. Anderson et al., 'Sublethal Behavioral and Physiological Effects of the Biomedical Bleeding Process on the American Horseshoe Crab, *Limulus Polyphemus*', *The Biological Bulletin*, 225.3 (2013), 137–151, DOI: 10.1086/BBLv225n3p137; Tom Maloney et al., 'Saving the Horseshoe Crab: A Synthetic Alternative to Horseshoe Crab Blood for Endotoxin Detection', *PLOS Biology*, 16.10 (2018), e2006607, DOI: 10.1371/journal.pbio.2006607.

4 Gail Davies et al., 'Science, Culture, and Care in Laboratory Animal Research: Interdisciplinary Perspectives on the History and Future of the 3Rs', *Science, Technology, & Human Values*, 43.4 (2018), 603–621, DOI: 10.1177/016224391875703.

5 Davies et al., 'Science, Culture, and Care in Laboratory Animal Research', p. 607.

6 Pru Hobson-West, 'What Kind of Animal is the "Three Rs"?' *Alternatives to Laboratory Animals: ATLA*, 37. Suppl 2 (2009), 95–99, DOI: 10.1177/026119290903702s11.

7 R. E. Banks, 'The 4th R of Research', *Contemporary Topics in Laboratory Animal Science*, 34.1 (1995), 50–51.

8 Carmen McLeod and Sarah Hartley, 'Responsibility and Laboratory Animal Research Governance', *Science, Technology, & Human Values*, 43.4 (2017), DOI: 10.1177/0162243917727866.

9 Richard Gorman, 'Atlantic Horseshoe Crabs and Endotoxin Testing: Perspectives on Alternatives, Sustainable Methods, & the 3Rs

(Replacement, Reduction and Refinement)', *Frontiers in Marine Science*, 7 (2020), DOI: 10.3389/fmars.2020.582132.

10 O. Westphal et al., "The Story of Bacterial Endotoxin', in *Advances in Immunopharmacology*, ed. by L. Chedid et al. (Oxford: Pergamon Press, 1985), pp. 13–34, DOI: 10.1016/B978-0-08-032008-3.50005-9.

11 Gram-negative bacteria are a type of bacteria found in many different environments. Some Gram-negative bacteria are harmless, but others can cause serious infections. Their cell wall structure makes Gram-negative bacteria more resistant to antibiotics and other treatments, meaning they can cause significant public health problems. If something disturbs the cell wall of Gram-negative bacteria, they can release endotoxins.

12 Jack Levin et al., 'Clotting Cells and Limulus Amebocyte Lysate: An Amazing Analytical Tool', in *The American Horseshoe Crab*, ed. by C. N. Shuster et al. (Cambridge, MA: Harvard University Press, 2003), pp. 154–188.

13 E. C. Hort and W. J. Penfold, 'Microorganisms and Their Relation to Fever', *Journal of Hygiene*, 12.3 (1912), 361–390, DOI: 10.1017/Fs0022172400005052.

14 Janet M. Bourn and Florence B. Seibert, 'The Cause of Many Febrile Reactions Following Intravenous Injections. I', *American Journal of Physiology*, 71.3 (1925), 621–651, DOI: 10.1152/ajplegacy.1925.71.3.652.

15 Caroline Vipond et al., 'Limitations of the Rabbit Pyrogen Test for Assessing Meningococcal OMV Based Vaccines', *ALTEX* (2016), 47–53, DOI: 10.14573/altex.1509291.

16 Louis Leibovitz and G. A. Lewbart, 'Diseases and Symbionts: Vulnerability despite Tough Shells', in *The American Horseshoe Crab*, ed. by C. N. Shuster et al. (Cambridge, MA: Harvard University Press, 2003), pp. 245–275.

17 Levin et al., 'Clotting Cells and Limulus Amebocyte Lysate', p. 311.

18 Levin et al., 'Clotting Cells and Limulus Amebocyte Lysate'.

19 Frederick B. Bang, 'The Toxic Effect of a Marine Bacterium on Limulus and the Formation of Blood Clots', *Biological Bulletin*, 105.2 (1953), 361–362.

20 Jack Levin, 'Discovery and Early Development of the Limulus Test', in *Endotoxin Detection and Control in Pharma, Limulus, and Mammalian Systems*, ed. by Kevin L. Williams (Cham: Springer International Publishing, 2019), p. 4.

21 Levin, 'Discovery and Early Development of the Limulus Test', p. 4.

22 Levin et al., 'Clotting Cells and Limulus Amebocyte Lysate'.

23 Levin et al., 'Clotting Cells and Limulus Amebocyte Lysate'.
24 For an incredibly rich history of the development and commercialisation of LAL, see Levin et al., 'Clotting Cells and Limulus Amebocyte Lysate'.
25 National Archives and Records Administration, 'Federal Register: 43 Fed. Reg. 35645 (Aug. 11, 1978)', Library of Congress, Washington, D.C. 20540 USA, 1978, pp. 35732–35733.
26 Atlantic States Marine Fisheries Commission, *Proceedings of the Atlantic States Marine Fisheries Commission Horseshoe Crab Management Board*, 2009, www.asmfc.org/uploads/file//may09proceedings.pdf [accessed 21 August 2021].
27 National Archives and Records Administration, 'Federal Register: 45 Fed. Reg. 32287 (May 16, 1980)', Library of Congress, Washington, D.C. 20540 USA, 1980.
28 National Archives and Records Administration, 'Federal Register: 45 Fed. Reg. 32287 (May 16, 1980)', p. 32297.
29 Davies et al., 'Animal Research Nexus'.
30 Richard Gorman, 'What Might Decapod Sentience Mean for Policy, Practice, and Public?' *Animal Sentience*, 7.32 (2022), DOI: 10.51291/2377-7478.1720.
31 Erica Burman, *Deconstructing Developmental Psychology* (London: Taylor & Francis, 2016).
32 Stuart Hall, *Representation: Cultural Representations and Signifying Practices* (London: Sage, 1997).
33 Maloney, Phelan, and Simmons, 'Saving the Horseshoe Crab'; Maike Piehler et al., 'Comparison of LAL and RFC Assays – Participation in a Proficiency Test Program between 2014 and 2019', *Microorganisms*, 8.3 (2020), DOI: 10.3390/microorganisms8030418.
34 John Dubczak, 'Standing Guard: Comparing the Established LAL Assay to Current Alternative Endotoxin Detection Methods', *PDA Letter*, 2018, www.pda.org/pda-letter-portal/home/full-article/standing-guard [accessed 21 August 2021]; Masakazu Tsuchiya, 'Innovative Mechanism of Limulus Amebocyte Lysate Activation to Achieve Specificity and Sensitivity to Endotoxin; Comparison With Recombinant Factor C Reagents', 10 (2020), 6, DOI: 10.37118/ijdr.19019.05.2020.
35 Richard Gorman, 'Horseshoe Crabs and the Pharmaceutical Industry: Challenges and Alternatives – Project Report' (Exeter: University of Exeter, 2020), DOI: 10.13140/RG.2.2.24616.60164/2.
36 Thomas Hartung, 'Three Rs Potential in the Development and Quality Control of Pharmaceuticals', *ALTEX*, 18 (2001), p. 6.
37 Oliver Flint, 'A Timetable for Replacing, Reducing and Refining Animal Use with the Help of in Vitro Tests: The Limulus Amebocyte

Lysate Test (LAL) as an Example', in *Alternatives to Animal Testing: New Ways in the Biomedical Sciences, Trends and Progress*, ed. by Christoph A. Reinhardt (Wiley Online Library, 1994), pp. 27–43.

38 Sarah Zhang, 'The Last Days of the Blue-Blood Harvest', *The Atlantic*, 2018, www.theatlantic.com/science/archive/2018/05/blood-in-the-water/559229/ [accessed 21 August 2021].

39 Levin et al., 'Clotting Cells and Limulus Amebocyte Lysate', p. 347.

40 Flint, 'A Timetable for Replacing, Reducing and Refining Animal Use', p. 31.

41 Charles River Laboratories, '3Rs: Sustainability in Endotoxin Testing', 2020, www.web.archive.org/web/20200617123912/https://www.criver.com/products-services/qc-microbial-solutions/endotoxin-testing/3Rs-achieving-sustainability-endotoxin-testing [accessed 21 August 2021].

42 National Archives and Records Administration, 'Federal Register: 42 Fed. Reg. 57683 (Nov. 4, 1977)', Library of Congress, Washington, D.C. 20540 USA, 1977.

43 *Animal Experimentation: Working Towards a Paradigm Change*, ed. by Kathrin Herrmann and Kimberley Jayne (Leiden: Brill, 2019), p. 659.

44 Gail Davies, 'Figuring it Out: Questions of Comparison, Culture, and Care in Animal Use Statistics', 2019, www.animalresearchnexus.org/blogs/figuring-it-out-questions-comparison-culture-and-care-animal-use-statistics [accessed 21 August 2021].

45 Alexandra Palmer and Beth Greenhough, 'Out of the Laboratory, into the Field: Perspectives on Social, Ethical and Regulatory Challenges in UK Wildlife Research', *Philosophical Transactions of the Royal Society B: Biological Sciences*, 376.1831 (2021), DOI: 10.1098/rstb.2020.0226.

46 Sarah Wolfensohn and P. Honess, 'Laboratory Animal, Pet Animal, Farm Animal, Wild Animal: Which Gets the Best Deal?' *Animal Welfare*, 16.2 (2007), 117–123.

47 Howard J. Curzer et al., 'The Ethics of Wildlife Research: A Nine R Theory', *ILAR Journal*, 54.1 (2013), 52–57, DOI: 10.1093/ilar/ilt012.

48 Eleanor Drinkwater et al., 'Keeping Invertebrate Research Ethical in a Landscape of Shifting Public Opinion', *Methods in Ecology and Evolution*, 10.8 (2019), 1265–1273, DOI: 10.1111/2041-210X.13208.

49 Beth Greenhough and Emma Roe, 'Attuning to Laboratory Animals and Telling Stories: Learning Animal Geography Research Skills from Animal Technologists', *Environment and Planning D: Society and Space*, 37.2 (2019), 367–384, DOI: 10.1177/0263775818807720.

50 Robert Loveland, 'The Life History of Horseshoe Crabs', in *Limulus in the Limelight: A Species 350 Million Years in the Making and in Peril?*, ed. by John T. Tanacredi (New York: Springer 2002), p. 99.
51 Moore, *Catch and Release.*
52 Brigitte Rusche, 'The 3Rs and Animal Welfare – Conflict or the Way Forward', *Altex*, 20. Suppl 1 (2003), 63–76.
53 Alexandra Palmer et al., 'Edge Cases in Animal Research Law: Constituting the Regulatory Borderlands of the UK's Animals (Scientific Procedures) Act', *Studies in History and Philosophy of Science Part A*, 90 (2021), 122–130, DOI: 10.1016/j.shpsa.2021.09.012.
54 A. C. Gray et al., 'Animal-Friendly Affinity Reagents: Replacing the Needless in the Haystack', *Trends in Biotechnology*, 34.12 (2016), 960–969 (p. 961), DOI: 10.1016/j.tibtech.2016.05.017.
55 Moore, *Catch and Release.*
56 Atlantic States Marine Fisheries Commission, *2019 Horseshoe Crab Benchmark Stock Assessment and Peer Review Report*, 2019, www.asmfc.org/uploads/file/5cd5d6f1HSCAssessment_PeerReviewReport_May2019.pdf [accessed 21 August 2021].
57 Josh Eagle, 'Issues and Approaches in Regulation of the Horseshoe Crab Fishery', in *Limulus in the Limelight*, ed. by Tanacredi, p. 86.
58 Priska Gisler and Mike Michael, 'Companions at a Distance: Technoscience, Blood, and the Horseshoe Crab', *Society & Animals*, 19.2 (2011), 115–136, DOI: 10.1163/156853011X562971.
59 Atlantic States Marine Fisheries Commission, *2019 Horseshoe Crab Benchmark Stock Assessment and Peer Review Report*, p. 250.
60 Charles River Laboratories, '3Rs: Sustainability in Endotoxin Testing'.
61 Gail Davies et al., 'Developing a Collaborative Agenda for Humanities and Social Scientific Research on Laboratory Animal Science and Welfare', *PLOS ONE*, 11.7 (2016), 1–12, DOI: 10.1371/journal.pone.0158791.
62 R. Blechová and D. Pivodová, 'Limulus Amoebocyte Lysate (LAL) Test – An Alternative Method for Detection of Bacterial Endotoxins', *Acta Veterinaria Brno*, 70.3 (2001), 291–296 (p. 291).

3

'The place for a dog is in the home': why does species matter when rehoming laboratory animals?

Tess Skidmore

Introduction

Some research animals are rehomed after life in the laboratory, with rehoming defined by the UK's Home Office as 'the movement of a relevant protected animal from an establishment to any other place that is not an establishment under A(SP)A [The Animals (Scientific Procedures) Act]'.[1] The 'place' referenced is typically a private home, farm, aquarium, or zoo.[2] Rehoming is seen as beneficial for numerous reasons, including promoting the ethical profile of animal research, boosting staff morale, and improving the lives of research animals.[3] However, in practice most laboratory animals are euthanised, with only 2,322 animals known to have been rehomed between 2015 and 2017.[4] Some research animals are ineligible for rehoming as the state of their health is too poor, as a result of the procedures they have undergone or their phenotype.[5] There are also other barriers to rehoming. For example, it is typically coordinated by laboratory staff, and can be time-consuming and resource-intensive. It can also be difficult to ensure that animals will be properly cared for post-rehoming, which may be one reason why the majority of rehomed animals are rehomed by laboratory staff, or their friends and family. In addition, different species are either under- or over-represented in rehoming statistics, with rodents representing 94.15% of animals kept in laboratories, but only 19.14% of those rehomed. In contrast, birds, cats, dogs, horses, amphibians, and agricultural animals make up 80.86% of those rehomed, despite making up just 5.84% of animals kept in laboratories.[6] This chapter considers this discrepancy, exploring why some species are viewed as more appropriate candidates for rehoming than others.

Research across multiple disciplines – including anthropology, geography, sociology, and animal welfare science – has sought to unpack why species are valued, and thus treated differently. The majority of this research concludes that such differences are due to perceptions of sentience.[7] Despite difficulties in measuring sentience,[8] the term generally refers to the capacity of individual animals to experience pleasurable states such as joy, as well as those aversive including pain and fear.[9] It is generally accepted that animals believed to be self-aware, solve problems, and possess 'higher' mental abilities, are sentient.[10] Scholars have acknowledged that the attribution of sentience commonly results in species having a higher ethical status[11] and being more likely to receive care, individualised attention, or even 'love' in the laboratory and other settings.[12] However, as this chapter will reveal, differences in the way species are categorised are more complex than perceptions of sentience, and are a product of wider political and cultural factors.

It is important to study differences in the way species are valued because, although a cultural phenomenon, these perceptions hold important political, ethical, and regulatory implications. For example, some species are specially protected in UK animal research legislation – namely dogs, cats, horses, and primates – as a product of the perceptions that these species are highly valued by the public, and hence there would be greater opposition to their use in research.[13] Public perceptions of species can therefore inform the practices and regulatory guidance of research involving animals.[14] Perceptions of species also shape pet-keeping practices, with some species more likely to be granted a privileged position in the home and loved as family members,[15] while others are demoted to tools for scientific research, their bodies amounting to data as the naturalistic animal and duties of care toward it are lost.[16] Of course, many species appear in both settings, therefore muddying understandings and categorisations. For example, a mouse can appear both in the home as a loved pet, and in the laboratory as a research tool.

This chapter will show that animals thought to be sentient *did* have a greater chance of being rehomed. However, understanding rehoming decisions involves a deeper analysis of socio-cultural practices. After describing the research methods, the chapter firstly explores how animal individuality and aesthetics shape the decisions of laboratory staff about which animals are the best rehoming candidates. It then turns to practical issues affecting the decisions of staff, including time needed to rehome and number of homes required. It then examines broader cultural considerations, such as domestication and long-evolved human–animal bonds and narratives of 'home' and which animals 'belong' there, before discussing the external consideration of public opinion. Finally, It discusses the resistance displayed by staff towards the dominant narratives presented throughout the chapter.

Methods

This research explored the rehoming of laboratory animals, to private homes, wildlife sanctuaries, zoos, aquariums, and farms. It involved 22 semi-structured interviews with 28 participants (four interviews were undertaken with two people) lasting between 30 minutes and two and a half hours with: animal research facility staff (17), such as researchers, Named Veterinary Surgeons (NVSs), Named Animal Care and Welfare Officers (NACWOs), and facility managers; rehoming organisation employees (8); and individuals who had rehomed laboratory animals (10). Some individuals belonged to more than one of these groups, such as facility staff who personally rehomed animals. The research additionally involved informal conversations and ethnographies as part of the interview process, including tours of six animal research facilities and observations of rehomed animals in their homes while interviewing owners. To protect participants, pseudonyms are used throughout. In addition to this qualitative work, the rehoming research involved an online questionnaire, which was completed by 41 facilities (response rate was around 25%) to better understand the numbers and types of animals rehomed from UK research facilities, and the main motivations for, and barriers to, rehoming.[17] The research was approved by the University of Southampton Ethics Committee (Submission Number: 32026). The following focuses on the outcomes of the qualitative analysis, exploring how animal individuality and aesthetics, facility and home spaces, and public concerns affect the implementation of efforts to rehome animals from the laboratory.

Animal individuality

Non-human individualisation impacts 'rehome-ability'. Once animals are considered individuals by laboratory staff, it becomes easier to foster a relationship with them.[18] The formation of a relationship in turn makes staff more likely to instigate the necessary procedures, such as training and socialisation, to prepare animals for rehoming, or even to consider rehoming animals themselves. Individualisation was viewed as shaped by several factors, one being the extent to which members of a species are typically viewed as 'one' or 'many'. Amy, who coordinates her facility's rehoming scheme, argues that laboratory staff do not tend to individualise fish (see also Message, Chapter 8 below):

> I think that people tend to see fish slightly differently – maybe because they're not seen as individual pets. So they would have them if they have a tank or a pond or something like that, whereas the dogs and cats, quite often, if the person works internally, they'll have a relationship with that animal. (Amy, facility employee, interview, 2019)

The 'relationship' discussed here is important, suggesting that not only do humans react and feel concern toward certain animals, but that these animals might display particular behaviours in response. Milton argues that animals found to react positively to human presence are more likely to trigger emotional concern, thereby also prompting objections to euthanasia.[19] Hillman, too, advocates that criteria for ethical consideration operate most strongly in animals that display reciprocity to human interaction.[20] Animals less responsive to human presence may therefore sit outside of protective ethical boundaries. While laboratory staff interviewed by Message (Chapter 8) often spoke of the challenge of bonding with fish in contrast to more relatable mice, even mice might be less interactive than other laboratory animals. According to Hillman, mice are 'autonomous animals' that resist domestication and thus are unlikely to interact with humans on a personal level.[21] As Richard, who works at a rodent facility, argues, mice too are thus difficult to socially interact with:

> I mean, I'd be ambivalent about it myself. I don't see [mice] as pet animals, and I wouldn't like to be seen to be breeding pet animals. I don't see reptiles or spiders as pets either for that matter.
>
> Q. And why is that?
>
> I think it's because you don't get the same level of social interaction that you do with a cat or a dog, or even a rat. (Richard, rodent facility employee, interview, 2018)

Temporality is also important for individualisation. Research finds that the longer staff spend with animals, the more likely a bond is to develop.[22] It is therefore important to consider the diverse lifespans of laboratory animals. For example, rhesus macaques can spend 20 to 30 years in a laboratory,[23] whereas mice live a maximum of four years.[24] As Chris, who works at a primate research facility, explains:

> [Bonds develop] even more so with primates because they're so long-lived. Our staff know their animals as individuals. I know a scientist who works in the unit, and she could tell which was which just from a photo of them all. And I think that happens too with dogs, especially when you're working closely with the ones that go into experiments. I used to work closely with the dogs and they all had names and personalities. (Chris, primate research facility employee, interview, 2018)

Chris highlights how individualisation is not only a product of properties inherent to the animals themselves, such as their lifespan and the degree to which they interact with humans, but also practices such as naming. Beck and Katcher assert that being named is the essence of being considered an individual.[25] The naming of animals involves attributing them with individuality and personhood, transforming anybodies into somebodies.[26] Milton discusses what she terms 'human extensionism', which facilitates the allocation of 'personhood' to individual non-humans and leads to a sense of care toward particular animals.[27] Naming is a practice typically deemed appropriate only for pets,[28] and thus is also characteristically associated with certain species, such as cats and dogs as Chris notes. Indeed, the fact that laboratory animals are rarely named highlights the significance of naming.[29]

Alongside naming, the practices of breeding and conducting procedures on animals can affect their individualisation. In 2017, the number of procedures involving mice totalled 2,781,685, and those involving zebrafish 422,138. Conversely, just 3,847 procedures were undertaken using dogs.[30] Research finds that when animals are kept in large numbers, the ability to care for the individual is compromised. While Levi-Strauss famously proposed that animals are 'good to think with',[31] Porcher and Despret warn that 'numbers help us to stop thinking'.[32] Buller, exploring human relations with farm animals, explains that livestock are both 'one' and 'many', and that their very number poses a challenge to individualisation: 'the more of many, the less of one'.[33] This thinking also holds true of laboratory animals kept in large numbers: intensive industrialised systems where thousands of mice are housed in cages stacked one on top of another results in difficulty seeing individual life, and thus also in viewing particular species as individuals.[34] This process of de-individualisation has important ramifications for the treatment of animals; Derrida suggests that sheer numbers mask individual lives, hence rendering them more 'killable'.[35]

Fish and mice, in addition to being kept in especially large numbers, are also more likely to be genetically altered, and hence easier to objectify. Genetically altered experimental animals are more likely to occupy space as a research tool, devoid of subjectivity. Indeed, the conception and patenting of animals like Oncomouse embody the suggestion that a mouse can be seen by some as just another research tool.[36] Therefore, some species such as fish and mice are less likely to be viewed as individuals by laboratory staff making rehoming decisions, and therefore less likely to be chosen as rehoming candidates or taken home by staff themselves. De-individualisation was related to factors such as these species' lack of interaction with humans, shorter lifespan (and therefore reduced possibility of bond formation), housing in larger numbers, and their unnamed and genetically altered status.

Animal aesthetics

It is not only aspects of an animal's perceived individuality or personality that are important in rehoming decisions; aspects of their physical appearance (or 'corporeal charisma')[37] also come into play. Animal aesthetics may result in increased ethical concern and human compassion. For example, when explaining why beagles are easily rehomed, Hannah (laboratory staff member) explains: 'Well, [beagles] look so cute don't they?' William, a researcher, also reflects on the aesthetic draw of beagles in explaining why his facility invited journalists to cover their beagle research programme, which included rehoming:

> We had to decide to be really open and up front and invite journalists in and positively say, 'Look, we've set up this colony, here are the dogs, this is why we're doing it, come and cuddle some cute puppies.' Because of their beagle background they really are very, very cute. (William, researcher, interview, 2018)

Freya, a facility manager, made a similar point: 'We used to have four or five litters of puppies at any one time and they were half beagle, half cavachon, so they were all adorable.'

As these quotes illustrate, dogs, and puppies in particular, were viewed by staff to be aesthetically appealing, or 'cute', increasing feelings of compassion. Ethologist Lorenz suggests that this concern arises from the physical similarity of puppies to human babies, including their big forehead, upright posture, flat face, and large eyes.[38] Lévinas's concept of the face also holds significance here, as the face is depicted as the medium through which all interaction takes place.[39] Non-humans in possession of a face that resembles that of a human are more likely to be subject to care, and, by extension, be considered for rehoming. Lorimer terms this phenomenon 'cuddly charisma': humans empathise more with those to whom they can relate.

However, this research suggested that 'cuddly charisma' in some cases was challenged, or viewed as less important than other factors. This was particularly evident in how facility staff spoke about rats. Despite being a small rodent, embedded in Western imaginations as vermin, a harbinger of disease,[40] or 'pests',[41] rats were described by staff as highly intelligent and social animals, and therefore excellent rehoming candidates. As Megan, an NVS, explained when asked which animals are more suited to rehoming:

> I suppose probably the rats, out of all of the animals, because they seem to form bonds with humans more easily. They seem less scared of interaction. Whereas the gerbils and the mice ... they're lovely but I don't think you ever quite get that same bond. I think they're just a bit more scared. (Megan, NVS, interview, 2018)

Freya, who manages an animal facility, agrees with Megan's notion of 'affection' displayed by rats toward humans, and suggests: 'Wild-type mice you could rehome, but then they're not like rats are they? They're not affectionate.' Louisa, an animal technician, concurs: 'Rats are very friendly. They love sleeping, they love sitting on you. They love being stroked.' Tully suggests that 'unlike mice, rats have excellent pet characteristics that include a charming personable disposition and extreme intelligence'.[42] Rats are described as social animals, increasing the possibility of bond development with humans, which, as discussed in the previous section, may lead to rehoming.[43] Thus, relationships evolve in the laboratory with species that may typically be viewed as unappealing from an aesthetic, 'cuddly charisma' perspective.[44]

However, facility staff did acknowledge the negative perception they felt the public may have of rats. As Sophie, a Named Information Officer, explains: 'If you know rats then you love rats, the technicians love rats, but I think lots of the general public don't like them. I think the general public need more education about rats because they make wonderful pets.' Isobel, an animal technician, has similar reflections on the appeal of rats as a species. She draws on their aesthetics both to explain why the public might not reflect positively toward them, but also why she felt empathy toward the species: 'Others were like, "Ew, no, just no." I think it's the tails. But I think they're really sweet. They have proper cute little faces.'

Many members of staff felt a certain affinity with rats. When interacting daily with rats, it was possible to overwrite cultural constructions of them as vermin and elevate them to the status of pet.[45]

Thus, while rats are common candidates for rehoming, rehoming of rats is typically on a much smaller and informal scale compared to species such as dogs, and rats are usually rehomed to staff with whom they had formed a personal bond.[46]

Facility resources

Practical and economic factors also shape decisions about which species can be rehomed. Facility staff argued that some species (generally smaller mammals and fish) are less resource intensive to rehome. As facility manager Freya explains:

> 'Are the animals fit for rehoming?' is always the question. It's easy with the – I hate to say lesser species, but the less emotional species, because they'll just adapt wherever. You know, fish will just go – oh there's a new tank, thank you very much. Rats pretty much the same, but when you've got like a dog … getting animals socialised is the big thing. (Freya, facility manager, interview, 2018)

Thus, a complex situation arises whereby, although less likely to be considered for rehoming, facility staff acknowledge the relative ease of rehoming fish and mice over dogs. This was largely due to perceptions that it would be easier to maintain the welfare of these species with less resource input from the facility, crucial if rehoming is to be attempted.[47] Freya suggests that the difficulty comes with ensuring large mammals, such as dogs, are adequately socialised. Similarly, Alice, a vet, explains the difficulties of providing additional enrichment to guarantee a good quality of life for primates post-rehoming:

> Higher [quality of life] for a primate is going to involve a lot more effort than higher for a mouse. Because with a mouse you can give them some tunnels, some stuff to chew, and another mouse that they know, bedding, and a little nesting area and a little running wheel, and that's your environment sorted. (Alice, vet, interview, 2018)

She implies that maintaining rodent welfare once rehomed is a relatively easy task, yet juxtaposes this with worries regarding rehoming dogs: 'With dogs it will depend on how they've been treated, what's been done to them, and their socialisation period. Mice would be fine. [...] They'll just get on with it.'

However, despite the acknowledged difficulty of rehoming larger mammal species, efforts to rehome them were not prevented. Alice reveals a construction of primates as 'deserving' of rehoming: 'With primates, I think their age doesn't matter, and they deserve a nice life regardless of their age.' Echoing this idea, Fluery states that retiring chimpanzees is time-consuming, but necessary in terms of its potential to enhance animal welfare.[48] Similarly, guidance from the Laboratory Animal Science Association recommends that the resources needed to successfully rehome should not be a deterrent, and that rehoming should still be undertaken 'for the sake of the dogs'.[49]

Conversely, Freya describes the difficulty in justifying the time it takes her staff to rehome one rabbit: 'As the facility manager there, I found it hard justifying my staff's time, and I'm being really brutal saying this, spending three or four days trying to rehome a rabbit. I know that sounds awful, but there's like a balance isn't there, between how much time you can spend on some things.' This reveals a complex situation whereby non-traditional companion animals are less likely to be rehomed, even if the process is acknowledged to be easier.

Exploring the rehoming of laboratory animals reveals complex economic and political factors that hold implications for the way laboratory animals are valued and treated.[50] Rehoming decisions are always situated within political-economic systems that serve both to restrain, and promote, rehoming possibilities.[51] Economic considerations that serve to impede rehoming do not affect all species universally, but instead manifest differently based upon the individual animal being considered for rehoming.[52]

Number of homes available

As previously discussed, animals kept in large numbers are less commonly individualised and attributed distinct personalities, reducing their chances of being selected for rehoming. In addition, practical issues present themselves in terms of the numbers of animals needing homes. Those working in facilities keeping thousands of mice or fish often viewed rehoming as impossible. Thus, it is crucial to consider factors external to the laboratory, including the demand for animals and number of homes available. Rose, who works for an animal welfare organisation, explains:

> It's also about the nature of the animal. There are likely to be more people who want dogs than who want rats, or mice. And there are far less dogs, and horses and cats than there are rats and mice. [...] It's the number of homes that are available that is the limiting factor I would say. (Rose, animal welfare organisation employee, interview, 2019)

William, whose facility has a comprehensive laboratory dog rehoming scheme, also explains: 'I don't think there's a huge number of people who keep pet rats and pet mice and therefore volume wise, there's no chance of a significant proportion being rehomed.' If a large proportion cannot be rehomed, a small subset would need to be selected. Jane, a researcher at a rodent facility, explains the difficulties in choosing which individual animals would warrant rehoming efforts: 'Yeah, I mean there's quite big numbers. You know, we breed over six figures every year. So it would be difficult to choose which to rehome.'

Jane's argument can be probed using theories of fairness, equity, and justice. Hay's research explores how Barry's concept of 'procedural fairness'[53] is conceived as a proper and complete adherence to rules, and, critically, the correct application of such rules to all cases (in this case, individual laboratory animals).[54] According to Hay, rules should be 'consistent, non-arbitrary and even-handed'.[55] In such a system, rehoming one animal over another would be unjust. This is heightened in the context of a 'multitude' or 'mass' of animals.[56] As rehoming one animal over another would violate procedural fairness, a policy of not rehoming is then applied consistently to all animals.

Facility staff consequently drew on moral and ethical arguments to justify their lack of rehoming. Such moral arguments rest upon the rational and logical. The scale of the numbers of animals needing homes is inconceivable to staff, and as such rehoming them was construed as impossible. This means that some species (those typically kept in larger numbers such as rats, mice, and fish), were less likely to be considered for rehoming due to the difficulties with either 1) finding them all homes, or 2) developing fair systems to select a subset of animals for rehoming.

Ideas of 'home'

Key cultural beliefs also impact species choice in rehoming decisions. Space – and the way it is defined and modified by objects, actions, and actors – is an important concept to explore.[57] For example, influential in shaping rehoming practices are cultural narratives of 'home' and where animals 'belong', which change across species and spaces. Western constructions and imaginations typically place the dog within the boundaries of the human home. This perspective was reflected by facility staff, who stressed the idea that dogs in particular 'belonged' in the home, not the research facility. As Ella, an NVS, explains:

> There was this feeling that the place for a dog is in the home. You know, and I don't necessarily think you would feel that so much for a rat or a mouse or a fish. And I think, apart from just society's perspective, there's a question about what the animals need. And I think for a dog, it's very difficult to, maybe, I'm not saying this for sure, but maybe for a fish or a mouse, you could give them a good quality of life outside of the home environment. (Ella, NVS, interview, 2018)

Staff reveal a perception that dogs possess the potential to enjoy a higher quality of life that can only be experienced once rehomed. It is crucial to unpick why such narratives have developed. Indeed, it is not simply access to the home space that facility staff see as important, but also cultural perceptions of the behaviours dogs are then able to engage in within these spaces. For example, Sophie, a Named Information Officer, explains: 'Obviously with the dogs and cats, because they're kept in [...] very kennelled environments which, you know, is typical for research animals, it was just lovely to see them laying on a sofa.'

Olivia also mentions the sofa, an object typically reserved for humans, as a space in which a loved family dog may reside: 'I think the general feeling is that people just feel really sorry for these dogs, and they think "oh my god they've never had that sofa, they've never had that one to one love from a family" and everything like that.' Amy, who works at a facility keeping cats and dogs, also explains the desire people feel for 'puppies to go and sit by a fire'.

These narratives disclose a complex understanding and ordering of species, and reveal inherent differences in human constructions of where species belong. Humans order animals spatially,[58] and construct how species are categorised and whether they are considered in or out of place.[59] Both human and non-human actions produce and re-produce spatial boundaries and designate meaning to those boundaries.[60] Power argues that pets are often afforded unrestricted access to 'human' spaces – including family rooms, bedrooms and furniture.[61] In fact, dog identity is entwined in human home narratives; domestication of the dog has changed the nature of both canine bodies and the image of the human home space.[62] Domestication has systematically worked both to integrate dogs into the home, and strengthen the human–dog relationship. As Alex (laboratory staff member) points out: 'I mean, since very early civilisation there's been a close link between dogs and humans. They're seen by most people as the most compatible other species to humans [...] There's a very long running bond between dogs and humans.'

This relationship does not only result in a human expectation that dogs should reside in the home, but extends to the belief that it is what dogs yearn for as they search for human interaction. As Chris, who previously worked at a facility housing dogs, describes: 'You know, dogs have had thousands of years of domestication so they're very human-orientated, and they'll look for a human to meet their needs. They like getting the appreciation of somebody saying well done, you know.' In contrast to the animal species described by staff as avoiding human companionship (such as mice and fish), dogs are framed as 'human-orientated'. Lorimer discusses animal atmospheres, and proposes that dogs can read 'human emotional cues, affective intensities and shared atmospheres', and that crucially this attunement flows bilaterally.[63]

Dogs have thus been trained over time to 'become pets'; they are conditioned to meet human expectations with regard to house training, walking calmly on a lead, and socialising suitably with other animals. In fact, it is now considered unacceptable for many that they be kept in the research laboratory. Thus, both culturally and spatially, dogs (and arguably other species such as cats)[64] are more likely to be considered for rehoming.[65]

The notion that some species belong in the home is in direct contrast to ideas about other animals relegated to spaces outside of it. As Richard, who works at a rodent facility, outlines:

> We very rarely get any trouble here, because people who have issues with animal rights around here go to [*another research facility*], because they have monkeys. I think people's perceptions of mice are slightly different. If you had mice in your house, you would just put out traps and kill them. (Richard, rodent facility employee, interview, 2018)

In explaining why the public may not be supportive of the rehoming of laboratory mice, Richard draws on a narrative that equates mice to vermin and pests.[66] He explains that, instead of loving a mouse in your home like you might a dog, the most common interaction people have with mice in the home space is viewing them as pests and disposing of them accordingly. Consequently, cultural constructions of species, and the spaces in which it is believed they should reside, have direct implications on the way in which animals are valued and treated,[67] and ultimately whether they are considered for rehoming.

Public concern

It is crucial to situate the impetus to rehome within the wider cultural context, including public perceptions surrounding animal research. Public perspective guides regulation and therefore daily practices in the laboratory space.[68] Many of those working in research facilities believe the public would feel positively about rehoming, largely because of perceptions that the inverse option (staying in the laboratory or euthanasia) is detrimental to animal welfare. As Megan, an NVS, discusses: 'I think that [the public] would think that [rehoming] was a good idea. Automatically they would assume that the life that an animal has in someone's house is better than in the lab.'

The world of animal research is notoriously ethically contentious,[69] worsened by public perceptions of secrecy and constructions of systematic animal harm for human benefit. Thus, rehoming can become romanticised in the public imaginary. This results both in pressure to consider rehoming as an option, but also in more complex choices surrounding which species should be rehomed. As Chloe, a research facility manager, discusses: 'Dogs and cats and horses, they're regarded as specially protected species, and there's not any reason as to why that is apart from public opinion. The public are more emotive about them.' In explaining why the facility he works for has developed a comprehensive laboratory dog rehoming scheme, William explains that: 'You know, obviously dogs are of great ethical concern to people.' Due to the engrained attachment toward dogs, there is also more emotional difficulty and strain embedded in their potential euthanasia. As Josh, who works at an amphibian facility, proposes: 'You know, would people feel much worse about putting a frog down, or a zebrafish down, at the end of its life, or the end of its working life, as opposed to a dog? Well I don't think they would.'

Staff draw on notions of what Hobson-West and Davies term 'societal sentience', described as 'imagined feelings of an abstract entity called the public or society'.[70] Imaginings of these societal concerns are generally assumed to be much stronger for certain mammalian species. This relates back to the feelings of empathy discussed earlier in the chapter. Humans tend to feel more ethical concern toward species that share human characteristics, and are less supportive of research that uses animals typically classed as 'companions'.[71] This leads to a greater effort to rehome these animals. The impact of public perception on animal research is also exemplified through those species that typically generate public concern (cats, dogs, and primates) being used less frequently in research.[72]

Choices regarding animal research thus extend beyond the most suitable animal model to also consider which animal is the most culturally and socially acceptable to use. In fact, European legislation states that, where appropriate, dogs and cats who have been used in experiments should be 're-homed in families as there is high public concern as to the fate of those animals'.[73] Indeed, scientific practice and public sensitivities both impact, and are impacted by, each other.[74] As Brown and Michael assert, the practice of rehoming is inherently entangled in 'idealised versions of what will count as public cultural acceptance'.[75] Consequently, rehoming involves complex negotiation between the purpose of the laboratory, the views and perceptions of the public, and ethical principles and regulations.

Objections to dominant narratives

However, social constructions of a species hierarchy as discussed throughout this chapter were not displayed by all working in research facilities. In fact, some actively voiced their objections to these dominant narratives, believing any form of speciesism was unjustified. For example, Peter, who manages an animal facility, explains:

> I don't really like the idea that we tend to differentiate between different species in terms of the rights and wrongs of research. I think because, at the end of the day, animals are animals, and we shouldn't assume that rats have any different rights to a dog. (Peter, animal facility manager, interview, 2018)

Such constructions feed into Lorimer's feral charisma,[76] a theory grounded in understanding and respect for all non-human others and their complexity, difference, wildness, and autonomy. Echoing the aquarists discussed by Message (Chapter 8), many research participants argued for an inherent respect for all non-human life, and not a privilege for those that aesthetically resemble humans, seek human attention and companionship, or are culturally kept in the home. Indeed, in Hobson-West and Davies' research, staff interviewed at animal facilities viewed the construction of a species hierarchy as an 'ethical trap'.[77] Of course, there is an argument that, especially in the context of being interviewed regarding personal beliefs on species value and treatment, participants did not want to express themselves as biased.

Others seemed to accept constructions of speciesism, but still believed such thinking to be unjust, criticising their own views and displaying reflexivity. As Josh, a researcher at an amphibian facility, explains:

> Dogs, cats and rats would be the three at the top I would imagine. The most suited to rehoming.
> Q: And why do you think that?
> Just because they're very common pets. They're big enough that the effort is worth making. Isn't that a terrible thing to say? (Josh, researcher, interview, 2018)

Here, Josh practises reflexivity through identifying that what he has implied is 'terrible'. He recognises that human understandings of species hierarchies, and differential displays of ethical concern, have little scientific grounding and are instead a reflection of cultural and social beliefs. Indeed, as Kirk explains, the wider social, cultural, and political context surrounding the scientific use of animals influences how humans perceive non-humans, and consequently allocate and act on the value attributed to them.[78]

Josh's account also meets Fox's demand for the need for humans to explore thoroughly their relations with non-humans and reflect on how perceptions of particular species manifest in their unequal treatment.[79] Fox criticises what she terms 'unreflective speciesism', and argues that humans must attend to species borders and conceptualisations in order to fully reveal complex multispecies relations.[80] Indeed, ethologist Bekoff argues that there are no high or low species in terms of sentience, and that instead humans make these distinctions because it serves them well in deciding who lives and dies, including which animals are rehomed and which euthanised.[81] According to Bekoff, it is only once humans have reflected on the unequal value allocated to species, and the differences in treatment that result, that they can fully address the imbalanced kinship with the non-human other.

Conclusion

This chapter has shown that cultural, economic, regulatory, and social factors influence species choice in rehoming decisions. Together, these reveal that despite those working with experimental animals being constructed at times as objective and rational,[82] they, too, are subject to cultural constructions of a species hierarchy. Competing constructions of laboratory animals exist, meaning that there are differences in how species are valued, treated, and ultimately whether they are considered for a life after the laboratory. The interactions between regulation, science, and public opinion result in complex ways of ordering and valuing animal life.[83] Exploring the diverse mechanisms through which rehoming decisions are made also demonstrates that although wider ethical frameworks guide human–animal relations in the laboratory, the manner in which ethical principles are practised is always a personal act, modified through researcher intimacies with their animal subjects.[84] Exploring species choice in rehoming decisions also reveals the processes by which laboratory workers' views about animal ranking are shaped by the views of imagined publics. Indeed, this symbolic ranking of animals has consequences for the interpretation of animal research regulation, and behaviours within the laboratory space. In summary, contested and competing meanings of ethical and moral value are attached to, and travel with, different species.[85]

However, some staff did object to these narratives, arguing that displaying any species preferences served only to reject wider ideals of the equality of species.[86] Others expressed shame that they consciously fed into these beliefs. Despite these displays of defiance, this chapter has demonstrated that, ultimately, key decisions surrounding rehoming are shaped by: engrained affections toward certain species; wider societal and cultural expectations regarding species treatment; and practical considerations, including resource output and number of homes available. It is difficult to separate these factors. For example, it is the aesthetic appeal and greater potential for interaction that means dogs have a long-evolved bond with humans. Similarly, it is because of this bond that the public feels more moral and ethical concern regarding their treatment and potential rehoming.

This chapter shows how values enter into decision-making but are not easily visible; crucially, their importance is not easily evaluated in shaping laboratory animal care. For example, the personal feelings about certain species held by laboratory staff are based on complex perceptions and cultural ideas of those animals, which might in turn shape their decisions about which animals to rehome. Given this complexity, it is difficult to standardise rehoming or ensure it is conducted equitably between species. Certain species are more likely than others to experience individual, tailored interaction and be subject to care practices that extend beyond required welfare legislation, meaning that they may be greater recipients of a 'culture of care' than other laboratory animals (Greenhough and Roe, Chapter 5). Investigating the rehoming of laboratory animals from a social scientific perspective can critically engage with these multifaceted issues and enable a greater understanding of the practical complexity surrounding decisions to enact new ethical policies and behaviours.

Notes

1 Animals in Science Regulation Unit, *Re-Homing and Setting Free of Animals Advice Note: 03/2015* (London: Home Office, 2015) available at https://assets.publishing.service.gov.uk/government/uploads/system/uploads/attachment_data/file/470146/Advice_Note_Rehoming_setting_free.pdf, p. 12 [accessed 14 June 2023], p. 10.

2 Tess Skidmore and Emma Roe, 'A Semi-Structured Questionnaire Survey of Laboratory Animal Rehoming Practice across 41 UK Animal Research Facilities', *PlOS One*, 15.6 (2020), e0234922, DOI: 10.1371/journal.pone.0234922.

3 Larry Carbone et al., 'Adoption Options for Laboratory Animals', *Laboratory Animals*, 39.9 (2003), 37–41, DOI: 10.1038/laban1003-37; Sarah Wolfensohn and Paul Honess, 'Laboratory Animal, Pet Animal, Farm Animal, Wild Animal: Which Gets the Best Deal?' *Animal Welfare*, 16.2 (2007), 117–123; Pauleen Bennett and Vanessa Rohlf, 'Perpetration-Induced Traumatic Stress in Persons Who Euthanize Nonhuman Animals in Surgeries, Animal Shelters, and Laboratories', *Society & Animals*, 13.3 (2005), 201–220, DOI: 10.1163/1568530054927753; Mark J. Prescott, 'Finding New Homes for Ex-Laboratory and Surplus Zoo Primates', *Laboratory Primate Newsletter*, 45 (2006), 5–8; Alexandra Palmer et al., 'When Research Animals Become Pets and Pets Become Research Animals: Care, Death, and Animal Classification', *Social & Cultural Geography* (2022), 1–19, DOI: 10.1080/14649365.2022.2073465; Tess Skidmore, 'A Life after

the Laboratory: Exploring the Policy and Practice of Laboratory Animal Rehoming' (unpublished doctoral thesis, University of Southampton School of Geography and the Environment, 2020).

4 Skidmore and Roe, 'A Semi-Structured Questionnaire Survey of Laboratory Animal Rehoming'.

5 Palmer et al., 'When Research Animals Become Pets and Pets Become Research Animals'.

6 Skidmore and Roe, 'A Semi-Structured Questionnaire Survey of Laboratory Animal Rehoming'.

7 Jonathan Birch, 'Degrees of Sentience?' *Animal Sentience*, 3.21 (2018), 11, DOI: 10.51291/2377-7478.1353; Lynne U. Sneddon et al., 'Fish Sentience Denial: Muddying the Waters', *Animal Sentience*, 3.21 (2018), 1, DOI: 10.51291/2377-7478.1317.

8 Helen S. Proctor, Gemma Carder, and Amelia R. Cornish, 'Searching for Animal Sentience: A Systematic Review of the Scientific Literature', *Animals*, 3.3 (2013), 882–906, DOI: 10.3390/ani3030882.

9 D. M. Broom, 'Quality of Life Means Welfare: How is it Related to Other Concepts and Assessed?' *Animal Welfare*, 16.S1 (2007), 45; Jonathan Birch, 'Animal Sentience and the Precautionary Principle', *Animal Sentience*, 2.16 (2017), 1, DOI: 10.51291/2377-7478.1200.

10 Harold A. Herzog and Shelley Galvin, 'Common Sense and the Mental Lives of Animals: An Empirical Approach' (Albany, NY: State University of New York Press, 1997).

11 Pru Hobson-West and Ashley Davies, 'Societal Sentience: Constructions of the Public in Animal Research Policy and Practice', *Science, Technology, & Human Values*, 43.4 (2018), 671–693, DOI: 10.1177/0162243917736138.

12 Fon T. Chang and Lynette A. Hard, 'Human–Animal Bonds in the Laboratory: How Animal Behavior Affects the Perspective of Caregivers', *ILAR Journal*, 43.1 (2002), 10–18, DOI: 10.1093/ilar.43.1.10.

13 Hobson-West and Davies, 'Societal Sentience'.

14 Elisabeth H. Ormandy and Catherine A. Schuppli, 'Public Attitudes toward Animal Research: A Review', *Animals*, 4.3 (2014), 391–408, DOI: 10.3390/ani4030391; Sneddon et al., 'Fish Sentience Denial'.

15 Susan Phillips Cohen, 'Can Pets Function as Family Members?' *Western Journal of Nursing Research*, 24.6 (2002), 621–638, DOI: 10.1177/019394502320555386; Rebekah Fox and Katie Walsh, 'Furry Belongings: Pets, Migration and Home', *Animal Movements, Moving Animals: Essays on Direction, Velocity and Agency in Humanimal Encounters*, ed. by J. Bull (Uppsala: University Printers,

Uppsala University, 2011), pp. 97–117; Ann Ottney Cain, 'Pets as Family Members', *Marriage & Family Review*, 8.3–4 (1985), 5–10.

16 Michael E. Lynch, 'Sacrifice and the Transformation of the Animal Body into a Scientific Object: Laboratory Culture and Ritual Practice in the Neurosciences', *Social Studies of Science*, 18.2 (1988), 265–289, DOI: 10.1177/030631288018002004.

17 Skidmore and Roe, 'A Semi-Structured Questionnaire Survey of Laboratory Animal Rehoming'.

18 Beth Greenhough and Emma Roe, 'From Ethical Principles to Response-Able Practice', *Environment and Planning D: Society and Space*, 28.1 (2010), 43–45.

19 Kay Milton, 'Anthropomorphism or Egomorphism? The Perception of Non-Human Persons by Human Ones', in *Animals in Person*, ed. by John Knight (London: Routledge, 2020), pp. 255–271.

20 J. Hillman, *Going Bugs* (Brilliance Audio, 1988).

21 Hillman, *Going Bugs*.

22 Kathryn Bayne, 'Development of the Human–Research Animal Bond and its Impact on Animal Well-Being', *ILAR Journal*, 43.1 (2002), 4–9, DOI: 10.1093/ilar.43.1.4.

23 Jeffrey D. Fortman et al., *The Laboratory Nonhuman Primate* (Chicago, IL: CRC Press, 2017).

24 Sulagna Dutta and Pallav Sengupta, 'Men and Mice: Relating Their Ages', *Life Sciences*, 152 (2016), 244–248, DOI: 10.1016/j.lfs.2015.10.025.

25 Alan M. Beck and Aaron Honori Katcher, *Between Pets and People: The Importance of Animal Companionship* (Lafayette, IN: Purdue University Press, 1996).

26 Katrina Holland, '"Biosensors in Fluffy Coats": Interspecies Relationships and Knowledge Production at the Nexus of Dog-Training and Scientific Research' (unpublished PhD thesis, University College London, 2019); Barbara Bodenhorn and Gabriele Vom Bruck, *The Anthropology of Names and Naming* (Cambridge University Press, 2006).

27 Milton, 'Anthropomorphism or Egomorphism?'

28 Milton, 'Anthropomorphism or Egomorphism?'

29 Mary T. Phillips, 'Proper Names and the Social Construction of Biography: The Negative Case of Laboratory Animals', *Qualitative Sociology*, 17.2 (1994), 119–142, DOI: 10.1007/BF02393497.

30 Home Office, 'Annual Statistics of Scientific Procedures on Living Animals in Great Britain 2019', 2019, www.assets.publishing.service.gov.uk/government/uploads/system/uploads/attachment_data/

file/901224/annual-statistics-scientific-procedures-living-animals-2019.pdf [accessed 1 February 2023].

31 C. Levi-Strauss, *Totemism* (Needham, Boston: Beacon Press, 1963).

32 Jocelyne Porcher and Vinciane Despret, 'Etre Bête', *Sciences Humaines* (Arles: Actes Sud, 2007), p. 36.

33 Henry Buller, 'Individuation, the Mass and Farm Animals', *Theory, Culture & Society*, 30.7–8 (2013), 162, DOI: 10.1177/0263276413501205.

34 Gail Davies, 'Caring for the Multiple and the Multitude: Assembling Animal Welfare and Enabling Ethical Critique', *Environment and Planning D: Society and Space*, 30.4 (2012), 623–638, DOI: 10.1068/d3211.

35 Buller, 'Individuation, the Mass and Farm Animals', p. 161.

36 Julie L. Urbanik, *Geography and Animal Biotechnology: How Place and Scale are Shaping the Public Debate* (Worcester, MA: Clark University, 2006); Hobson-West and Davies, 'Societal Sentience'.

37 Jamie Lorimer, 'Nonhuman Charisma', *Environment and Planning D: Society and Space*, 25.5 (2007), 911–932, DOI: 10.1068/d71j.

38 Konrad Lorenz, *The Foundations of Ethology* (New York: Springer, 1981).

39 Cary Wolfe, *Zoontologies: The Question of the Animal* (Minneapolis, MN: University of Minnesota Press, 2003).

40 Jacques Derrida, *On Cosmopolitanism and Forgiveness* (London: Routledge, 2003).

41 Lynda Birke, 'Animal Bodies in the Production of Scientific Knowledge: Modelling Medicine', *Body & Society*, 18.3–4 (2012), 156–178, DOI: 10.1177/1357034X12446379.

42 Thomas N. Tully Jr, 'Mice and Rats', in *Manual of Exotic Pet Practice* (London: Elsevier, 2009), p. 335.

43 Bayne, 'Development of the Human–Research Animal Bond'; Birke et al., *The Sacrifice*.

44 Beth Greenhough and Emma Roe, 'Attuning to Laboratory Animals and Telling Stories: Learning Animal Geography Research Skills from Animal Technologists', *Environment and Planning D: Society and Space*, 37.2 (2019), 367–384, DOI: 10.1177/0263775818807720.

45 Charlotte Robin et al., 'Pets, Purity and Pollution: Why Conventional Models of Disease Transmission do not Work for Pet Rat Owners', *International Journal of Environmental Research and Public Health*, 14.12 (2017), 1526, DOI: 10.3390/ijerph14121526.

46 Skidmore and Roe, 'A Semi-Structured Questionnaire Survey of Laboratory Animal Rehoming'.

47 Animals in Science Regulation Unit, *Re-Homing and Setting Free of Animals Advice Note: 03/2015* (London: Home Office, 2015) available at https://assets.publishing.service.gov.uk/government/uploads/system/uploads/attachment_data/file/470146/Advice_Note_Rehoming_setting_free.pdf, p. 12 [accessed 14 June 2023].

48 Erika Fleury, 'Money for Monkeys, and More: Ensuring Sanctuary Retirement of Nonhuman Primates', *Animal Studies Journal*, 6.2 (2017), 30–54, https://ro.uow.edu.au/asj/vol6/iss2/4.

49 LASA, 'A Report Based on a LASA Working Party and LASA Meeting on Rehoming Laboratory Animals', 2002, p. 24, www.lasa.co.uk/wp-content/uploads/2018/05/LASA-Guidance-on-the-Rehoming-of-Laboratory-Dogs.pdf [accessed 1 February 2023].

50 Greenhough and Roe, 'Attuning to Laboratory Animals and Telling Stories'.

51 Beth Greenhough, 'Citizenship, Care and Companionship: Approaching Geographies of Health and Bioscience', *Progress in Human Geography*, 35.2 (2011), 153–171, DOI: 10.1177/0309132510376258.

52 Gail Davies et al., 'Animal Research Nexus: A New Approach to the Connections between Science, Health and Animal Welfare', *Medical Humanities*, 46.4 (2020), 499–511, DOI: 10.1136/medhum-2019-011778.

53 Brian Barry, *Political Argument (Routledge Revivals)* (Oxford: Routledge, 2010).

54 Alan M. Hay, 'Concepts of Equity, Fairness and Justice in Geographical Studies', *Transactions of the Institute of British Geographers*, 1995, 500–508, DOI: 10.2307/622979.

55 Hay, 'Concepts of Equity, Fairness and Justice in Geographical Studies', p. 501.

56 Buller, 'Individuation, the Mass and Farm Animals'.

57 Emma Power, 'Furry Families: Making a Human–Dog Family through Home', *Social & Cultural Geography*, 9.5 (2008), 535–555, DOI: 10.1080/14649360802217790; Emma R. Power, 'Domestication and the Dog: Embodying Home', *Area*, 44.3 (2012), 371–378, DOI: 10.1111/j.1475-4762.2012.01098.x.

58 Chris Philo and Chris Wilbert, *Animal Spaces, Beastly Places* (Routledge, 2004).

59 Lauren E. Van Patter and Alice J. Hovorka, '"Of Place" or "of People": Exploring the Animal Spaces and Beastly Places of Feral Cats in Southern Ontario', *Social & Cultural Geography*, 19.2 (2018), 275–295, DOI: 10.1080/14649365.2016.1275754.

60 Tora Holmberg, *Urban Animals: Crowding in Zoocities* (London; New York: Routledge, 2015).

61 Emma R. Power, 'Domestication and the Dog'.
62 Emma R. Power, 'Domestication and the Dog'.
63 Jamie Lorimer et al., 'Animals' Atmospheres', *Progress in Human Geography*, 43.1 (2019), p. 35, DOI: 10.1177/0309132517731254.
64 Penny L. Bernstein, 'The Human–Cat Relationship', in *The Welfare of Cats*, ed. by I. Rochlitz (Dordrecht: Springer, 2007), pp. 47–89.
65 Holmberg, *Urban Animals*; Skidmore and Roe, 'A Semi-Structured Questionnaire Survey of Laboratory Animal Rehoming'.
66 Lynda Birke, 'Who – or What – Are the Rats (and Mice) in the Laboratory', *Society & Animals*, 11.3 (2003), 207–224, DOI: 10.1163/156853003322773023; Birke et al., *The Sacrifice*.
67 Philo and Wilbert, *Animal Spaces, Beastly Places*.
68 Davies et al., 'Animal Research Nexus'.
69 Carol Kilkenny et al., 'Animal Research: Reporting In Vivo Experiments: The ARRIVE Guidelines', *British Journal of Pharmacology*, 160.7 (2010), 1577–1579, DOI: 10.1111/j.1476-5381.2010.00872.x.
70 Hobson-West and Davies, 'Societal Sentience', p. 685.
71 Innes C. Cuthill, 'Ethical Regulation and Animal Science: Why Animal Behaviour is not so Special', *Animal Behaviour*, 74.1 (2007), 15–22, DOI: 10.1016/j.anbehav.2007.04.003; Ormandy and Schuppli, 'Public Attitudes toward Animal Research'.
72 Ipsos Mori, 'Public Attitudes to Animal Research in 2018', 2018, www.ipsos.com/sites/default/files/ct/news/documents/2019-05/18-040753-01_ols_public_attitudes_to_animal_research_report_v3_191118_public.pdf [accessed 1 February 2023].
73 Commission of the European Communities, Directive 2010/63/EU of the European Parliament and of the Council of 22 September 2010 'on the protection of animals used for scientific purposes', p. 3.
74 Kay Peggs, 'Nonhuman Animal Experiments in the European Community: Human Values and Rational Choice', *Society & Animals*, 18.1 (2010), 1–20, DOI: 10.1163/106311110X12586086158367.
75 Nik Brown and Mike Michael, 'Switching between Science and Culture in Transpecies Transplantation', *Science, Technology, & Human Values*, 26.1 (2001), p. 14, DOI: 10.1177/016224390102600101.
76 Lorimer, 'Nonhuman Charisma'.
77 Hobson-West and Davies, 'Societal Sentience', p. 680.
78 Robert G. W. Kirk, 'Recovering the Principles of Humane Experimental Technique: The 3Rs and the Human Essence of Animal Research', *Science, Technology, & Human Values*, 43.4 (2018), 622–648, DOI: 10.1177/0162243917726579.

79 Rebekah Fox, 'Animal Behaviours, Post-Human Lives: Everyday Negotiations of the Animal–Human Divide in Pet-Keeping', *Social & Cultural Geography*, 7.4 (2006), p. 149, DOI: 10.1080/14649360600825679.
80 Fox, 'Animal Behaviours, Post-Human Lives'.
81 Marc Bekoff and Carron A. Meaney, *Encyclopedia of Animal Rights and Animal Welfare* (London: Routledge, 2013).
82 Kuno Lorenz, 'Science, a Rational Enterprise?' in *Constructivism and Science*, ed. by Robert E. Butts and James Robert Brown (Dordrecht: Springer, 1989), pp. 3–18.
83 Tone Druglitrø, '"Skilled Care" and the Making of Good Science', *Science, Technology, & Human Values*, 43.4 (2018), 649–670, DOI: 10.1177/0162243916688093.
84 Martyn Pickersgill, 'The Co-Production of Science, Ethics, and Emotion', *Science, Technology, & Human Values*, 37.6 (2012), 579–603, DOI: 10.1177/0162243911433057.
85 Lene Koch and Mette N. Svendsen, 'Negotiating Moral Value: A Story of Danish Research Monkeys and Their Humans', *Science, Technology, & Human Values*, 40.3 (2015), 368–388, DOI: 10.1177/0162243914553223.
86 Peter Singer, 'Speciesism and Moral Status', *Metaphilosophy*, 40.3–4 (2009): 567–581.

4

Commentaries on changing and implementing regulation

Edited by Robert G. W. Kirk

This chapter offers an overview and three-part response to the first section of the book, and looks at the changing relationships between regulation and animal research. Opening the conversation is a commentary from Liz Tyson, an academic and animal advocate currently working within the animal protection and conservation not-for-profit sector. She offers an insightful critical analysis identifying the limitations of our work on the 'Animal Research Nexus', limits which result from our choices made in structuring research activity. Tyson invites reflection on how the structuring of research includes some perspectives and excludes others, prompting readers to consider not only what had been said but what had been left unsaid, and why. Amy Hinterberger, academic and social scientist, similarly highlights how the Animal Research Nexus productively focuses attention on the institutional lives of animals but in doing so may leave little space for considering other approaches to animal research. In closing, Robert G. W. Kirk, academic and historian, considers the Animal Research Nexus as a historically constituted object, reflecting how past trends may inform future change while inviting speculation on how the lay public may be better engaged with the ever-evolving relationship of regulation to animal research.

4.1

Accentuate the positive ... silence the negative

Liz Tyson

The three authors in this section provide insights into various parts of the animal experiment process – from the origination of the current legislation governing animal testing in England and Wales (Myelnikov, Chapter 1), to the experiences of some of the animals subjected to testing (or used in associated intrusive procedures) (Gorman, Chapter 2), and finally to the few animals who may be rehomed once they are no longer of use to the laboratories (Skidmore, Chapter 3).

I define myself as an animal rights activist and scholar. As part of my day-to-day work as Director of one of the largest primate sanctuaries in the US, I care for monkeys formerly used in animal experimentation. My position on animal testing comes then, not from direct experience of working in laboratories, but from my overall ethical opposition to the use of animals for human benefit and my experience of supporting some of the very few animals who make it out of laboratories alive as they work to overcome the trauma of animals exploited in this way.

What struck me throughout the three articles was the way in which the carefully constructed narratives of animal protection within the 'vivisection' industry begin to unravel when we scratch beneath the surface. Constructed narratives, such as that around the use of horseshoe crabs, pay lip service to the popular ideals of replacement, reduction, and refinement (the 3Rs) while the reality of using horseshoe crabs for blood extraction fails in both reduction and replacement of animals in procedures. Gorman notes that this makes the horseshoe crabs 'invisible'; their exploitation not only hidden from statistics on animal testing, but the entire invasive process to which they are subjected and, as a result of which a significant percentage of these animals die, being referred to as an 'alternative' to animal testing. In this, they become *non-animals*. They fall through the gaps left by lack of regulatory protection, and are, in effect, erased.

Horseshoe crabs become *non-animals* because the legal parameters within which the use of animals is regulated were constructed not (only) to mitigate harm to animals based on animal welfare science or ethics, but also as Myelnikov references: 'to enable scientists to get on with their work in peace' and to 'clip the wings of the increasingly violent extreme anti-vivisectionist movement by isolating them from moderate opinion' (p. 45). As Myelnikov clearly demonstrates, political manoeuvring played a significant role in the development of ASPA, arguably to the detriment of the animals themselves.

In the consideration of ASPA, the theoretically neutral civil servants explicitly excluded those who were calling for an end to vivisection. The use of terms such as 'radical' to describe those who oppose animal testing pitches them against the 'moderate' animal welfare advocates and the scientists, combined with mention of lab workers' fears of 'terrorist' attacks by 'animal rights' activists, serves to paint the civil servants as the rational mediators seeking to come to a conclusion that works for everyone. But by excluding the voices who question the ethical, welfare, and scientific foundations of animal testing on the basis that these views are *too radical* is to erase an essential part of the conversation. In silencing these voices, who seek to speak on behalf of the animals, like the horseshoe crabs, the constructed narrative that animal welfare, animal testing, and science, can coexist and thrive together is perpetuated without challenge.

Finally, Skidmore's discussion of rehoming animals after the lab provides a perfect example of constructed narrative in the description of the lab that invited journalists to see the 'cute puppies'. In this arguably cynical public relations exercise, the suffering of the other dogs exploited in labs is invisible while the puppies who are 'very, very cute' (p. 86) are put front and centre for the world to see. There is reference made to primates being rehomed by laboratories in some of the interview commentaries in Skidmore's piece, but no stats are provided that compare the number of animals who find homes, and those who live and die in the lab.

While a handful of dogs, and an even smaller number of primates, may be rehomed when they are no longer of use to the lab, the vast, vast majority of animals used in labs will be killed by the lab; either as part of the experiment itself, or once they are no longer deemed of use. And, as someone who runs one of the largest primate sanctuaries in the US, I can confidently say that the rehoming of primates, at least, does not represent the 'happily ever after' that members of the public might think it does. While a dog may be able to acclimatise to life in someone's home and – to a greater or lesser extent, dependent on the individual – overcome some of their past trauma, when it comes to primates, those who are not killed outright carry the trauma of the laboratory with them for the rest of their lives. One of the workers in Skidmore's chapter on rehoming refers to decisions on rehoming being made 'at the end of [the animal's] working life'. But we must remember that this is not 'work' for the individuals used in this way. No animal chooses this; they are non-consenting innocents, not consenting workers. This is not retirement with the proverbial gold watch. My concern here, as I will elaborate in the case studies that follow, is that we must guard against believing – and perpetuating the associated narratives – that the harms of the lab are reversed or remedied by the provision of post-lab care for the animals involved. One cannot justify the other.

Of the many individuals used in labs previously who are now cared for at the sanctuary I work at, there are a few whose stories demonstrate the ongoing harm that vivisection causes, even to those who are given the chance of life after the laboratory. There was Theo – a Rhesus macaque who was used in a lab for years. He developed conjunctivitis, which was not resolved by standard treatment, so he was taken to an eye specialist. Under anaesthetic, several inches of conductive wire was found in his skull, behind his eye, and was removed. It had been left in there from whatever experiment was conducted on him previously. He was almost blinded by this negligence. Notwithstanding the unnecessary pain he was subjected to, he struggled socially, too. His aggression was so severe that the sanctuary he was housed at before ours removed his canines in a desperate attempt to prevent him injuring others, and himself. When he came to us, we worked with him for years before we were able to see him settle into a social group that worked for him. He passed away peacefully in 2020. I am proud of the care we gave him, but I don't pretend he lived a happy life, either in the lab or after.

Dawkins (see Figure 4.1), another Rhesus macaque, arrived with us on a regimen of sertraline – an anti-depressant. In the three years that he has lived with us, he has struggled to socialise with other monkeys – something that would have come naturally to a monkey who has been raised in his familial troop and in his natural environment. Instead, Dawkins has spent limited time with other monkeys and, when he is able to cope with the company of others, he soon becomes stressed and aggressive. Due to his large size, when he becomes aggressive and begins to show stereotypical behaviour – often self-biting or grabbing fur – this puts both him and other monkeys at risk. As a result, he has spent a large part of his years with us alone.

Theo and Dawkins are two of dozens of monkeys who have passed through our doors, but their stories are representative. For wild animals, such as monkeys, life after the lab can never be a completely fulfilling life – even with the highest standard of care. Skidmore's point that public perception believes that dogs belong in the home is an important one. Rightly or wrongly, because we have subjected dogs to tens of thousands of years of domestication, they have the potential to go on to have a 'good' life after the lab (though this does not negate the trauma they have been subjected to). For non-human primates, whose captivity is a fundamental part of their suffering, even the best sanctuary cannot provide them with the life that they deserve. They will continue to live caged; they will continue to live in unnatural social groups (as even the best sanctuaries cannot safely recreate the complex social hierarchies that exist in primate troops in the wild), they live in the wrong country, the wrong climate, and are fed a diet that may be nutritionally complete but will be unlikely to properly resemble their diet in their natural home. Importantly, their worlds are so small and without the challenges of their natural environment. Their lives, even with the best care, are a shadow of the lives that they would live in freedom.

There is a tendency to believe that those who fundamentally oppose animal experiments do so from a position of ignorance. There is a narrative that has been perpetuated by industries that exploit animals, and by governments, that those who oppose animal testing are radical, anti-science, uninformed, and that their

Figure 4.1 Rhesus macaque Dawkins (Source: Born Free USA/ Ruth Montiel Arias).

demands are unrealistic. But when those voices are excluded from the conversation altogether – and those voices belong to the people charged with caring for the victims of vivisection – not just in the laboratories themselves, but in the years that follow, a vital part of the conversation is missing. When we have no one speaking for the animals, and only for the animals, we have a system that was deliberately designed to make invisible those who are harmed the most.

So, while I agree that horseshoe crabs should not be made *non-animals* by virtue of their exclusion from legal protection, and while rehoming more dogs, or more monkeys, from labs is generally a 'good' thing, these details seem to be distracting from the bigger issue. If horseshoe crabs *were* included within ASPA, it is likely that the procedures that they are submitted to would be the same, their suffering the same, their deaths the same. If more dogs or monkeys were rehomed, it still would not even scratch the surface when we consider the millions of animals who suffer and are killed each year in labs. Those who do get out would still go into their new lives traumatised – sometimes irrevocably so. Small changes and minor reforms may play out well in the public domain, much like the PR exercise of inviting journalists to meet cute puppies, but if we are *truly* concerned about the welfare of animals in labs, then conversations that make visible the individual animals and their lived experiences, and that allow space at the table for those who seek to speak up for their interests, are essential.

4.2

The institutional life of animals

Amy Hinterberger

When we study the institutional and legal life of animals, what do we study? The history of twentieth-century science and policy shows us that controversies about the use of animals in research are conduits for reassessing the relationship between humans and other animals. Within arguments about animal care, welfare, husbandry practices, reductions, and bans lie new imaginings for the political, economic, and social management of animal life in our societies. When I reflect on these new imaginings, however, I am also led to consider how, as social researchers, we actually study the institutional life of animals in a practical sense. In my own research, I have explored how cell-based biotechnologies are influencing and changing our ideas of human health and wellbeing. The goal of cutting-edge biomedical research is targeted at improving human lives, but animals are often the objects of research before it reaches the clinical stage (indeed if it ever does). To study biomedicine as a social and cultural practice means to also study the institutional treatment and management of animals. In tracing out how cell biology may be transforming human health, I have encountered the institutional management of animal life in many places: from primate research centres in Europe, to committee meeting room, to the 8th floor of a tower block in a busy North American town which housed genetically altered pigs. Accessing these sites as a social scientist is not easy, and researchers have been more excited to show me a dish of reprogrammed human heart cells that beat under the microscope than the animals housed down the hall, or in a nearby facility, who will be receiving these cells in experimental procedures.

So what about these many different animals and institutional processes that often remain hidden from view, unable to gain traction, not only in the social studies of biomedicine, but also in public or alternative visions of our relationship with animals? As these chapters make clear, our understanding and appreciation of animals in research requires expanded attention and vision. From these chapters we learn that there is much to be discovered about the institutional life of animals. By institutional life, I mean the forms of governance that are part of the general ecosystem of activities, laws and regulations relating to the use of animals in research involving both state-led proscriptions for actions, along with forms of governance that are initiated by actors and groups beyond the state.

From the politics of horseshoe crab blood to rehoming lab animals after experiments have finished, to the tracing out of the UK's legislative history of animal use, these chapters illuminate the lesser-known species and spaces of animal life. What can we glean from these chapters, then, about the evolution of our institutional and legal relationship with animals? The following are some lines of thought initiated by the many revelations and insights that cut across this section.

The institution's animal

A striking leading theme of the chapters in this section is that what counts as an animal in law is not self-evident, nor are the processes that shape institutional definitions of animal life. When it comes to the institutional management of animal life, animals are classified in numerous ways with some receiving forms of protection and others not. What we may consider to be an animal in everyday life, does not necessarily count as an animal under the law. These chapters lay bare many of the qualifications and technical complexities that accompany animals as they become the subjects of bureaucratic management. Gorman (Chapter 2) shows us how the horseshoe crab, an invertebrate whose blood forms part of the supply chains of contemporary biomedicine, does not count as an animal within formal animal welfare legislation. Animals that do not have a backbone have traditionally not counted as animals in welfare regulation, whereas animals that do have a backbone are viewed as sentient and entitled to some forms of protection. Recent changes in UK law, for example, granting sentient status to invertebrates such as crabs, octopuses, and lobsters, highlight how becoming animal in the law is a political and social process, subject to change. Indeed, in the US, for the purposes of the Animal Welfare Act, mice are not considered animals.

To this end, the institution's animal is specific and prescribed. Yet we also learn from these chapters about the arbitrariness of institutional animal regulation, along with in Skidmore's account of rehoming policies for lab animals (Chapter 3), how social and cultural relations to animals shape research practices. Myelnikov (Chapter 1) writes an illuminating history of the Animals (Scientific Procedures) Act 1986 in the UK and the bureaucratic Codes of Practice that accompanied it. He explains how some regulations around husbandry practices such as temperature and humidity were drawn from ethological knowledge, but others, such as pen size for dogs were enshrined simply by measuring the pens that dogs were already kept in.

Institutional personality

A great variety of animal species are put into motion, within different domains, across the chapters. Here it is useful to read across the section to learn about the significance of species when considering the ways in which our institutions care, control, create, and destroy animal life. A theme that emerges from doing this is that animals often require specific forms of personality to qualify for institutional protection. In an examination of rehoming practices for animals used in research, Skidmore explains how an animal higher on the species hierarchy is more likely to be considered an individual and receive care practices beyond required welfare legislation. Animals used in research, she explains, are not just ordered in relation to their proximity to humans but also in relation to each other, with monkeys, dogs, and cats often being viewed as individuals with personalities in need of care, as opposed to mice and rats.

The accounts of institutional animal life here point to a problem of care – if we can't connect, we can't care. This perhaps says less about animals and more about our relationship to the concept of care itself. Where care is not warm and fuzzy, but more troublesome, cold and quiet, we struggle to cultivate care. Gorman's horseshoe crab is beset by a number of challenges, and for those who want to reduce the half a million bled each year, how to cultivate care is a top priority. However, Gorman also shows us that a focus on care can obscure the significance of the technical infrastructures that shape the use of horseshoe crab blood – how it is seen in the research community as an *in vitro* procedure, how it is viewed as an alternative to using other animals. These are dilemmas that need attention to advance the care of horseshoe crabs. Such dilemmas about how to intervene or transform the long-standing research infrastructures that shape biomedical research and experimentation require understanding of our historical present. This is one reason why the kind of analysis provided by Myelnikov, in a detailed historical reconstruction of how the Thatcher government came to revise the UK's one-hundred-year-old animal laws, is so crucial. We may exhibit social care to some animals used in biomedical research we attach personality to, or connect culturally to some of the famous animals produced by bioscience (for instance, Dolly the cloned sheep), but meaningful change requires understanding the research infrastructures that shape current practice.

Institutional habitats

Finally, these chapters are significant because they ask us to consider: what is the proper place of animals within our institutions? What kind of institutional habitats should we be cultivating for animals used in research? The slogan discussed by Myelnikov, 'Put Animals into Politics!' used by animal welfare campaigners in the 1970s highlights the role and influence of animal advocacy movements. The message here is clear: formal regulation through legislation is needed. Though as we can see in both Gorman and Skidmore's studies, formal regulation covers only a small portion of animals used in research – the ecosystem is complex with a variety of governance structures both formal and informal.

For me, these questions of proximity and space should lead us also to look at where the regulation of animal life sits in relation to human life. Existing regulatory structures have been set up over the years to govern research on humans or animals through separate mechanisms and forms of oversight in bioscience. In most countries there exist two separate ethical and regulatory streams, one for biomedical research on humans and one for animals. These regulatory structures are part of deeper and underlying logics that have shaped assumptions about how to divide and organise research on human health between human beings and the animals that are used as substitutes or proxies for human subjects.

However, such regulatory structures come under strain as the techniques and materials of bioscience change and transform. For example, developments in stem cell science and gene editing techniques enable researchers to integrate human cells more accurately and comprehensively into non-human animals at different stages of development. These new forms of biomedical research require social scientists and humanities scholars to work across human and animal boundaries in biomedicine and ethics.

Biomedical researchers have historically had limited options to create models of human disease and development in laboratory settings. Animal models are commonly used but are subject to both experimental and ethical problems. However, new kinds of research tools that are made of human cells have significantly increased the ability researchers have to study and model human disease. New cell-based technologies are opening up and transforming the pursuit of human-specific models of disease and development. We do not yet know the outcome of what these changes mean for animals in research. However, what is clear is that assessing the place of animal use in bioscience will require engaging with our changing understandings of not only animal, but also human biology.

Conclusion

Euro-American culture is not disinterested in animals – quite the opposite. If you walk into any bookshop and survey the new titles section, you will no doubt find an abundance of books about animals. Far from the old Cartesian idea of animals as unfeeling machines, our current orientation to animals demonstrates an insatiable fascination with the social life of animals. There is a magnetic attraction of the human imagination to find alliances with animal life but how might our understanding of animal life be enriched if we turned this curiosity towards the institutional life of animals? Or towards the ways our own institutions manage and classify animal life? The chapters in this section, and this collection as whole, open up the meaningful possibilities offered by such investigations. Charting the institutional life of animals offers an opportunity to reassess how we approach animals, along with the political, cultural, and ethical stakes of contemporary biomedical research.

4.3

Regulatory connections and challenges

Robert G. W. Kirk

Writing as a historian and a member of the Animal Research Nexus Programme, if I was asked to identify a single theme that characterises this volume, I would choose connections. How different elements relate, become entangled, and reshape each other to drive historical change in what we refer to as the 'animal research nexus'. Tracing connections focuses attention upon the complex processes that simultaneously create, connect, and transform the very elements that make up our object of study: the scientific use of animals. From this perspective, animal research is approached as a historically constituted object, a moving object, made up of many parts, each changing the other and the whole over time. But what holds the whole together? What gives form to the animal research nexus? Many answers are possible, but perhaps the most prominent is regulation. Today, regulation is one of the primary functions of the modern state. The way regulatory authority is applied and administered creates and governs connections between state, industry and other stakeholders, and the public. Regulation is therefore a political arena and unsurprisingly humanities and social sciences research has much to say on the subject. Conventionally framed as the state operationalising an obligation to protect the public, regulation is nonetheless vulnerable to 'regulatory capture' where the work of regulation gradually becomes inverted toward protecting industry, stakeholders and other interests against those of the public.[1] Various social, political, and economic arguments lend themselves toward processes of 'deregulation'. Regulation can be construed to place unwarranted limits upon innovation and productivity, while a tendency toward centralising power within an unelected, invisible, and seemingly unaccountable administration – as well as the alternative approach that diminishes centralised authority in favour of cooperative structures of public-private governance – might each be perceived as exhibiting a 'democratic deficit' that falls short of the expectations of a modern liberal democracy.[2]

Within animal research, regulation is neither a simple nor a stable object. Nevertheless, examining regulation reveals the most prominent and influential concerns that make up the 'animal research nexus' upon which most of society might be thought to agree upon. In Britain, the Cruelty to Animals Act 1876 was introduced to reassure the 'public' that animals were not subjected to unnecessary suffering in the pursuit of science. As Myelnikov (Chapter 1) shows, the push toward public reassurance made 'consensus' the guiding strategy of the reforms that produced its successor the Animals (Scientific Procedures) Act 1986 (ASPA). Inevitably, such an approach empowered some voices over others. In pursuing consensus, arguments at the polar end of debate lost influence and in this example those who favoured abolition (such as 'anti-vivisectionists') were gradually excluded from shaping reform. Nevertheless, building consensus on issues such as pain (where the interests of animal welfare gave ground), and recognition of the value of veterinary expertise on laboratory animals (where, eventually, historical resistance to veterinary involvement in regulation gave way), among many others, was far from simple. Yet deals could be done on the basis of the need to diffuse public concern. The resultant regulation gave shape to the animal research nexus and that in turn shaped the scope of the research that makes up the content of this volume. Yet, as Tyson's commentary (Chapter 4) makes clear, our work has not done as much as it might have to make visible the interests and arguments of those that are excluded from regulation. In a similar critique later in this volume, Giraud (Chapter 8) argues that the form of animal welfare and care examined here is specific to and operates through the logic of a situated institution: the laboratory. Hinterberger's commentary (Chapter 4), too, recognises that this volume is tightly focused on the 'institutional life of animals', recognising the value of such work in revealing the political, cultural, and ethical stakes of contemporary biomedical research. As useful as this may be, given that the laws of the laboratory are determined primarily (though not exclusively) by regulation, Tyson and Giraud make a cautionary and necessary point in drawing attention to the limits of our analysis of the animal research nexus. Gorman's (Chapter 2) provocative exploration of the horseshoe crab reveals regulation to be central to contemporary biomedical practice, broadly invisible to the public imagination and almost impossible to locate within regulatory concern as it straddles animal research, ecology, and conservation while transcending national boundaries. Through the horseshoe crab Gorman reveals the limits of regulatory imagination. Yet, as Tyson suggests, the inclusion of the horseshoe crab within regulation (however that might be achieved) would do little to assuage the concerns of those who believe that no level of animal welfare can justify the practice of animal research. How, then, might we better include those interests and voices that hitherto have marginal influence upon animal research?

Perhaps greater attention to perceptions of the public. The horseshoe crab is a marvellously interesting animal, but can hardly be said to possess what Lorimer describes as a 'non-human charisma' in the contemporary imagination.[3] In 1876 dogs, cats, and equines received privileged regulatory status due to a perception that the presence of these species in human society made them of most concern to the public and, at least some believed, a heightened capacity for suffering. These species continued to be privileged in ASPA, no doubt in part because of the conservative nature of reform but also, as Myelnikov describes, in light of controversy about the use of dogs in tobacco research that can only have emphasised their prominence within public concern. Perception of charisma in animals, as Skidmore (Chapter 3) demonstrates, can shape not just what is included within regulation but how regulation is enacted. But such perceptions are malleable, shaped by experience and by the accumulation of knowledge about a species (thus animal technicians who work with rats develop a different understanding of the species to that of 'vermin', which tends to dominate public imagination). In this way, animal research itself can be seen to drive shifts in how society perceives non-human animal charisma. The reforms of 1986 extended privileged protection to non-human primates, but neglected to extend regulatory scope to include *octopus vulgaris* despite a clear consensus that this species had near equivalent sentience to higher vertebrates. A significant difference between non-human primates and *octopus vulgaris* was their location within the public imagination. Popular science, ethological study, and environmental concerns had propelled the non-human primate to the forefront of public concern whereas *octopus vulgaris* had enjoyed no comparable transformation and so had to wait until 1993 before recognition as a regulated species and perhaps as late as *My Octopus Teacher* (Netflix, 2020) before public imagination fully embraced the species.[4] If regulation changes in lockstep with scientific knowledge and public concern, what steps could we take to facilitate and systematise this process? What might bring forward recognition of the non-human charisma of the horseshoe crab and make visible further excluded voices, interests, and species?

Animal research regulation has, historically, served to reassure the lay public without involving lay persons within its governance structure. The Cruelty to Animals Act 1876 made no provision for lay involvement, although having a non-scientist chair of the Advisory Committee served a bureaucratic purpose in that it encouraged scientists to render the discussion and conclusions in language more easily understood by the Home Office.[5] Under ASPA, lay involvement was slightly expanded in the reformed Animal Procedures Committee, though it was not until the late 1990s and the development of the Animal Welfare Ethical Review Board that a role for lay opinion was gradually institutionalised. The involvement of lay persons can be understood as a response to the democratic deficit within regulatory decision-making but it is also more than this in that it brings diverse perspectives to a problem that is at once scientific and societal. Perhaps we can build on that through a mechanism that would allow more systematic, and regular, public participation in the making as well as the application of regulation. This mechanism could include those voices and interests that are as yet too frequently excluded from the consensus upon which regulation and the animal research nexus rest. One possibility might be to adapt the model of the citizen assembly.[6] Consisting of a representative group of randomly selected citizens, the model of the assembly is specifically intended to address divisive and complex issues by providing time for engaged discussion and informed decisions to serve long-term societal interests. Given that perceptions of public concern appear to play a significant role in shaping regulation, a mechanism like a citizen assembly may facilitate more accurate tracking of the former through the latter, by including presently excluded interests such as abolitionist arguments. It would also empower a representative group of citizens to deliberate and inform the future direction or regulatory change within animal research.

Notes

1 Science and knowledge production can play a role in regulatory capture; see for instance, Andrea Saltelli et al., 'Science, the Endless Frontier of

Regulatory Capture', *Futures*, 135 (2022), 102860, DOI: 10.1016/j.futures.2021.102860.

2 Victor Bekkers et al., *Governance and the Democratic Deficit: Assessing the Democratic Legitimacy of Governance Practices* (Aldershot: Ashgate 2007); David Levi-Faur, *Handbook on the Politics of Regulation* (Cheltenham: Edward Elgar Publishing, 2011).

3 J. Lorimer, 'Nonhuman Charisma', *Environment and Planning D: Society and Space*, 25.5 (2007), 911–932, DOI: 10.1068/d71j.

4 See also Peter Godfrey-Smith, *Other Minds: The Octopus, the Sea, and the Deep Origins of Consciousness* (London: William Collins, 2017).

5 Robert G. W. Kirk and Dmitriy Myelnikov, 'Governance, Expertise, and the "Culture of Care": The Changing Constitutions of Laboratory Animal Research in Britain, 1876–2000', *Studies in History and Philosophy of Science*, 93 (2022), 107–122, DOI: 10.1016/j.shpsa.2022.03.004.

6 https://citizensassembly.co.uk/ [accessed 1 February 2023].

Part II

Culturing and sustaining care

5

Subjugated love: aligning care with science in the history of laboratory animal research

Robert G. W. Kirk

Introduction

> We are heirs to an ancient tradition that opposes the life of the mind to the life of the heart, and to a more recent one that opposes facts to values. Because science in our culture has come to exemplify rationality and facticity, to suggest that science depends in essential ways upon highly specific constellations of emotions and values has the air of proposing a paradox ... The ideal of scientific objectivity, as currently avowed, insists upon the existence and impenetrability of these boundaries. I will nonetheless claim that not only does science have what I will call a moral economy ... certain forms of empiricism, quantification, and objectivity itself are not simply compatible with moral economies; they require moral economies.[1]

Where do laboratory animals come from? Where do our contemporary practices of laboratory animal care come from? These questions are not meant in a practical or literal sense. Rather, they mean to ask where the *concept* of laboratory animals came from; where the *ideas* underpinning our approaches to animal care came from. Or more properly, when. This chapter takes a historical approach to answer these questions. In doing so, it argues that the social and material infrastructural framework that makes up and sustains the contemporary work of laboratory animal research must be thought of as historically situated: a product of its time and place. As such, the laboratory animal is, to a greater or lesser degree, an invention. An invention that has changed, can change, and will be subject to further change. From this perspective, the history of animal research can, therefore, not only deepen our understanding of why current practices are as they are but also why they remain so and how, potentially, they may be changed.

Historical work is notoriously equivocal in its theoretical commitments. History is not a science, but nor is it entirely an art. The historian's impossible task is to understand the past on its own terms. Accordingly, historical research is imagined as being driven by empirical evidence from the past as opposed to theoretical hypotheses from the present. In practice, however, the past can only be read from the perspective of the present. Nothing is so antagonistic to cordial debate, or corrosive to the production of historical work, than asking historians to define precisely what – epistemologically speaking – historical knowledge is.[2] Nevertheless, Jill Lepore summarises history well in writing that it 'is the art of making an argument about the past by telling a story accountable to evidence'.[3] One of the many challenges of collaborating across disciplines is building common ground; how does a discipline that often runs fast and loose with its theoretical commitments work with the social sciences where theory seemingly has a more prominent role in driving research agendas? One approach is to insist on the historicity of knowledge. An example is Donna Haraway, whose work is shaped by a long career traversing and transforming the humanities and social sciences in unexpected and innovative ways, ways that are first and foremost attentive to history yet remain shaped and to a large part motivated by her early training in the biological sciences and later theoretical critical commitments.[4] Haraway writes that 'Beings do not preexist their relatings' as '[t]here are no pre-constituted subjects and objects, and no single sources, unitary actors, or final ends'.[5] Starting from this observation, and without discounting that the scientific use of animals has a longer history, this chapter argues that the 'laboratory animal' might best be understood as a product of the mid-twentieth century. Prior to the laboratory animal, animals were used for all manner of scientific purposes, but these animals were more often than not obtained from random sources, purchased from commercial markets with little interest in science and its needs.[6] In contrast, the laboratory animal was conceived as a new form of life created for and of the laboratory, produced through new ways of relating embodied in new fields of knowledge ('Laboratory Animal Science and Medicine'), new social and material infrastructures and practices (requiring the reconfiguration and elevation of the status of the 'animal house' relative to the 'laboratory'), and new roles (such as the 'animal technician'). The first and second sections below sketch the contours of these historical developments with a focus on the 'invention' of the 'laboratory animal' and the 'animal technician' respectively. Fundamental to the success of this endeavour was the transformation of animal care from undeveloped and inexpert activity delegated to unskilled labour, to a new profession grounded in new scientific knowledge and technical practice. Animal care became animal technology.

Yet scientific objectivity, as the historian and philosopher Lorraine Daston reminds us, exemplifies 'rationality and facticity' in working to exclude emotions and other subjective values.[7] The third, and final section, charts how the invention of the 'laboratory animal' and the 'animal technician' considerably expanded the elements that made up the animal research nexus and attempted to align the new whole to the service of science. As a technical practice, animal care was, and remains, premised on prioritising science as the objective framework for ordering the management of laboratory animals. In recent years, the social sciences have turned to investigating how animal care operates within the biomedical sciences, and how care for the animal has been connected to care for science, as well as how animal care in the laboratory can become entangled with care for the patient in the clinic.[8] This work has begun to map how everyday working practices in animal laboratories operate through collective moral decisions that exceed a straightforward adherence to regulatory expectation.[9] At the same time, work within the history and philosophy of science has examined how the 'animal model' and/or 'model organism' operates within experimental epistemology.[10] In charting the historical origins of laboratory animals and animal technology, this chapter explores attempts to align animal care and husbandry with the perceived needs of the epistemology of experimental science. It examines the challenge of managing subjective elements such as a feeling of love for animals, which simultaneously appeared essential to components of care, yet in resisting quantitative measurement and scientific standardisation seemingly had to be subjugated within the rational discourse of animal technology. Subjugated, but never entirely erased. As such, laboratory animal care and animal research might therefore be productively understood as possessing what Daston describes as a moral economy of science, wherein objectivity is not simply compatible with subjective factors such as emotions and affect saturated relations, but dependent upon them. Laboratory animal care, as much as the laboratory animal itself, is therefore shown to be of and for the laboratory.

Inventing the 'laboratory animal'

From the 1940s, a new field of expertise, 'Laboratory Animal Science and Medicine' (LASM), developed independently at multiple locations for initially localised reasons, before cohering at the national level and eventually becoming an internationally coordinated programme. Conceived as an auxiliary specialism in the service of animal dependent biomedical research, proponents of LASM aimed to improve the quality and quantity of animals available to biomedical research.[11] Conjoining 'laboratory' and 'animal' within the title of this new field emphasised the intent to create new and 'sophisticated' forms of life for, and of, the laboratory described at the time as being:

> [d]eprived of their primitive naturalness; deprived, that is, as far as practicable of generic imprecision, disease, infestation, nutritional deficiency, and other stressful influences. They are animals whose sophistication implies removal of adulteration, corruption, perversion and falsification.[12]

'Science' and 'medicine' were equally meaningful, indicating the methods and epistemological concerns by which these new sophisticated animals were to be created. LASM sought to align the diagnostic and experimental use of animals (serving the needs of science) with a more clinical approach drawing on human and veterinary therapeutic practice (serving the needs of animal welfare) whilst simultaneously rationalising the logistics of animal provision for biomedical research worldwide. Though closely related and possessive of the same end goal, scientific research and clinical medicine operate in distinctive ways, which can prove challenging to successfully align.[13] By drawing on heterogeneous and otherwise independent forms of expertise, including genetics, ethology, pathology, microbiology, epidemiology, nutrition, veterinary medicine, and others, LASM positioned itself as a new auxiliary field managing the interface of the needs of experimental epistemology and those of animal welfare, ultimately working to improve the production, supply, and above all quality of laboratory animals. Now understood as unique products of and for the laboratory, LASM tasked itself with establishing new, instrumentally orientated ways of relating to animals by making the experimental animal itself the object of scientific study. Ultimately this work invented a new form of life, the 'laboratory animal', by transforming spaces such as the 'animal house' from a place of little consideration or status to one governed by the same logics as the laboratory and therefore deserving of equal respect, prestige, and reward.

The Second World War provided a power stimulus to the formation of LASM by simultaneously disrupting the supply of animals for experimental research and heightening the importance of scientific work through its contribution to the war effort. In Britain, for example, an unprecedented coalition of scientific organisations formed in 1942 under the banner of the 'Conference on the Supply of Experimental Animals' (CSEA) to call for state intervention to improve the quantity and quality of animals available.[14] A. L. Bacharach, who chaired the CSEA and went on to oversee the implementation of its recommendations by the British Medical Research Council, believed post-war reconstruction should establish a national supply of animals for scientific research. Bacharach was an early British proponent of the importance of using genetically uniform laboratory animal populations that could serve as 'standard' tools equivalent to the chemist's litmus paper.[15] By the 1940s, however, Bacharach and others had realised that environmental considerations were equally important to experimental outcomes.

Accordingly, one of the motivating concerns of LASM was to ensure that only purpose-bred animals of known environmental background were used for biomedical research. Reflecting on the challenge of bringing this about in 1947, Bacharach mused:

> The more I think of the phrase 'experimental animals', the less I like it ... because the animals are not strictly speaking experimental, except in so far as they carry out experiments on the unfortunate people who try to produce and tend them! ... [I]s not the phrase 'laboratory animals' more accurate than the intended one?[16]

This was reflecting on more than a question of nomenclature. The preference for *laboratory* over *experimental* marked a symbolic break from the past. Whereas *experimental* defined the animal instrumentally, *laboratory* prioritised the animal's material environment and social relationships. In naming this new animal, and thereby the field that was to sustain it, Bacharach sought to emphasise its unique identity distinct from animals found in nature. Subsequently, as LASM established itself as a specialist, auxiliary, scientific field, the 'laboratory animal' crystallised as a historically and ontologically distinct form of life. The laboratory animal was simultaneously the product and object of scientific knowledge. The construction, production, maintenance, and care of laboratory animals was not, however, conducted primarily in the laboratory. Rather, it was the work of a newly re-imagined 'animal house', elevated from sheds, basements, or similar otherwise inconsequential places to a space designed for and governed by science.

In the early twentieth century, inbreeding mice and rats to create animals exhibiting similar responses to similar stimuluses began to be used to create 'standardised' experimental tools.[17] However, using such purpose-bred animals was adhered to mostly in disciplines where it could not be avoided (such as genetics) and within institutions that were particularly well resourced (such as the National Institute of Medical Research in the UK). As late as the 1950s, most animals used for experiential purposes were sourced from a commercial pet trade that had little understanding of, or interest in, science.[18] Breeding animals in-house was difficult, laborious, expensive, and risky due to the frequency of disease wiping out stocks, as well as the general dearth of knowledge as to best practice. Yet purchasing purpose-bred animals was near impossible due to the absence of scientifically literate commercial breeders. At the time, the culture of sourcing animals for experimental use was focused on obtaining animals in sufficient number because demand exceeded supply. In a commercial market where demand outstripped production, the quality of animals was rarely considered because power lay in the hands of the supplier. Recognising that laboratories required large numbers of animals and were seemingly unconcerned by their quality, commercial breeders disposed of animals that previously had no viable sale value (so-called 'wasters') to the new and seemingly insatiable laboratory consumer. Laboratories had little choice but to prioritise quantity over the quality of their animals.

Reliance on a dysfunctional commercial market created numerous problems, not least the common practice of having to source animals from anywhere they could be found for sale. Mixing animals from different breeders meant that the animal house brought together populations with varied and unknown microbial backgrounds. Infections tolerated in one population frequently caused illness in another, making disease outbreaks common. The consequent wasteful loss of animals to illness or controlled slaughter placed further pressure on an already unfit for purpose commercial market. In sum, early twentieth-century laboratory animal houses struggled to maintain adequate stocks of animals, and standards of animal health were far from what desirable.

Widespread recognition of an urgent and growing problem with the provision of animals for science did not mean individual researchers were receptive to changing their ways of working. On the contrary, this was the norm and had come to be accepted. As late as the 1960s one commentator complained that, despite acknowledging that uncontrolled variables such as ectoparasites, endoparasites, protozoa, bacteria, and viruses posed challenges to the reliability of experiment, and the unexpected loss of animals prior to or during experiment due to ill-health frequently frustrated experiments, researchers thought of this problem as 'of little moment', preferring to work in 'something of a fool's paradise' than to take action to improve the situation.[19] Improving the quality of laboratory animals required a greater acknowledgement of the significance that animal provision, as well as animal husbandry and care, had for experimental research. Hitherto, the work of the animal house had not previously been imagined as having any importance beyond making available animals as and when experimental work required them. Reimaging the work of the animal house as being integral to the work of the experimental laboratory required greater financial investment in animal houses as well as improved recognition and a change of social status for those who worked within them. Continuing to live in a 'fool's paradise' was far preferable to facing the cultural and practical implications that bringing about such changes entailed. Few were willing to accept without argument the transfer of already limited resources from the experimental laboratory to animal house.

To overcome this and other barriers to change, much of the early work of LASM focused on establishing that reliable experimental science required animals of high 'quality' (quality being defined as a known standard of health).[20] Ectoparasites, endoparasites, protozoa, bacteria, and viruses introduced uncontrolled variables into the experimental design that would, in today's language, lead to problems with the validity and replicability of any outcomes. Epistemological concerns, however, were less persuasive than economic arguments. Unexpected loss of animals during experiment from disease or otherwise inexplicable death had led to the common practice of using more animals for a given purpose than was otherwise statistically necessary to obtain the desired results. This not only placed greater pressure on an already overstrained commercial market but had demonstrable economic consequences. The use of seemingly affordable randomly sourced animals of unknown backgrounds was not merely unscientific, it was grounded in a false economy. This approach made experimental work dependent on a commercial market where costs fluctuated uncontrollably, and action to improve the animal quality was impossible to enact. On the other hand, investment in purpose-bred animals, of known background, might appear more expensive at first but would contribute to better controlled experiments. Moreover, improving quality and standards of health would require fewer animals to be used to obtain the same results. LASM argued that investment in the production of higher quality laboratory animals might prove costly in the short term but promised better science and economic savings in the long-term.

This argument gained momentum once placed with the wider context of mid-twentieth-century shifts in thinking about the role of the nation state. LASM was keen to include the provision of laboratory animals within wider reforms that resulted from governments embracing an obligation to deliver improved social health and welfare for their citizens. In Britain, the founding of the National Health Service in 1948 established a new and significant economic burden on an already stretched economy. In the view of LASM, the national health and economy would be best served by state investment in laboratory animal provision because the 'cost of laboratory animals, the cost of medical research, and the cost of maintaining the health of the population are in steeply ascending order of magnitude' meaning it would be wise to begin by improving 'the cheapest end of the scale'.[21] For most laboratories at the time, however, purpose-bred animals were not just prohibitively expensive but logistically beyond reach. Little systematic investigation had been done to establish how to produce and maintain large numbers of small animals, and even less was known as to how animal genetics, health, and other experience shaped the responses of animals under experiment. Accordingly, the invention of what one early pioneer of LASM named the 'sophisticated' laboratory animal required more than just investment in the animal house. It necessitated research into and the development of new knowledge, technologies, practices, skills, and approaches to breeding, maintaining, and caring for animals. From a social and cultural perspective, it also called for the creation of a new profession, akin to a new human identity for those who worked in the animal house.

Inventing the animal technician

In 1952, Edward Joseph Fitzgerald, a member of the medical staff of the British Home Office tasked with the regulation of animal experimentation, reflected:

> In the old days 'Sarah Gamps' were the only nurses in hospitals, and today trained nurses are indeed skilled people. My own personal view is that a similar raising of the standards of animal house attendants is highly desirable, not only in the interests of research but in meeting any criticism that may arise from the general public on the care of experimental animals.[22]

Sarah Gamp, imagined by Charles Dickens and featuring in the novel *Martin Chuzzlewit*, was a well-known inept, untrained, and disinterested caricature of a nurse prior to the late nineteenth-century transformation of this role into a skilled profession. Portrayed as untidy, slapdash, and overly fond of gin, Gamp was as much a moral indictment of the lower classes as she was a criticism of past approaches to the care of the sick.[23] Under the Cruelty to Animals Act (1876), British law regulated scientific experimentation providing no mandate for intervening into the animal house. Nevertheless, Home Office inspectors frequently made recommendations on the proper approach to husbandry and care in the animal house, having become informal experts on best practice due to their ability to contrast approaches taken in different institutions. In drawing on the unflattering character of Sarah Gamp, Fitzgerald was tacitly critiquing existing standards of care within the animal house and encouraging the professionalisation of animal care on the model of modern nursing.

Prior to the 1940s, animal care was an unimportant role delegated to whomever happened to have time available, often 'unskilled workers such as porters and janitors'.[24] Reform would be challenging, not least because it necessitated rebalancing the relative standing of the animal house and laboratory and improving the status of those who worked within the former to something similar to that of the latter. It would also require the development and introduction of new skillsets grounded in new knowledge to underpin a new profession:

> The training and status of animal attendants require attention. At present they are usually drawn from relatively low grade labour and their prospect for advancement is small. If the quality and quantity of experimental animals are to be improved, this state of affairs cannot continue. In the course of their training all biological laboratory technicians should spend a proportion of their time in the animal houses, so they acquire familiarity with the rules of good animal husbandry and are competent to care for animals under experiment in their departments. The status and therefore pay of regular animal attendants should be comparable with those of laboratory technicians.[25]

As with the invention of the laboratory animal, naming this new profession was considered critically important. Hitherto, there had been no attempt to standardise animal house labour; workers were variously referred to as 'attendants', 'keepers', 'caretakers', and 'curators' (smaller institutions sometimes included the work within the remit of the janitors). In an effort to mark a break from the past and emphasise the professional and skilled nature of the new profession, these terms were abandoned in favour a new title: the 'animal technician'. The language of technician evoked the intended transformation of animal care from inexpert largely affective labour to a skilled technical profession grounded in science on a par with the laboratory technician who assisted experimental work. As one contemporary explained, this was the 'day of the specialist' and therefore '[p]roperly cared for animals were attendant upon properly cared for and trained Technicians who were *not* just muckers out of animal cages'.[26] Animal technology was imagined as a field, membership of which would bring the pride and status accorded to any professional occupation. Accordingly, professional organisations were established. In 1950, the British Animal Technicians Association formed to advance 'the modern view of the status of the animal house and its staff' and to 'ensure that the enhanced prestige of animal technicians is justified by their professional standards'.[27]

Professionalising animal care required 'a system of training men and women for the job', capable of embedding new skilled and technical forms of labour into biomedical working cultures.[28] Establishing animal technology as a profession, and ensuring 'the care of animals on a real understanding of their needs', was an essential component of the wider trend to make the laboratory animal itself an object of scientific study.[29] The goal was to standardise all activity within the animal house to ensure the production and maintenance of healthy animals suitable for controlled scientific experiment. Aligning labour in the animal house to better serve that of the laboratory assumed that animal care should be grounded in objective standards derived from the scientific study of laboratory animals. This approach lent itself to education, qualification through examination, and ultimately centred concern on the material practice and infrastructure of the animal house. Care became a matter of learning to properly manage matter. For example, one early guide to animal technology emphasised that a 'clean and tidy animal house is the hall-mark of a good animal technician' because it reflected an ordered approach governed by due attention to animal health.[30] Where animals were sourced externally, new regimes of stock management were introduced to minimise the risk of latent infection entering the animal house. Specialist technologies such as disinfectant footbaths and the routine use of autoclaves were developed, alongside new standards of personal and environment cleanliness. The prominence of attention to cage design, animal foodstuff, bedding, and environmental controls illustrate how 'skilled' animal care was institutionalised through practices and infrastructure which, in being material, could be rendered into and represented as objective standards.

Efforts to create and maintain animals of known pathogenic background, or Specific Pathogen Free (SPF) animals, are indicative of the trend to transform animal care into a technical practice. New hygiene regimes, coupled with rigorous decontamination practices and architectural innovation, remodelled the boundaries of the animal house to serve as robust physical barriers. The known microbial ecology within the animal house was thereby separated from the unknown pathogen filled environment without. Innovative surgical techniques combined biology with engineering to enable the routine production of 'germ free' animals – organisms entirely separated from microscopic and parasitic life. The womb of pregnant females was surgically removed before birth, decontaminated and transferred into a sterilised mechanical isolator, within which progeny were surgically released and reared by hand (if first generation) or by foster parents (if second generation or later). Animals could be maintained in a germ-free state or exposed to an environment of known microbial ecology to create 'clean' animals free of specified pathogens. Producing and maintaining clean, microbially defined animals transformed the skillsets required by the animal technician as well as the material infrastructure that made up the animal house, reshaping animal care into a technical practice.

Work to professionalise animal care, improve the laboratory animal and raise the status of the animal house did not go uncontested. In 1968 Peter Medawar, Nobel laureate and early advocate of improved standards of laboratory animal care, reflected on how critics and supporters alike recognised the new animals as marking a significant break from the past.

> Before its principles could be adopted, elder statesmen in the world of medical biology had to be re-assured that pathogen-free animals were not in some way abnormal or unrepresentative, and therefore liable to give misleading results. Today we are more inclined to think that it was the animals they worked on that were abnormal, and (what is worse) abnormal in unknown ways ... the SPF revolution is simply raising the general level of hygiene from nineteenth- to twentieth-century standard ... we are now in fact doing for laboratory animals what has been done for human beings over the past 100 years.[31]

In subsequent years, as the use of SPF animals became established as the expected standard of health, attention turned to the ecological interdependence of all aspects of the animal house. The work of maintaining secure microbial environments was a collective endeavour, which placed ever greater emphasis on connections and interdependence. Maintaining the personal hygiene of the individual animal technician was as important as policing the health of animals, producing an increasingly ecological framing of the animal house.

Alongside the biological and material considerations, social and cultural considerations formed part of the emerging ecological vision of the animal house and its working relations. Early advocates of LASM were keen to elevate the status of animal care to be considered equal in prestige and importance to the experimental work of the laboratory, arguing that the

> animal house may be considered *either* as a subdivision of the laboratory *or* as a complementary department. In the first case, there is often a tendency for the most junior laboratory technician to be posted to the animal house. Those who show no particular promise remain there indefinitely ... The more ambitious technician looks forward to his return to the laboratory, where alone he sees a reasonable future. Thus the animal house attracts nobody and retains only the less ambitious or competent members of staff. In the second case, the animal house is regarded as a complementary department, where duties and promotion run parallel with the laboratory. Hence animal technology offers a career alternative to laboratory technology, and fully as satisfying ... If it is true that good experiments demand good animals (and who can doubt this?) then the latter arrangement offers every advantage over the former ... The old idea of staffing the

> animal house with men and women who are not considered fit to be trained as laboratory technicians should be abandoned as wasteful, inefficient and obsolete.[32]

As Medawar alluded to above, work to convince the 'elder statesmen in the world of medical biology' that new approaches to laboratory animals, their care, and the status of their careers, was appropriate, took time. Reasons for resistance were many and varied. While few doubted that 'good experiments demand good animals', the meaning of 'good' was open to dispute. Any argument that the cheap animals and ad hoc approach to animal care that had been used for years were inadequate for experimental science going forwards risked casting doubt on the validity of the work of the past. There was also the economic cost of new practices, which was often substantial. A significant challenge to change, however, was epistemological, relating to a lingering concern that the work of the animal house was not, and could not be, entirely transformed into a scientifically grounded technical practice. Some aspects of animal care appeared to resist codification into rational, objective, and measurable standards, posing a more conceptual barrier to fully integrating the work of the animal house with that of the laboratory.

Beyond objectivity: care and the moral economy of animal technology

Animal health could be rendered tangible, measurable. Health, as the marker of quality in laboratory animals allowed the standardisation of labour in the animal house. Bacteria, viruses, and other pathogens and parasites could be identified, rendered visible, and thereby controlled though new technologies and ways of working. The physical and infrastructural environment of the animal house could be similarly studied and adjusted to provide optimal temperature, humidity, and comfort to enable laboratory animals to thrive. Studies of laboratory animals' nutritional needs produced standardised diets available all-year round that were no longer dependent on seasonal production of foodstuffs. These and other new ways of working allowed animal care to be formalised, communicated, taught, learned, and assessed. In this way, labour in the animal house could be aligned with experimental science and its commitment to objectivity. Nevertheless, some aspects of animal care resisted efforts to objectify them as technical practices because elements of their practice evaded representation in language. Methods of handling animals, for example, struggled to rationalise the immaterial, affective considerations that were essential to avoid causing distress.

Unlike practices related to improving health and hygiene, which mapped relatively easily onto technical and material factors such as cage design and cleanliness, physical interaction between animal technicians and laboratory animals was challenging to standardise. As one commentator explained, the 'proper method of handling animals could not be taught by lecturing or learned by reading … the most satisfactory method of teaching is by practical demonstration, and of learning by constant practice'.[33] Success relied on an emotional attunement of human to animal, performed through shared responsiveness and learned in large part by shared experience built over time. In a sense, it was an intimate relationship that could only be understood in the doing. While species-appropriate approaches to handling animals could be communicated in writing and through illustration, some aspects nonetheless escaped these forms of communication.[34] Animal handling necessitated a tactile and subjective feeling for the animal that was equally attentive and responsive to the feeling of the animal toward the handler. Handling revealed the immaterial, affective factors that underpinned good animal care and consistently supervened in attempts to standardise and rationalise the work of the animal house.

The use of physical tools such as '[f]orceps or leather gloves' to minimise the subjective factors involved in handling were found to undermine the relationship upon which animal care depended.[35] In part, this was for practical reasons as gloves reduced physical sensitivity and thus the ability to sense whether one was using too much or too little pressure when holding an animal. However, forceps, gloves and other mechanical implements were also found to agitate the animal in a way that human touch appropriately deployed would not. Moreover, feeling for the animal that developed through the shared experience of intimate handling was productive. Regular handling was essential to creating 'docile' laboratory animals as familiarity was found to discourage feelings of 'insecurity' in animals, providing that the animal technician was 'relaxed' and 'approached the animal with quiet confidence'.[36] On the other hand: 'Improper handling may result in injury to the animal, to the technician himself, or, most of all, to the animal-man relationship' that was vital to animal care.[37] Arguably, more than activity within the animal house, handling revealed highly specific, albeit illusive, constellations of affective experience and emotion, which resisted alignment with the scientific ideal of objectivity. Furthermore, as one influential guide to animal technology explained, the 'importance of handling animals in the correct manner cannot be overrated' because it sustained the positive human–animal relationship upon which good animal care depended.[38] Handling was crucial to animal care and in being so, demonstrated the critical role that subjective elements played within the moral economy of animal research.

In contrast to the laboratory, where the subjective factors were expunged from published reports of experimental science, the affect and the human–animal relationship haunted animal technology literature. Animal technology discourse adopted technical language whenever possible, yet subjective elements nonetheless supervened because affective and often unsayable experience was recognised to be an essential component of good animal care. Animal technology's aspiration to objective representation was thereby diluted through the necessity of acknowledging its dependence on subjective experience. One consequence of this tension was greater prominence of the language of emotion alongside anecdotal and ill-defined knowledge within animal technology literature to capture the role of affective relationships and other qualities that could not be easily measured, rendered tangible, or made certain.

A prominent example was reports of suspected correlations between changes in human personnel and the health or behaviour of animals within the animal house. In one instance, the absence of an animal technician led to an otherwise healthy colony experiencing numerous sudden animal deaths that could not be explained biologically but ceased on the return of the usual animal technician from their holiday.[39] In another, routine monitoring of growth rates revealed unexplained yet consistent cycles where weekly growth slowed or reversed over the weekend. Such observations saw animal technology discourse switch registers from quantitative to qualitative evidence:

> If one visits the animal house in the evening, and one stands outside the door of the mouse room, the commotion going on inside is considerable, much greater than during the day ... If one enters the room, the commotion dies down ... it seems that the presence of human activity depresses murine activity, so that the mice keep quiet and grow fat.[40]

The author was apologetic in offering such a 'rash' observation and acknowledged that 'presence' pushed at the limit of what could be accepted as a tangible scientific object. Yet it was the fact that such observations were difficult to align with science that indicated their importance. Anecdote, alongside ill-defined reference to emotion and experience, provided a partial explanation for phenomena that would otherwise be left unknown and invisible yet nevertheless were essential to animal care. In this way, the human–animal relationship marked the limits of technical discourse, indicating that some elements of animal care lay beyond the calculable.

Animal technology aimed to professionalise care by making the activities of the animal house 'more of a science and less of a craft'.[41] Yet animal care required qualitative elements that were akin to craft-like practices not easily reduced to standardised, technical, and quantitative discourse. Handling was one amongst several areas where subjective qualities supervened within otherwise objective language. For instance, a 1945 report recommended improvements to laboratory animal provision in Britain by increasing the 'employment of women, who have a particular aptitude for the care of animals and are in some ways more reliable than men for this work'.[42] Why women were more 'reliable' and in what form their superior 'aptitude' presented itself was unexplained and therefore may be taken to be widely accepted. Perhaps, the perceived association was less an essence possessed by women than an imagined orientation akin to a 'feeling for the organism'.[43] Aptitude, in this example, would therefore involve adopting a situated positioning of remaining open to the non-human other, of listening and nurturing shared understanding through affect, attention, and familiarity. Arguably, by leaving the nature of this aptitude undefined, the 1945 recommendations asserted its importance while protecting its integrity as a subjective component of animal care.

Haraway's claim that 'response-able' relating requires a 'feeling for the organism' resonates with the 1945 association of a 'particular aptitude for the care of animals' with being 'more reliable'. For Haraway, 'response-able' relating operates in registers others than those of calculation because:

> mattering is always inside connections that demand and enable response, not bare calculations or ranking. Response ... grows with the capacity to respond, that is responsibility. Such a capacity can be shaped only in and for multidirectional relationships, in which always more than one responsive entity is in the process of becoming. That means ... animals as workers in labs, animals in all their worlds, are response-able in the same sense as people are; that is, responsibility is a relationship crafted in intra-action through which entities, subjects and objects come into being. People and animals in labs are both subjects and objects to each other in ongoing intra-action.[44]

Haraway's use of intra-action highlights a critical presupposition: it is not that things pre-exist relationships, rather they emerge through performative interactions – things made in the doing and therefore situated within specific historical times and places. From this perspective, the performance of care becomes an act of shared becoming. Within the animal house, boundaries are indeterminate and new forms of life, human and non-human, in relating creatively constitute one another. Such relations involve radically different positions of power and varying degrees to influence. However, inequalities of power and other disparities are not barriers to responsible relating. Rather, they are the condition of possibility for reasonable-able care. In the absence of difference, there would be no imperative to care.[45]

The suggestion that animal technologists and laboratory animals shape each other in their relating appears in early guides to management of the animal house. For instance, the second and much enlarged 1957 edition of *The UFAW Handbook on the Care and Management of Laboratory Animals* explained that:

> With every species the human attendant who is prepared to lavish care on his charges and makes determined efforts to make pets of them is an essential ingredient to success. An unsympathetic man will drive the best of animals into a vicious circle of suspicion and moroseness.[46]

In this way, efforts to define the desirable characteristics of the ideal 'animal technician' contended with the limits of the ability of language to capture and express qualities that chafed with the objective aspiration of scientific discourse. When animal technology approached such points, the calculative, rational, and objective language of science gave way to anecdotal description and sometimes, as in this example, direct reference to emotion:

> One little recognised factor in the care of animals is a genuine love for them. This should not be confused with the maudlin sentimentality of the antivivisection faction. Neither should it be assumed that an unintelligent love is sufficient; it is rather an understanding of animals to the extent that one recognises their happiness under certain conditions ... so important is this love of animals factor that it might well be placed at the head of the list of requirements for a good animal man.[47]

Here, the capacity for 'intelligent love' took precedence over technical competence in determining the ability of an individual to perform good animal care. The unasked and therefore unanswered question as to how one could meaningfully assess intelligent love, or how it could be distinguished from other less intelligent forms of love, illustrates the challenges that laboratory animal care posed to scientific discourse. Such was the importance of this affective orientation that had qualities such as 'intelligent love' proved compatible with scientific epistemology then the role of the animal 'technician' could well have been 'animal carer' or perhaps even 'animal lover'. One approach at translating the affective qualities of animal care into an objective language involved a quantitative approach to a taxonomy of human personality. Drawing on recent psychological innovations such as personality testing, animal care was distilled into fifteen core behavioural and psychological traits ranging from 'self-confidence' through 'quality of work' to 'communication with others', 'respect for authority', 'sense of humour' and the ever illusive 'attitude to animals'.[48] Despite all intents, scoring each trait on a scale of one to seven still turned on subjective judgement. In practice, the assessment of 'aptitude' for animal care was left to the discretion of the senior animal technician. Those with experience, through virtue of possession of an aptitude for animal care, could be trusted to recognise it in others. If one had it, one knew it when one saw it in others.

Explicit attempts to translate the subjective to objective within animal care, to measure the immeasurable, reveal the invisible or make material the immaterial were unusual. More typically, the subjective elements of animal care supervened within animal technology in the form of being referenced yet not explained. The author of the discussion of human presence quoted earlier, for instance, almost shamefully diminished their own observation by dismissing it as being 'not particularly original' and moving quickly on to:

> introduce the next topic; one which is of urgent interest to all users of laboratory animals. This is the problem of cage design which ... can only be satisfactorily solved by properly conducted studies of behaviour.[49]

Such a shift in register to studying the 'physical environment and the physical response of the animal' was characteristic of animal technology in reaching for the certainty of the calculable. Cage design could be studied scientifically, the consequences assessed and species specific standards established and communicated though the emerging professional journals of animal technology. The Animal Technicians Association's early training syllabus expected examinees to understand the 'merits and limitations of conventional and modern materials for cages and equipment' and qualifications were created that enhanced the status of the new profession.[50] In this way animal care could be rendered calculable and presented in a technical discourse that seemingly aligned the ways of working in the animal house with the scientific expectations of the experimental laboratory. Consequently, the emotional and affective components of animals were increasingly subjugated within animal technology discourse – but never entirely erased. Animal care relied on a moral economy and its early practitioners knew this. One might subjugate the role played by love in the day-to-day work of animal technology but as one commentator reminded the reader: 'Emotion is bound to creep into any discussion of relations between animals and man … To ignore it is unrealistic, and to be ashamed of it unnecessary; but it should not be allowed to cloud our reason, any more than any other emotion'.[51]

Conclusion

Today, all animal research within an institution is generally co-located within a Biological Services Facility, which integrates the former animal house and laboratory under one standardised organisational system. To an extent this transition realises the vision of mid-twentieth-century LASM, albeit largely in the form of material infrastructure and institutional organisation. The historical legacy of animal house and laboratory haunts the modern Biological Services Facility, wherein the animal technician (responsible for care and routine management of animals) and scientist form part of the whole – colleagues whose status is equally valued yet still sometimes hierarchically divided in status, reward, and visibility. Work to further improve these 'intra-actions', particularly that orientated toward overcoming remaining barriers to communication and understanding across different roles, is a key driver of early twenty-first-century emphasis on the 'culture of care'.[52]

This chapter provides a historical outline of the contours of mid-twentieth-century animal research at a moment of significant change. Following the disruption of the Second World War, efforts to improve the quantity and quality of animals available to scientific research established what can be thought of as the conditions that made the world of twenty-first-century animal research possible. Improving the provision of animals for experimental work was achieved by transitioning animal research to the use of purpose-bred 'sophisticated' laboratory animals. These new and unique forms of life marked a rupture with the past as they possessed a known genetic and environment history, making them a more reliable tool for controlled experiment. This involved the formation of a new multi-disciplinary scientific specialism, LASM, that fostered the new knowledge on the production, provision, management, and care of laboratory animals, as well as the professionalisation and transformation of animal care into animal technology. This work was as much an outcome of political, social, and cultural interventions as it was the product of a new, applied scientific specialism. The invention of the animal technician, as an example, illustrates how LASM contributed to change across the interconnected registers of knowledge, material infrastructure, and social relations that made up the animal research nexus.

Aligning the work and status of the animal house with that of the laboratory required the former to reimage its activities to reflect the epistemological expectations of the latter. This involved imposing rationality and facticity upon animal care whilst erasing subjective elements in line with what Daston has described as the 'tradition that opposes the life of the mind to the life of the heart and ... opposes facts to values'.[53] Animal care as technical practice tended toward separating human from animal. Animal technology sought to standardise the relationship about objective controls to ensure that the animal technician and laboratory animal behaved the same regardless of the individual human or animal occupying the role. This was the objective ideal at the heart of the vision of animal technology. Yet working relations within the animal house repeatedly challenged this goal. Few who worked with animals could deny the importance of subjective factors in the delivery of good animal care. One way to think of this tension is to understand the performance of animal care as 'implosive' in the sense that it tended to collapse boundaries.[54] Which is to suggest that thinking about animal research as nexus highlights how the doing of working with animals persistently challenges the very categories, boundaries, and orderings that are assumed to be necessary for success. Or, to think with Daston, we can recognise animal research as an example par excellence of a moral economy of science wherein objectivity is not simply compatible with subjective values but could not succeed without them. Perhaps if we found ways to engage more openly with the role of subjective elements within animal research this would serve to further integrate the two interdependent areas of work whilst also transforming professional and public understanding and enhancing what we now call the culture of care. Haraway reminds us that subjugated knowledge is 'savvy to the modes of denial through repression, forgetting, and disappearing acts' and therefore engaging with it can 'promise more adequate, sustained, objective accounts of the world'.[55] The question asked here, but not answered, is whether love for animals should continue to be subjugated within the moral economy of animal dependent scientific research.

Notes

1 Lorraine Daston, 'The Moral Economy of Science', *Osiris*, 10 (1995), 2–24, DOI: 10.1086/368740, p. 3.
2 A useful, relatively impartial and current introduction to historical method is John Tosh, *The Pursuit of History: Aims, Methods and New Directions in the Study of History* (London: Routledge, 2021).
3 Jill Lapore, *The Story of America* (Princeton, NJ: Princeton University Press, 2013) p. 15.
4 T. H. Goodeve, *How Like a Leaf: An Interview with Donna Haraway* (London: Routledge, 1999).
5 Donna J. Haraway, *The Companion Species Manifesto: Dogs, People, and Significant Otherness* (Chicago, IL: Prickly Paradigm Press, 2003) p. 6.
6 Non-human species have long served as sources of biological knowledge. See, for example, Anita Guerrini, *Experimenting with Humans and Animals: From Aristotle to CRISPR* (Baltimore, MD: Johns Hopkins University Press, 2022).
7 Daston, 'The Moral Economy of Science', p. 3.
8 Gail Davies, 'Caring for the Multiple and the Multitude: Assembling Animal Welfare and Enabling Ethical Critique', *Environment and Planning D: Society and Space*, 30.4 (2012), 623–638, DOI: 10.1068/d3211; Carrie Friese, 'Realizing Potential in Translational Medicine: The Uncanny Emergence of Care as Science', *Current Anthropology*, 54:S7 (2013), S129–S138, DOI: 10.1086/670805; Mette N. Svendsen, *Near Human Border Zones of Species, Life, and Belonging* (New Brunswick, NJ: Rutgers University Press, 2021) and Afterword in this volume.
9 B. Greenhough and E. Roe, 'Ethics, Space, and Somatic Sensibilities: Comparing Relationships between Scientific Researchers and Their Human and Animal Experimental Subjects', *Environment and Planning*

D: Society and Space, 29.1 (2011), 47–66, DOI: 10.1068/d17109; Lesley A. Sharpe, *Animal Ethos: The Morality of Human–Animal Encounters in Experimental Lab Science* (Berkley, CA: University of California Press, 2018).

10 'Animal model' or 'model organism' should not be conflated with 'laboratory animal': the former are highly situated forms of life embedded within specific experimental systems and research trajectories. While the 'laboratory animal' can be this, it was (and is) also conceived as a more generic 'tool'. For studies of 'model organisms' see Angela N. H. Creager et al., Wise *Science without Laws: Model Systems, Cases, Exemplary Narratives* (Durham, NC: Duke University Press, 2007); Nicole C. Nelson, *Model Behaviour* (Chicago, IL: Chicago University Press, 2018). The history of the animal model concept has not yet been fully charted, but see C. Logan, 'Before There Were Standards: The Role of Test Animals in the Production of Scientific Generality in Physiology', *Journal of the History of Biology*, 35.2 (2002), 329–363, DOI: 10.1023/A:1016036223348. For an insightful framework distinguishing 'model' from other 'experimental' uses of animals, see Rebecca A. Ankeny and Sabina Leonelli, 'What's so Special About Model Organisms?' *Studies in History and Philosophy of Science Part A*, 42.2 (2011), 313–323, DOI: 10.1016/j.shpsa.2010.11.039.

11 Robert G. W. Kirk, 'A Brave New Animal for a Brave New World: The British Laboratory Animals Bureau and the Constitution of International Standards of Laboratory Animal Production and Use, circa 1947–1968', *Isis*, 101.1 (2010), 62–94, DOI: 10.1086/652689; Tone Druglitrø, '"Skilled Care" and the Making of Good Science', *Science, Technology, & Human Values*, 43.4 (2017), 549–670, DOI: 10.1177/0162243916688093.

12 W. Lane-Petter, 'Sophisticated Laboratory Animals', in *British Postgraduate Medical Federation Scientific Basis of Medicine Annual Reviews*, 60 (London: Athlone University Press, 1966) pp. 54–70, p. 54.

13 S. Sturdy, 'Knowing Cases: Biomedicine in Edinburgh, 1887–1920', *Social Studies of Science* 37.5 (2007) 659–689, DOI: 10.1177/0306312707076597; S. Sturdy, 'Looking for Trouble: Medical Science and Clinical Practice in the Historiography of Modern Medicine', *Social History of Medicine* 24.3 (2011), 739–757, DOI: 10.1093/shm/hkq106.

14 Standing Committee of the Conference on the Supply of Experimental Animals, *Memorandum on Conference on the Supply of Experimental Animals* (London: privately published, 1945).

15 A. L. Bacharach, 'The Albino Rat in Biochemical Investigation', *Pharmaceutical Journal and Pharmacist*, 62 (1926), 629–630; Karen Rader, *Making Mice: Standardizing Animals for American Biomedical Research 1900–1955* (Princeton, NJ: Princeton University Press, 2004).
16 The National Archives of the UK (TNA): Letter A. L. Bacharach to A. S. Parkes, 8 April 1947, p. 1, FD1/383.
17 Bonnie T. Clause, 'The Wistar Rat as a Right Choice: Establishing Mammalian Standards and the Ideal of a Standardized Mammal', *Journal of the History of Biology* 26.2 (1990), 329–349, DOI: 10.1007/BF01061973; Adele E. Clarke and Joan H. Fujimura, *The Right Tools for the Job: At Work in the Twentieth-Century Life Sciences* (Princeton, NJ: Princeton University Press, 1992); Logan, 'Before There Were Standards'; C. Logan, 'Are Norway Rats ... Things?' Diversity Versus Generality in the use of Albino Rats in Experiments on Development and Sexuality', *Journal of the History of Biology*, 34.2 (2001), 287–314, DOI: 10.1023/A:1010398116188; Rader, *Making Mice*.
18 Robert G. W. Kirk, '"Wanted – Standard Guinea Pigs": Standardisation and the Experimental Animal Market in Britain ca. 1919–1947', *Stud Hist Philos Biol Biomed Sci.*, 39.3 (2008), 280–291, DOI: 10.1016/j.shpsc.2008.06.002.
19 W. Lane-Petter, 'The Provision and Use of Pathogen-Free Laboratory Animals', *Proceedings of the Royal Society of Medicine*, 55.4 (1962), 253–263, p. 253, DOI: 10.1177/003591576205500402.
20 W. Lane-Petter, 'A Policy for Breeding Laboratory Animals', *Veterinary Record*, 66 (1955), 833–836; Lane-Petter 'Sophisticated Laboratory Animals'.
21 W. Lane-Petter, 'The Place of Laboratory Animals in the Scientific Life of a Country', *Impact of Science on Society* 9.4 (1959), 178–196, p. 180.
22 TNA: Letter E. J. Fitzgerald to Sir Bryan Mathews, 13 August 1952, HO 285/24.
23 A. Summers, 'The Mysterious Demise of Sarah Gamp: The Domiciliary Nurse and Her Detractors, c. 1830–1860', *Victorian Studies* 32.3 (1989), 365–386, www.jstor.org/stable/3828497 [accessed 1 February 2023].
24 Anon., 'Animals for Research', *The Lancet*, 294.7620 (1969), 582.
25 Standing Committee of the Conference on the Supply of Experimental Animals, *Memorandum*, p. 4.
26 Anon., 'Report on Proceedings of Congress of Animal Technician', *Journal of the Animal Technicians Association* 1.1 (1950), 5–8, p. 7. Capitalisation and emphasis as in original.

27 A. S. Parkes, 'Foreword', *Journal of the Animal Technicians Association* 1.1 (1950), p. ii.
28 Anon., 'The Training of Animal Technicians', *Laboratory Animals Bureau Newsletter No. 3 Supplementary Memorandum* (London: Medical Research Council, 1952).
29 P. Medawar, 'Foreword', in *The UFAW Handbook on the Care and Management of Laboratory Animals*, ed. by W. Lane-Petter et al. (London: E. & S. Livingstone, 1967), p. v.
30 D. J. Short and D. P. Woodnott, *The IAT Manual of Laboratory Animal Practice and Techniques* (London: Crosby, Lockwood & Son, 1963), pp. 68–72.
31 P. Medawar, 'Foreword', in *The IAT Manual of Laboratory Animal Practice and Techniques*, ed. by D. J. Short and D. P. Woodnott, 2nd edn (London: Crosby, Lockwood & Son, 1969), pp. v–vi, p. v.
32 Anon., 'The Training of Animal Technicians', p. iii (emphasis in original).
33 Short and Woodnott, *The IAT Manual*, 1st edn, p. 47.
34 See, for instance, Short and Woodnott, *The IAT Manual*, 1st edn, p. 49.
35 Short and Woodnott, *The IAT Manual*, 1st edn, p. 147.
36 Short and Woodnott, *The IAT Manual*, 1st edn, p. 147.
37 D. J. Short, 'Handling Laboratory Animals', in *The UFAW Handbook on the Care and Management of Laboratory Animals*, ed. by A. N. Worden and W. Lane-Petter (London: UFAW, 1957), pp. 141–150, p. 141.
38 Short, 'Handling Laboratory Animals', p. 141.
39 W. Lane-Petter, 'Some Behavioural Problems in Common Laboratory Animals', *British Journal of Animal Behaviour*, 1.4 (1953), 124–127.
40 Lane-Petter, 'Some Behavioural Problems', p. 126.
41 W. Lane-Petter, 'The Place of Laboratory Animals in the Scientific Life of a Country', *Impact of Science on Society*, 9.4 (1959), 178–196, p. 182.
42 Standing Committee of the Conference on the Supply of Experimental Animals, *Memorandum*, p. 4.
43 E. Fox Keller, *A Feeling for the Organism: The Life and Work of Barbara McClintock* (New York: St. Martin's Press, 1983), pp. 197–207.
44 Donna J. Haraway, *When Species Meet* (Minneapolis, MN: University of Minnesota, 2007), pp. 70–71.
45 Haraway explains that the 'parties in intra-action do not admit of preset taxonomic calculation; responders are themselves co-constituted in the responding and do not have in advance a proper checklist of properties. Further, the capacity to respond, and so to be responsible,

should not be expected to take on symmetrical shapes and textures for all the parties. Response cannot emerge within relationships of self-similarity' (*When Species Meet*, p. 71).

46 R. E. Rewell, 'The Choice of Experimental Animal', in *The UFAW Handbook*, ed. by Worden and Lane-Petter, p. 167.

47 C. N. W. Cummings and F. G. Carnochan, 'The Importance of the Trained Laboratory Animal Attendant', *The American Journal of Medical Technology*, 17.6 (1951), pp. 298–301, p. 300.

48 D. McKelvie and A. K. Solarz, 'Personnel Problems', *Journal of the Institute of Animal Technicians*, 17.3 (1966), 109–115, p. 112.

49 Lane-Petter, 'Some Behavioural Problems', p. 126.

50 D. J. Short, and D. P. Woodnott, *The IAT Manual of Laboratory Animal Practice and Techniques*, 2nd edn (London: Crosby, Lockwood & Son, 1969), p. 232.

51 W. Lane-Petter, 'Science, Animals and Humanity', *Nature*, 174.4429 (1954), 532–534, p. 532.

52 Carrie Friese, 'Cultures of Care? Animals and Science in Britain', *British Journal of Sociology* 70.5 (2019), 2042–2069, DOI: 10.1111/1468-4446.12706.

53 Daston, 'The Moral Economy of Science', p. 3.

54 cf. Donna J. Haraway, *Modest_Witness@Second_Millenium.Female Man©_Meets_OncoMouse™* (London: Routledge, 1997).

55 Donna J. Haraway, 'Situated Knowledges: The Science Question in Feminism and the Privilege of Partial Perspective', in Donna J. Haraway, *Simians, Cyborgs, and Women* (New York: Routledge, 1991), pp. 183–201, p. 191.

6

Culturing care in animal research

Beth Greenhough and Emma Roe

Introduction

Since 2015 the concept of a 'culture of care' has become increasingly prominent within animal research. It is promoted by regulators of animal research in the UK[1] and widely recognised as being instrumental to improving the welfare of both staff and animals in animal research facilities, and to the quality of the science produced.[2] At the same time it is a complex and multifaceted phenomenon, and uncertainties remain as to '[h]ow can a *culture of care* be defined, what does it look like in institutions where it is functioning well, and what factors enable or constrain its development?'[3] Furthermore, animal research regulation and guidance emphasises the importance of caring for the animal and identifies strategies to support this (e.g., training, distributing responsibilities to those with competence), but often overlooks the care of facility staff – a growing concern for many who work in this sector.[4] Alongside internal tensions and uncertainties around what constitutes a good culture of care within animal facilities are further tensions between those who accept animal research as necessary for the advancement of medical and scientific research, and those outside these organisations for whom the very idea of cultures of care in animal research is an anathema (see Giraud, Chapter 8).

Many of those writing on a 'culture of care', ourselves included, have sought to emphasise the complex and multi-faceted nature of care in animal research: it can include care for the animals used in the research (often seen as synonymous with animal welfare), but also the need to care for those working with them (care for staff and colleagues), a commitment to the broader scientific and institutional objectives and standards they pursue (care for the science), and, occasionally, care for those who may benefit from the research in the future.[5] Concurrently, animal research professionals and advocacy organisations are developing initiatives aimed at better defining the culture of care and advising research settings and establishments on how best to promote a culture of care. For some, the institutional ethical review board, or as it is known in the UK, the Animal Welfare and Ethics Board (AWERB) can play a key role.[6] For others, guidelines and benchmarks offer insight into the multiple dimensions constituting a culture of care, from institutional level initiatives to interpersonal relations,[7] while training programmes offer an opportunity to promote the need for a culture of care amongst those with a licence to practice animal research.[8]

This chapter provides an overview of the different ways in which care finds meaning within the practises of regulation, institutional management, and daily animal caretaking within animal research. It seeks to understand what care looks like in practice from the perspective of those working in animal research facilities. How do different individuals interpret their responsibilities to support and develop a culture of care under the Animals (Scientific Procedures) Act 1986 (ASPA), which regulates UK animal research (see also Chapters 1 and 10)? How is this reflected in the ways these individuals talk about their work, the way they feel about that work, and the things they do to try to provide good care? How is the capacity of those occupying different roles within an animal facility to deliver good care both enabled and restricted by broader institutional infrastructures and governance practices, as well as by the social relations between members of staff, and between staff and the animals they work with?

The chapter begins by outlining our methodology, and then turns to locate the increasing interest in the 'culture of care' in animal research in the UK and internationally, before examining the roles played by physical infrastructure, governance, and human–human and human–animal relations in facilitating and/or restricting a culture of care. Finally, by way of conclusion, it suggests that in addition to the harm–benefit analysis that informs the formal licensing of animal research procedures, there is also a form of harm–care analysis within animal research facilities, through which those working there negotiate tensions and pressures in their day-to-day work.

Methodology

Our evidence comes primarily from a series of in-depth interviews with seven junior animal technologists[9] (those who provide day-to-day care for laboratory animals), conducted between 2013 and 2015.[10] We focused on this group as we wanted to understand how, at the beginning of their careers, our participants learned to care for, and cared about, the animals they worked with. More specifically we were able to chart how these individuals adapted to their institutions' culture of care, as their narratives shifted from elucidating a broadly felt 'love of animals' towards recounting specific actions (derived from their growing knowledge of the species they work with and institutional animal welfare protocols) as illustrative of providing particular forms of care. We became attuned to the stories they told of us of relationships they built up with the animals in their care, and reflected on what these in turn might tell us about how care takes place in animal research facilities. These interviews were then analysed using the NVivo12 coding software to extract key themes and patterns in the data through allocating specific codes, of which the most frequently used (highest number of occurrences) concerned care, emotions, animal suffering, and communication. Throughout this chapter we use quotations from these interviews and extracts from our fieldwork diaries to illustrate how these key themes emerged within our data, arguing these in turn evidence different aspects and understandings of a culture of care. We also spent time as participant observers in animal research facilities and at professional meetings and events, and we interviewed twelve other key stakeholders in the animal research community to better understand the wider context within which our junior animal technologists were working. All interviews were conducted with the informed consent of participants, and all the names referred to in the text are pseudonyms to protect the identity of interviewees.[11]

We supplement the above with interview and participant observation material from the wider Animal Research Nexus Programme (2017–23), and with material from our readings of regulations, guidelines, and academic and professional publications (identified through our ongoing engagement with stakeholders) that discuss the culture of care, in particular where these speak to changes in the understanding of the culture of care over time. Before examining our findings in more detail, we will next set the scene by exploring the ways in which the culture of care is being defined in professional guidance within the animal research community, and the extent to which this is echoed by emerging social science research on how care is conceived of and practised in animal research.

Culturing care in animal research

Today, within the animal research community, the culture of care is increasingly recognised as a complex and multi-faceted concept. Reviewing recent guidelines as part of an exercise to develop a new culture of care training resource, we identified a range of different qualities that stakeholders suggest are key to developing and sustaining a good culture of care (see Table 6.1). These take into account the need to develop a shared institutional vision of a culture of care, strong leadership that effectively communicates, supports, and promotes it, as well as the need for all staff to respect different roles and understand how their actions shape others' working lives, and therefore to be able to take and share responsibility for facility-wide animal and staff welfare.

These lists and accounting exercises provide a good sense of what a culture of care might involve, and go some way towards suggesting how this might be achieved in practice.[12] As social scientists, however, our key concern in this chapter is to reflect on how these measures shape and respond to how care takes place in the animal facility, enabling, but also sometimes constraining, different forms and expressions of care, with implications for animal and staff wellbeing. We are interested in how normative understandings of care, welfare, and wellbeing emerge: who is expected to do emotional care-work, who is expected to cope, and who is expected to be

Table 6.1 What makes a good culture of care?

Strong leadership: Senior management are committed to developing and promoting a culture of care, responsible animal use, the 3Rs, and animal welfare. An expectation of high standards with respect to the legal, welfare, 3Rs, and ethical aspects of the use of animals, operated, endorsed, and resourced at all levels throughout the establishment. Senior management and leadership champion their institution's culture of care values and recognise and support caring practices.
A shared institutional culture of care: A common set of values and standards, which is communicated, understood, and implemented across all parts of the establishment, and that is reflected in the condition of the animals, working environment, and all relevant documentation.
Commitment to animal care and welfare and the 3Rs: A commitment to and proactive implementation of good experimental design, good care, and the 3Rs. Dedication to a learning culture and the regular review and improvement of policies and processes to strive towards higher standards of animal welfare.
Commitment to staff care and welfare: A commitment to fostering a culture of inclusivity and mutual support where staff demonstrate empathy and understanding towards each other and appropriate mechanisms are in place to support staff well-being. Demonstrable respect for differing ethical perspectives on animal use.
Recognition of both shared and individual responsibility: An effective operational structure with clear roles, responsibilities, and tasks in which animal technologists and care staff, named persons (NVS, NACWO, NIO, NTCO),* trainers, and assessors are listened to and their work supported throughout the establishment. Roles and responsibilities with respect to developing a culture of care are clearly defined and visible. There is a recognition of shared responsibility (without loss of individual responsibility) towards animal care, welfare, and use.
Training, competence, and continuing professional development: A robust framework for training on aspects of animal care and use, plus assessment of competence, together with recognition of the importance of continuing professional development for all staff, and with adequate opportunities and resources provided. Engagement with the latest developments in animal welfare science and experimental design. The importance of compliance is understood and effected.
Recognising and rewarding good practice: Programmes recognise achievements in the 3Rs and care excellence.

Table 6.1 (continued)

Empowered staff: Creation of an environment where staff at all levels throughout the organisation are respected, listened to, and feel empowered to come forward with any concerns or suggestions they have to improve animal care.
Empowerment of animal welfare oversight committees: Effective and well-supported ethical review of scientific work undertaken with a thoughtful and rational approach.
Good communication: Mechanisms to support open communication and collaboration between different research programmes, teams, and staff at all levels.
Commitment to openness and honesty about animal use both internally and in the public domain.
Commitment to take a culture of care into account when working with those outside the organisation. Mechanisms to ensure that standards at animal suppliers, contracted organisations, couriers, and research partners nationally and internationally are consistent with the good practice that is implemented in-house.

* NVS (Named Veterinary Surgeon); NACWO (Named Animal Care and Welfare Officer); NIO (Named Information Officer); NTCO (Named Training and Competency Officer)

Sources: This table is adapted from Sally Robinson et al., 'The European Federation of the Pharmaceutical Industry and Associations' Research and Animal Welfare Group: Assessing and Benchmarking "Culture of Care" in the Context of Using Animals for Scientific Purpose', *Laboratory Animals*, 54.5 (2019), 421–432, DOI: 10.1177/0023677219887998; European Commission, *A Working Document on Animal Welfare Bodies and National Committees to Fulfil the Requirements under the Directive* (Brussels: European Commission, 2014); LASA and RSPCA, *Guiding Principles on Good Practice for Animal Welfare and Ethical Review Bodies*, 3rd edn, 2015; Penny Hawkins and Thomas Bertelsen, '3Rs-Related and Objective Indicators to Help Assess the Culture of Care', *Animals*, 9.11 (2019), 969, DOI: /10.3390/ani9110969; M Brown et al., 'Culture of Care: Organizational Responsibilities', in *Management of Animal Care and Use Programs in Research, Education, and Testing*, ed. by Robert H. Weichbrod, Gail A. (Heidbrink) Thompson, and John N. Norton, 2nd edn (Boca Raton: Taylor & Francis, 2018).

able to handle the suffering of both themselves and the humans and animals they work with? We can see in the developments within nursing and medical education a growing sense that staff care, and conversations around care, coping, and suffering are starting to be handled differently, reflecting an awareness of the hidden culture that lies behind medical education practices.[13] To what extent is there also a hidden culture within animal research?

In opening up the question of how care is conceptualised as a practice in animal research, we build on a wider body of scholarship within science and technology studies and cognate disciplines where care has been a central theme. This trend is exemplified by the work of scholars such as Maria Puig de la Bellacasa, Annemarie Mol and colleagues,[14] which draws attention to care as iterative, relational practice; something which involves constant reflection and tinkering in response to the changing environments and relations within which the subjects of care are embedded. Building on the legacy of writing on care from Hochschild, Gilligan, and Tronto,[15] scholars from the social sciences and humanities working on laboratory animal research, as well as colleagues writing from within animal research, offer further nuance and specificity to this understanding of care as complex and multi-dimensional. Their work highlights how cultures of care in animal research are: mandated through regulation;[16] practised though skilled labour;[17] entangled through human–animal relations;[18] felt as emotional labour and cognitive dissonance[19]; embedded in infrastructure[20]; shaped by national cultures and contexts;[21] shared as stories;[22] balanced as complex obligations to patients, publics, and research subjects[23]; or enacted as a counter to unavoidable harm, violence, and suffering.[24] This chapter brings this literature into conversation with the emerging practice of caring in animal research experienced amongst the junior animal technologists we worked with. Furthermore, by focusing on the period between 2013 and 2015, before the culture of care became a widely established buzzword within UK animal research (and increasingly internationally), we can explore how the labour culture of animal care significantly precedes the label. The next section begins this exploration by examining how particular forms of care-work are – at times literally – built into the infrastructure of animal research facilities.[25]

Care as technological infrastructure

Firstly, we might suggest the entire structure of animal research facilities is designed to provide a particular kind of care (husbandry) for the animals, alongside a care for scientific progress through peer-reviewed experimentation. These are the elements of a culture of care as set out in institutional visions and in culture of care strategies which state that 'all establishments should ensure that they have a clear vision of what a culture of care means for them',[26] but which are also embedded in the technological, bureaucratic, and physical architectures of animal research facilities. Care becomes closely specified as a series of benchmarks that those tasked with providing animal care are mandated to provide. For example, in the 2014 *UK Animals in Science Regulation Unit* guidelines[27] the word care most frequently appears in conjunction with issues of animal welfare and animal accommodation (section 7, 'Code of Practice on the care and accommodation of protected animals'), which in turn focuses on requirements to meet animals' needs for freedom of movement, food and water, to check animals daily, and to minimise harms and suffering.

For some scholars, notably those working in critical animal studies, this conjuncture of caring for animals and caring about experimental set-ups leads to an instrumentalisation of care,[28] synonymous with the 'cold' forms of care described by Hochschild,[29] a reading arguably not helped by the now fairly well-established mantra within animal research that the poorly cared for animals leads to poor quality scientific data. However, such a reading sits awkwardly with animal technologists' professional expertise in offering skilled care[30] to the animals in their charge, for whom good husbandry practice – through, for example, the provision of environmental enrichment, as well as conscientious adherence to welfare protocols[31] – goes hand in hand with their emotional warmth towards the animals they care for, through to individual-specific consideration for mixing animal personalities in group-housed settings. For example, junior animal technologist Carrie (interview, 2015) speaks of how she is motivated by the fact that she cares for animals which require care for whatever reason: 'that was one of the questions they asked me, "Would I have a problem in this sort of environment?" [...] And I don't think I do because [...] I do it because I care for the animals'. At other times, though, conflicts can arise where offering care for one subject, human or animal, can lead to direct harm being imposed on another. This applies, for example, to the tendency to separate and singly-house aggressive male mice (who may harm both each other, and in so doing also impede the progress of a particular research protocol), despite the suffering those mice may experience from lack of social contact with their kin.

Furthermore, alternative forms of caring for the animals can lead to uneasy resolutions where there appears a hierarchy around what matters most. For example, innovations in husbandry practice can seek to improve both staff and animal welfare, but they do not always sit easily alongside a junior technician's desire to interact with and handle the animals they care for. Here is Claire talking about the introduction of the new Individually Ventilated Cages (IVCs):

> I do prefer conventional [cages] just because in an IVC they're in boxes, like a show box basically with an air ventilation. And you look at the mouse in that box and it's just doing what it would do–, anyways it's just going round and eating and playing with his friends or it's chewing on something or it's going in and out of its hide. Whereas like, mice that are kept in the conventional box because you can pull out the racks and look in behind [...] They do this thing where they put their noses up and they can smell you and they know the technician that looks after them every day. So, you kind of have a sense of more–, they know who you are so you're more part of their little world, which is quite nice really. [...] But then I guess if they get a cleaner life in an IVC, so they're less susceptible to bugs, then an IVC is better for them. So, I think it's swings and roundabouts really. (Claire, junior animal technologist, interview, 2013)

Claire's description captures very effectively how IVCs both enable and hinder care. They provide a healthier, germ-free life for the mice, which could improve their health status and welfare, and they reduce human welfare risks from exposure to animal allergens. At the same time, they hinder animal technologists' direct interactions with animals, which can be seen as detrimental to the care skills known to develop through close bonding with animals. Furthermore, direct interaction with animals is also a coping strategy for animal technologists, who seek out those intimate animal interactions – such as going to cuddle the rabbits – to counteract more challenging aspects of their work. In other words, physical infrastructure both enables the provision of some forms of care whilst simultaneously limiting others – a prioritisation where some kinds of (often more easily measured and evidenced) care are chosen over others. While this example of care-in-practice focuses on the importance of being in proximity to the animals concerned, the next section considers in more depth how care also takes place at a distance through regulation, governance, and the allocation of responsibility. This dimension also feeds into the processes of prioritisation, by focusing on some aspects of care more than others.

Care as governance

A second way in which cultures of care are enacted is through governance and regulation. As a controversial sector, regulation and associated oversight provide a key means through which animal research institutions are held accountable to wider society for the care they provide. For example, the EU directive 2010/63/EU (recital 31) states that an institution's animal welfare body should 'foster a climate of care', and further guidance[32] serves to assign responsibility for delivering this to: (i) regulatory officials and inspectors; (ii) those 'named' as having a specific role, such as Named Animal Care and Welfare Officers in the UK; and (iii) Named Veterinary Surgeons, as well as recommending that the culture of care is embedded in Education and Training Frameworks. An additional working document (National Competent Authorities for the implementation of Directive 2010/63/EU, 2014b) further emphasises the role to be played by National Committees promoting the importance and relevance of a good culture of care for good scientific and animal welfare outcomes, as well as Animal Welfare and Ethical Review Boards (AWERBs) which are seen as central to fostering a culture of care. All of these bring attention to the ways in which responsibility for providing particular kinds of care is assigned to key individuals and processes within research animal facilities. For example, overall responsibility for regulatory compliance rests with the Establishment Licence Holder and senior management, who experience care failings primarily as moments when systems flag a failure of the institution to comply with the conditions of their animal research licence. Project Licence Holders are accountable for work conducted under their individual licences, and this responsibility in turn may be delegated to named persons such as Named Animal Care and Welfare Officers, Named Training and Competency Officers, and to specific staff who are responsible for carrying out regulated procedures (such as surgeries or administering substances or behavioural tests) or daily care-tasks (such as checking water supplies or cleaning cages). The way in which these individuals understand and respond to these legal responsibilities in turn shapes the ways in which they may approach the culture of care.

Indeed, regulation is also open to interpretation and is shaped by national constitutions, global competition, and local cultures.[33] For example, EU regulation is reflected in UK Home Office guidance, but interestingly this guidance emphasises firstly the importance of regulatory compliance:[34] 'non-compliances' serve as a key performance indicator for a facility, and addressing these is central to retaining the establishment licence. For facility manager David (interview, 2012), who had experience in both the private and public sectors, compliance was a key indicator of a good culture of care, but also, importantly, a good culture of care is key to helping ensure compliance: 'How you implement a culture of care [...] shows through in compliance, service level'. By this method, good care would be signified by a low number of non-compliances, or moments where something goes wrong and the conditions of the licence are infringed (for example a failure to provide water). However, David continues to set out how care is not solely about compliance, and also that it extends beyond care for the animals, to encompass care for facilities, personnel, and customers: 'Because the culture of care isn't just about the animal in your hand, it's about caring for your facilities, caring for your people first and foremost, then caring for the animals, caring for your customers, developing a customer service ethos'. There is a strong tone of working within a customer-service economy in what he says, which begs the question about how customers (those who may source animals or animal testing services from a private sector facility) further down the supply chain may also shape emerging cultures of care. But there is also an interconnecting vision of the multiple aspects which collectively might be drawn on to nurture a culture of care.

This emphasis on how a culture of care encompasses human–human relations as well as human–animal ones is echoed in the guidance provided by professional bodies and other non-governmental organisations. For example, in the UK, the Laboratory Animal Science Association (LASA) and the Royal Society for the Prevention of Cruelty to Animals (RSPCA) have produced a guidance document on the *Guiding Principles on Good Practice for Animal Welfare and Ethical Review Bodies* (3rd edn, 2015). In contrast to the UK Home Office documentation, this guide places good communication at the heart of governing a culture of care and includes a key role for AWERBs in promoting a culture of care through two-way communication with senior management. This is echoed in recent research which suggests AWERBs could go further in recognising the care needs of staff as well as animals.[35] This emphasis on communication is also noted by Natalie Nuyts and Carrie Friese, whose research showed how the ways in which scientists and animal technologists communicate 'shapes if and how a culture of care takes shape within the organizations and institution of science'.[36]

Broadly, across much of the documentation on the topic is a sense that a culture of care needs to strive to go beyond complying with regulation. Robinson and fellow members of the European Federation of Pharmaceutical Industries and Associations' Research and Animal Welfare Group explicitly assert that:

> A Culture of Care goes beyond adhering to legal requirements. It refers to an organizational culture that supports and values caring and respectful behaviour towards animals and co-workers. A Culture of Care is the responsibility of everyone involved with animal studies, from those directly working on the studies and beyond to include animal facility management, sample analysts, study planners, engineers, biologists, chemists, statisticians, project leaders, managers and senior leaders. The culture should instil responsibility

> and accountability in those planning and implementing research programmes and those caring for animals, so they do the right thing ethically and strive for continuous improvement.[37]

Similarly, Hawkins and Jennings define a culture of care as exceeding minimum requirements, but place greater emphasis on values and attitudes than we have seen elsewhere, adding texture and nuance to the work that communication and interconnectedness can achieve:

> The culture of an organisation relates to the beliefs, values and attitudes of its staff and the development of processes that determine how they behave and work together. A Culture of Care is one that demonstrates caring and respectful attitudes and behaviour towards animals and encourages acceptance of responsibility and accountability in all aspects of animal care and use. This should go beyond simply having animal facilities and resources that meet the minimum requirements of the legislation.[38]

Such accounts recognise that while regulation and guidelines can provide a resource for developing a culture of care, and indeed constitute a form of care in and of themselves,[39] on their own they are insufficient.

The guidelines cited above offer a vision of a broad distribution of responsibility and accountability across all aspects of animal care and use, which extends beyond, and to some extent may even be hindered by, an over-emphasis on compliance and offering a good service. What is less visible in these guidelines, but became very apparent in the course of our research, is how putting such a vision into practice in and of itself requires a culture of not only taking care but accepting and sharing responsibility. For one of the junior animal technologists we spoke to, this became very visible when seeking a resolution to an animal problem that would be satisfactory for all those involved:

> I guess it's just the way–, the way things are dealt with. So, like, if there's ever an issue with someone or an issue between like–, with someone and someone else and the animals are in the middle it's always–, it will always be resolved and it's – the resolution is – always the right resolution for the animals. So, I guess that's how I know. (Claire, junior animal technologist, interview, 2015)

While Claire emphasised the wellbeing of the animal as being the primary concern, the capacity of a team to take responsibility, communicate, and find negotiated compromises to deal with issues between people that affected the animals was described as evidence of the presence of a culture of care. This emphasis on 'taking responsibility' in turn brings us to a third dimension we want to explore, as we consider how human–animal and human–human relations also play a key role in shaping cultures of care.

Care as relationships

Some of the earliest ethnographic work looking at animal research was quick to recognise that many of those who work in animal research, and animal technologists in particular, often form close emotional attachments with some of the animals in their care,[40] experiencing emotional harm and distress when those animals are used or killed as part of the research process.[41] This ability to care about as well as for the animals they work with, to attune to their needs, is often seen as a key ethical resource, and a quality which distinguishes 'good' from 'bad' animal technologists.[42] These ideas came to the fore strongly when we spoke to senior managers about the recruitment process for animal technologists. The managers we spoke to stressed the importance of getting the right kind of person,[43] someone who showed empathy towards animals and evidence of a strong work ethic and high standards.

The qualities sought out by managers and valued by others who work in animal research show how care is also a property of individuals, albeit one which can be cultivated. The managers we spoke to often asked new recruits about their experiences working with animals, while junior animal technologists narrated their journey towards working in the sector as part of a much longer history of wanting to work with animals, beginning on farms or in pet stores. As we have argued elsewhere, such experiences are key to developing not only practical skills in animal care and handling, but emotional experience in dealing with the more challenging side of living and working with animals, including when they become ill or die.[44] Yet at the same time it can leave animal technologists – as care providers – vulnerable to psychological and emotional harm.[45] As one manager put it, when you're recruiting new staff you need to remind them not to 'forget why the dogs are here' (interview with facility manager David, 2012). Here a 'culture of care' describes not only the care new animal technologists show towards the animals, but also the care offered to them to support them through the more challenging aspects of their work.

> We have a very, very detailed induction and probation period which we adopt for our animal technicians, and that includes being sympathetic and empathetic at all stages. You know, the first time an animal technician kills, like a new recruit. You know, they will watch it, we've got a video we show, they get lectures, they get all sorts of things, very much like a researcher. But the impact [of the first time they kill] is something that we are sensitive to. (David, facility manager, interview, 2012)

Furthermore, it is arguably these more intangible dimensions of care-work which are placed at greatest risk at times of increased pressure within a facility's working environment. Sources of such pressure may vary between sites, from funding and publication pressures in academic establishments, to the pressures for a fast turnaround or requests for the use of particular procedures from clients in commercial settings. The licensing process and paperwork needed to comply with animal research regulation provides a form of care for the animals by aiming to minimise animal suffering and harm, and care for the researchers in terms of legal protection and approval of their work. However, for some, especially junior researchers, these are also stressful and time-consuming processes, especially when they lack adequate support.[46] Animal technologists also faced pressures from staff shortages. The impacts of the COVID-19 pandemic placed considerable burdens on those working in animal research, with staff at all levels having to work longer hours, to live away from home or to kill larger numbers of animals as work was delayed or suspended. These pressures can result in staff at all levels, from technicians to named people and researchers, having to make difficult decisions and perform complex, careful, and vital work whilst being increasingly physically and emotionally exhausted.

This came across in our third interview with Debbie and Fiona, who had by that time both been working in their facility for nearly two years and had taken on more responsibility:

> I think it all like comes around like staffing, staffing is majorly important, because all that stress that is being put on you then affects you and then you're making mistakes, do you know, we're all human, we all make mistakes, do you know, and then you feel bad for it and then you're like, oh if I wasn't as stressed as I was, or if I wasn't thinking about doing this, and concentrating on what I was doing I wouldn't have made that mistake. (Debbie and Fiona, junior animal technologists, interview, 2015)

While without exception all those we spoke to saw the welfare of the animals as the primary concern, Debbie and Fiona's experiences speak eloquently to the complex intersection between human and animal care. As Williams explored in greater depth in her recent examination of cultures of care in animal research, how 'human carers work with and interpret the needs of animals depends to a significant extent on how human–human interactions address human, as well as animal, needs, wants, feelings, resources, and responsibilities'.[47]

An additional source of pressure comes from interactions with stakeholders and publics outside of the animal research facility. It is interesting to note that in the UK, where we undertook our research, the main regulator, the Animals in Science Regulation Unit,[48] emphasises the role of public opinion or 'societal expectations' in informing a culture of care: 'A good culture of care is an environment which is informed by societal expectations of respectful and humane attitudes towards animals used in research.' Here wider publics are seen as a key driving force for the culture of care in animal research, and yet beyond animal activism there is little visibility as to what laboratory animal welfare means in a lab context, in contrast to media and retailer attention to food animal welfare, which has shaped market segmentation on welfare standards. Nevertheless many of those we spoke to, working at all levels, spoke of the importance of external critique and scrutiny in driving reductions and refinements in animal use and the development of alternatives to animal testing: 'I think animal rights protestors are a good thing [...] I think it's very important they have a voice and they continue to have a voice in society because it does keep us aware of what we're doing' (group interview with animal technologists, 2013).

The dialectic between the 'culture of care' in animal research and anti-vivisectionist critique is therefore arguably a productive one. As Eva Giraud writes in her careful and nuanced account of activist practice, 'different approaches need to remain in fraught dialogue with one another in order to recognise that the contradictions inherent in each approach mark imperfect responses to an equally messy and contradictory ethico-political terrain'.[49] At the same time, this is something which needs to be handled with care. The history of direct-action animal rights protests in the UK, particularly during the 1990s, has left a legacy of concern which makes many of those working in the sector cautious about talking about what they do outside of the workplace. Being open to alternative perspectives brings with it its own emotional burdens.

> Well, when the Animal Rights [activists] were at their peak, you know, and some of their opinions based on what was going on I can understand where they were coming from, nothing is perfect in any sphere. But [following] the threats and the intimidation, we ended with very much a bunker mentality, you know, and people became very introverted. There was a lot of, I don't know, it was just a very negative place to be. (David, facility manager, interview, 2012)

Here care might be seen in the training provided by animal research facilities in how to handle protests, as well as more pragmatic measures to protect staff identity and provide a secure working environment. In the longer term though, such measures cannot perhaps alleviate the emotional toll of facing those criticisms, nor of being unable to discuss what you do outside of work or with a limited few trusted friends and relatives. We might ask what a more careful approach to such an encounter might look like, one which builds on these productive tensions and contradictions, while remaining respectful of fundamentally different perspectives and mindful of the emotional toll such encounters can have for all involved.

Conclusion

Each of the three different dimensions of 'cultures of care' set out above is characterised by a central tension between the need and desire to deliver good care and the often unavoidable harms that the practice of animal research imposes on animals and the humans who work with them. This juxtaposition of care and harm is also highlighted in recent work in conservation[50] and animal sacrifice.[51] We suggest this tension between care and harm is a productive area for future work at the interface of science and technology studies and work in anthropology and animal studies. Recent research in these fields has been productive in examining the consequences of animal research in terms of its impact on both human–animal relations and scientific praxis;[52] our work adds to this by drawing attention to the harms experienced by humans as well as the animals in these relations.

As we have shown, the physical infrastructures of animal research facilities are often a negotiated compromise between meeting the care needs of animals and the humans who work with them. For example, we noted how new IVC technologies reduced some forms of harm, offering improved animal welfare, improved biosecurity, and reduced risk of humans developing animal allergy, but also limited some kinds of care, such as the affective bonds technicians developed with animals through regular handling. Similarly, while regulations, guidelines, and training provide a top-down means of mandating towards good care, they cannot force people to care about the animals and people they work with. We saw how the ability of regulations and guidelines to enforce good care was tied up in the ways in which that regulation and guidance was interpreted by those tasked with implementing it, always shaped by cultural norms, and the capacity they had to take responsibility for not only delivering but striving to exceed those requirements. We then saw how this ability to take responsibility for and to deliver good care for both animals and colleagues was also conditioned through individual relationships and encounters, and noted how both internal and external factors, including project deadlines, financial and resource constraints, the impact of the pandemic, and the critiques offered by anti-vivisectionist movements can compromise and limit capacities to care and be cared for. In practice, caring about animals and people you work with is also constituted through diffuse affective emotional elements, but these, too, come with their own risks and vulnerabilities, for example, when the demands of maintaining standards of care during periods of staff shortage place physical and emotional burdens on animal care staff.

We therefore argue that cultures of care in animal research facilities are characterised not only by reasoned harm–benefit analysis[53] – the formal assessment process which informs the decision about whether or not to grant a licence for animal work – but by a constant ongoing renegotiation with scientific designs and institutional pressures. These scientific practices and pressures can impose harm, and this necessitates the provision of good care for both the animals and humans within these spaces (and possibly beyond them), or what we might term the harm–care nexus. Here the use of the term nexus signals a need to understand the process of negotiating the tension between giving care and imposing harm as one of being attuned to how animal research structures feelings and generates meanings both inside and outside research facilities. Institutions, managers, and regulators seeking to promote a culture of care need to be mindful of different, sometimes conflicting, understandings of what constitutes good care, and the roles played by (a) infrastructure (b) governance and (c) both human–animal and human–human relations in facilitating and/or restricting care in practice. Here the work of AnNex and other social science and humanities scholarship might play an important role in sharing stories about the experiences of those who work in animal research and in creating spaces for different kinds of conversations between different stakeholder groups.[54] Our work arguably offers a resource for building empathy and encouraging communication between those with different roles and responsibilities within animal research, and even beyond them,[55] thereby encouraging recognition of the many different cultures of care which emerge around animal research and the tensions but also potentially productive points of collaboration and synergy which emerge between them.

Notes

1 Animals in Science Regulation Unit, *Animals in Science Regulation Unit Compliance Policy* (London: Home Office, 2017), www.gov.uk/guidance/animal-testing-and-research-compliance-with-aspa [accessed 28 November 2019].
2 Tania Boden and Penny Hawkins, 'Communicating the Culture of Care – How to Win Friends and Influence People', *Animal Technology and Welfare* (2016), 151–156.
3 G. Davies et al., 'Developing a Collaborative Agenda for Humanities and Social Scientific Research on Laboratory Animal Science and Welfare', *PLOS ONE*, 11.7 (2016), 1–12, DOI: 10.1371/journal.pone.0158791. See also Sally Robinson et al., 'The European Federation of the Pharmaceutical Industry and Associations' Research and Animal Welfare Group: Assessing and Benchmarking "Culture of Care" in the Context of Using Animals for Scientific Purpose', *Laboratory Animals*, 54.5 (2019), 421–432, DOI: 10.1177/0023677219887998; Boden and Hawkins, 'Communicating the Culture of Care'.
4 Keith Davies and Duncan Lewis, 'Can Caring for Laboratory Animals Be Classified as Emotional Labour?' *Animal Technology and Welfare*, 9.1 (2010), 1–6; Carrie Friese and Joanna Latimer, 'Entanglements in Health and Wellbeing: Working with Model Organisms in Biomedicine and Bioscience', *Medical Anthropology Quarterly* (2018), www.anthrosource.onlinelibrary.wiley.com/journal/15481387 [accessed 4 July 2018]; Jordi López Tremoleda and Angela Kerton, 'Teaching a Culture of Care: Why It Matters', *Revista de Bioética y Derecho*, 15 (2021).
5 G. Davies, R. Gorman, and B. Crudgington, 'Which Patient Takes Centre Stage? Placing Patient Voices in Animal Research', in *GeoHumanities and Health*, ed. by Sarah Atkinson and R. Hunt (Cham: Springer, 2020), pp. 141–155; Emma Roe and Beth Greenhough, 'A Good Life? A Good Death? Reconciling Care and Harm in Animal Research', *Social & Cultural Geography*, 24.1 (2021), 49–66, DOI: 10.1080/14649365.2021.1901977; Annabella Williams, 'Caring for Those Who Care: Towards a More Expansive Understanding of "Cultures of Care" in Laboratory Animal Facilities', *Social & Cultural Geography*, 24.1 (2021), 31–48; Carrie Friese, 'Intimate Entanglements in the Animal House: Caring for and about Mice', *The Sociological Review*, 67.2 (2019), 287–298, DOI: 10.1177/0038026119829753; Nathalie Nuyts and Carrie Friese, 'Communicative Patterns and Social

Networks between Scientists and Technicians in a Culture of Care: Discussing Morality across a Hierarchy of Occupational Spaces', *Social and Cultural Geography* (2021), http://eprints.lse.ac.uk/108472/ [accessed 29 January 2021].

6 Mike King and Hazem Zohny, 'Animal Researchers Shoulder a Psychological Burden That Animal Ethics Committees Ought to Address', *Journal of Medical Ethics* 48.5 (2022) 299–303, DOI: 10.1136/medethics-2020-106945. See also Kirk (Chapter 5).

7 Penny Hawkins and Rebecca Thomas, *Assessing the Culture of Care: A Survey of Network Members*, July 2017, www.researchgate.net/publication/336253708_Assessing_the_Culture_of_Care_a_survey_of_network_members [accessed 26 November 2022]; Robinson et al., 'The European Federation of the Pharmaceutical Industry and Associations' Research and Animal Welfare Group'.

8 Tremoleda and Kerton, 'Teaching a Culture of Care'.

9 Throughout this chapter the term animal technologists is used to describe those who work in animal research facilities providing care and animal husbandry, as well as sometimes assisting in scientific work. Animal technologist is the preferred term used in the UK and by the Institute of Animal Technologists, a professional organisation for those providing care for research animals. Other terms often used are animal technician or tech and animal care staff.

10 These total eighteen interviews in all, five interviewees were each interviewed three times with roughly six months between interviews, the other two interviewees were also interviewed three times, but preferred to be interviewed together.

11 This study has been reviewed and approved through the Queen Mary University of London Research Ethics Committee, 8 April 2013 (QMREC2012/76), and through the Oxford University Central University Research Ethics Committee (CUREC) process, 24 October 2014.

12 For a good example of this see Robinson et al., 'The European Federation of the Pharmaceutical Industry and Associations' Research and Animal Welfare Group'.

13 Hanneke Mulder et al., 'Addressing the Hidden Curriculum in the Clinical Workplace: A Practical Tool for Trainees and Faculty', *Medical Teacher*, 41.1 (2019), 36–43, DOI: 10.1080/0142159X.2018.1436760.

14 María Puig de la Bellacasa, *Matters of Care: Speculative Ethics in More than Human Worlds* (Minneapolis, MN: University of Minnesota Press, 2017); María Puig de la Bellacasa, '"Nothing Comes Without its

World": Thinking with Care', *The Sociological Review*, 60.2 (2012), 197–216; *Care in Practice: On Tinkering in Clinics, Homes and Farms*, ed. by Annemarie Mol et al. (Verlag, Bielefeld: transcript publishing, 2010), p. 326.

15 A. R. Hochschild, *The Managed Heart: Commercialization of Human Feeling* (Berkley, CA: University of California Press, 2003); Carol Gilligan, *In A Different Voice* (Cambridge, MA: Harvard University Press, 1982); Joan C. Tronto, *Moral Boundaries: A Political Argument for an Ethic of Care* (New York: Routledge, 1993).

16 Brown et al., 'Culture of Care: Organizational Responsibilities', in *Management of Animal Care and Use Programs in Research, Education, and Testing*, ed. by Robert H. Weichbrod et al., 2nd edn (Boca Raton: Taylor & Francis, 2018).

17 Friese, 'Intimate Entanglements in the Animal House'.

18 Friese, 'Intimate Entanglements in the Animal House'.

19 Davies and Lewis, 'Can Caring for Laboratory Animals Be Classified as Emotional Labour?'; Angela Kerton and Jordi L. Tremoleda, 'Emotional Challenges in Our Work with Laboratory Animals: Tools That Support Caring for Others and Yourself', *Animal Technology and Welfare* 20.1 (2021), 43–60.

20 Robert G. W. Kirk, 'Care in the Cage: Materializing Moral Economies of Animal Care in the Biomedical Sciences, c.1945–', in *Animal Housing and Human–Animal Relations: Politics, Practices and Infrastructures*, ed. by Kristian Bjørkdahl and Tone Druglitrø, Routledge Human–Animal Studies Series (London; New York: Routledge, 2016), pp. 167–185.

21 Gail Davies, 'Locating the "Culture Wars" in Laboratory Animal Research: National Constitutions and Global Competition', *Studies in History and Philosophy of Science Part A*, 89 (2021), 177–187, DOI: 10.1016/j.shpsa.2021.08.010; Wakana Suzuki, 'Improvising Care: Managing Experimental Animals at a Japanese Laboratory', *Social Studies of* Science, 51.5 (2021), 729–749, DOI: 10.1177/03063127211010223.

22 Beth Greenhough and Emma Roe, 'Attuning to Laboratory Animals and Telling Stories: Learning Animal Geography Research Skills from Animal Technologists', *Environment and Planning D: Society and Space*, 37.2 (2019), 367–384, DOI: 10.1177/0263775818807720.

23 Richard Gorman and Gail Davies, 'When "Cultures of Care" Meet: Entanglements and Accountabilities at the Intersection of Animal Research and Patient Involvement in the UK', *Social & Cultural Geography* (2020), 1–19, DOI: 10.1080/14649365.2020.1814850.

24 Roe and Greenhough, 'A Good Life? A Good Death?'

25 Kirk, 'Care in the Cage'; see also Kirk (Chapter 5).

26 Robinson et al., 'The European Federation of the Pharmaceutical Industry and Associations' Research and Animal Welfare Group', p. 56.
27 Home Office, *Guidance on the Operation of the Animals (Scientific Procedures) Act 1986* (London: Home Office, 2014).
28 Tone Druglitrø and Robert G. W. Kirk, 'Building Transnational Bodies: Norway and the International Development of Laboratory Animal Science, ca. 1956–1980', *Science in Context*, 27.2 (2014), 333–357, DOI: 10.1017/S026988971400009X; Carrie Friese, 'Realizing Potential in Translational Medicine: The Uncanny Emergence of Care as Science', *Current Anthropology*, 54.S7 (2013), S129–138, DOI: 10.1086/670805; Eva Giraud and Gregory Hollin, 'Care, Laboratory Beagles and Affective Utopia', *Theory, Culture & Society*, 33.4 (2016), 27–49, DOI: 10.1177/0263276415619685.
29 Hochschild, *The Managed Heart*.
30 Tone Druglitrø, '"Skilled Care" and the Making of Good Science', *Science, Technology, & Human Values*, 43.3 (2018), 649–670, DOI: 10.1177/0162243916688093.
31 Beth Greenhough and Emma Roe, 'Exploring the Role of Animal Technologists in Implementing the 3Rs: An Ethnographic Investigation of the UK University Sector', *Science, Technology, & Human Values*, 43.4 (2017), DOI: 10.1177/0162243917718066.
32 European Commission, Directorate-General for Environment, *Caring for Animals Aiming for Better Science – Directive 2010/63/EU on Protection of Animals Used for Scientific Purposes* (Publications Office), 2019, https://data.europa.eu/doi/10.2779/26419 [accessed 1 February 2023]. European Commission, Directorate-General for Environment, *Caring For Animals Aiming for Better Science – Directive 2010/63/EU on Protection of Animals Used for Scientific Purposes: Animal Welfare Bodies and National Committees* (Publications Office), 2019, https://data.europa.eu/doi/10.2779/059998 [accessed 1 February 2023].
33 Gail Davies, 'Locating the "Culture Wars" in Laboratory Animal Research'.
34 Animals in Science Regulation Unit, *Animals in Science Regulation Unit Compliance Policy*.
35 King and Zohny, 'Animal Researchers Shoulder a Psychological Burden'.
36 Nuyts and Friese, 'Communicative Patterns and Social Networks'.
37 Robinson et al., 'The European Federation of the Pharmaceutical Industry and Associations' Research and Animal Welfare Group', p. 2.

38 Penny Hawkins and Maggie Jennings, 'The Culture of Care – a Working Concept', https://norecopa.no/media/7711/culture-of-care-working-concept.pdf [accessed 26 November 2022].
39 Tone Druglitrø, 'Procedural Care: Licensing Practices in Animal Research', *Science as Culture* (2022), 1–21, DOI: 10.1080/09505431.2021.2025215.
40 Lynda Birke et al., *The Sacrifice: How Scientific Experiments Transform Animals and People* (Lafayette, IN: Purdue University Press, 2007).
41 Robinson et al., 'The European Federation of the Pharmaceutical Industry and Associations' Research and Animal Welfare Group', p. 56.
42 Greenhough and Roe, 'Attuning to Laboratory Animals and Telling Stories'; Greenhough and Roe, 'Exploring the Role of Animal Technologists in Implementing the 3Rs'.
43 Roe and Greenhough, 'A Good Life? A Good Death?'
44 Roe and Greenhough, 'A Good Life? A Good Death?'
45 Davies and Lewis, 'Can Caring for Laboratory Animals Be Classified as Emotional Labour?'; American Association for Laboratory Animal Science, 'Cost of Caring' (American Association for Laboratory Animal Science, 2013).
46 See Gabrielle King, 'Towards a Culture of Care for Ethical Review: Connections and Frictions in Institutional and Individual Practices of Social Research Ethics', *Social & Cultural Geography* (2021), 1–17, DOI: 10.1080/14649365.2021.1939122.
47 Williams, 'Caring for Those Who Care', p. 15.
48 Animals in Science Regulation Unit, *Identification and Management of Patterns of Low-Level Concerns at Licensed Establishments* (London: Home Office, 2015), p. 4.
49 Eva Haifa Giraud, *What Comes after Entanglement? Activism, Anthropocentricism and an Ethics of Exclusion* (Durham, NC: Duke University Press, 2019), p. 138.
50 Thom Van Dooren, *Flight Ways: Life and Loss at the Edge of Extinction* (New York: Columbia University Press, 2014).
51 Radhika Govindrajan, *Animal Intimacies: Interspecies Relatedness in India's Central Himalayas* (Chicago, IL; London: University of Chicago Press, 2018).
52 See Matei Candea, 'Habituating Meerkats and Redescribing Animal Behaviour Science', *Theory, Culture & Society*, 30.7–8 (2013), 105–128, DOI: 10.1177/0263276413501204; Vinciane Despret (trans. Brett Buchanan), *What Would Animals Say If We Asked the Right Questions?* (Minneapolis, MN: University of Minnesota Press, 2016).

53 Animals in Science Committee, *Review of Harm–Benefit Analysis in the Use of Animals in Research* (London: Home Office, November 2017), www.gov.uk/government/uploads/system/uploads/attachment_data/file/662098/Review_of_harm_benefit_analysis_in_the_use_of_animals_in_research.pdf [accessed 1 February 2023]; G. Davies, 'Harm–Benefit Analysis: Opportunities for Enhancing Ethical Review in Animal Research', *Laboratory Animals*, 47.3 (2018), 57–58, DOI: 10.1038/s41684-018-0002-2.

54 Beth Greenhough and Hibba Mazhary, *Care-Full Stories: Innovating a New Resource for Teaching a Culture of Care in Animal Research Facilities*, 2021.

55 For an insightful commentary on the risks of marginalising more critical activist perspectives see Giraud (Chapter 8).

7

The good aquarist: morality, emotions, and expectations of care in zebrafish aquariums

Reuben Message

Introduction

'There is no substitute for the empathetic animal technologist', I once heard a speaker at a workshop on the standardisation of laboratory animal welfare terminology say. Empathy, I think he meant, cannot be standardised into a laboratory protocol, and yet it is essential to animal care-work. The claim connects to a common feature of animal welfare discourse: the assumption that failures of empathy underpin cruelty. Consequently, animal technologists should not merely oversee animal welfare: they are expected to care about, as well as for, the animals, to be able to imagine their suffering, and, consequently, to experience guilt at their demise.[1] Indeed, empathy and related 'moral emotions' are even spoken of as though they are kinds of 'soft skills' desirable for good animal technologists.[2] As industry insiders, Kerton and Tremoleda note that animal technologists today are increasingly expected to 'carry' emotional bonds as a part of their professional responsibility.[3] Outside of animal technology, of course, ambient culture has long coded emotions like empathy as appropriate in dealing with animals under human care. Now, in lab animal welfare circles, where it may once have been frowned upon as a source of bias, human–animal bonding is increasingly seen as not only inevitable but actually desirable for human and animal wellbeing.[4]

These claims about the singular importance of empathetic relations hint at prevailing ideas regarding how good technologists *should* feel. Of course, if competence in the job is understood to demand empathy, it stands to reason that being competent would serve as proof of empathetic or moral character.[5] Moral and technical competence are closely linked in this way. But, equally, between the proof and the experience, there is ample room for self-doubt. This chapter therefore explores what such expectations, and the gap between them and reported experience, might mean for aquarists – those animal technologists who specialise in working with fish. In particular, it focuses on those who care for those prototypically 'uncharismatic', non-cuddly lab denizens: zebrafish, a small, relatively short-lived species of subtropical fish who shoal in vast numbers in laboratories around the world.[6] These animals are of such vastly different appearance, physiology, and habitat to their human carers, that talk of 'bonding', 'relating', or 'empathy' seems strained.

The 'paradoxes' whereby humans both care for animals and deliberately cause them suffering or kill them is a staple theme in the social study of laboratory animal care.[7] The resulting pressures structure what Lesley Sharp calls the 'moral dimension' of working with animals in science.[8] In this context, social scientists have described various hacks, rationalisations, and improvisions that assist technologists to cope with moral and emotional strain in different contexts.[9] Species is one notable variable influencing the forms these take.[10] However, most existing research has centred on inter-mammalian relations, most often involving rodents, primates, or companion animals. It is the presence (and sometimes abundance) of feeling-attachments to other mammals that fuels the tensions experienced by technologists, adds to the high 'moral cost' of their work, and prompts the entry of a vocabulary of 'emotional labour' into animal technology.[11] As Sharp wrote, 'warm-blooded' animals have a power of 'sentimental leverage' that is capable of 'reconfiguring moral thinking in science'. But, as Sharp acknowledged, 'myopic understandings of which creatures matter most in the day-to-day of laboratory life and death' may also cause one to neglect the moral work being done around other species, including fish.[12] Certainly, much of this 'work' involves regulators, animal welfare scientists, animal advocates, and philosophers: the collective effect of their efforts has transformed fish, at least in some contexts, into sentient beings with legitimate moral claims.[13] This chapter takes a different approach to filling the fish-shaped gap in the literature by focusing attention on aquarists, who, as a group, may develop distinctive identities and outlooks connected to a perception that they work at the margins of a profession dominated by mammalian (especially rodent) concerns. Thus, it investigates some quotidian activities, attitudes, and modes of speaking adopted by a particular category of technologist, but who, like all technologists, wishes to do good and be perceived as doing good. In particular, it explores ways aquarists try to pursue the good in the (relative) absence of sentimental leverage: that is, without the guide to actions, sign of character, or source of personal consolations that an emotional attachment to animals can represent.

This research is based on qualitative interviews and ethnographic experience of aquarium life, including two one-week stints of participant observation at different facilities in the UK and a non-random sample consisting of 27 in-depth interviews with individuals involved with different aspects of fish-based science and regulation. In addition, I attended various industry events and made numerous shorter visits to different facilities. The interviews were recorded, transcribed, and analysed thematically with Nvivo. Fieldnotes were kept and similarly analysed. Due to the sensitive nature of the topic, place names have been changed and pseudonyms are applied throughout.[14] This research was granted ethical approval by the Central University Research Ethics Committee of the University of Oxford (Reference Number: SOGE 18A-7).

My approach lends itself to interpretive accounts of ideas co-created in the situated interaction between researcher and research participant – its strength is in generating ideas and exploring possibilities, rather than generalising about states of affairs. Of course, any researcher – especially one billed as interested in the social relations of animal research – comes freighted with associations, and these in turn generate specific performances of moral identity, which demand reflexivity on their part. For example, an aquarist named Erica told me that 'it's because their faces are different [laughs], so you can't really empathise with something that looks different from you I think. Not that I'm saying that's the right thing, but–' (interview, 2018). First the laugh, then the other qualification: the work of image management in the face of an anticipated reaction is palpable. However, why specific renditions of moral identity are elicited cannot be explained by exclusive reference to the situated interaction alone. Performances are shaped by wider communicative contexts, in this case connected to perceptions of what it is right for technologists (or people generally) to feel about animals.

Thus, in what follows I explore some ways in which aquarists construct their moral-emotional identities in relation to other people and groups. I examine how they may, simultaneously, claim the mantle of objective, hard-headed professionals and sensitive animal lovers; I look at how aquarists adopt the role of animal advocates; and finally I discuss exceptional cases in which aquarists appear to develop, against expectations, specific relationships with individual fish, and speculate on the significance of this. First, however, I make an introductory foray into the 'changing moral landscape' of zebrafish aquariums in the UK – a key communicative context for all that follows.[15]

The zebrafish aquarium in a changing moral landscape

There are many reasons, practical and biological, why zebrafish have become such useful model organisms. One of the advantages is that their use tends to attract little uncomfortable public attention. Jim Endersby noted that early adopters of the model quickly recognised the advantages that came with animals whose use passed with a 'relative lack of ethical concerns'.[16] Illustratively, Endersby quoted a research scientist's satisfaction in finding that, whenever he discussed his work with members of the public, they readily understood that 'all things being equal, you'd rather do that [use a fish] than in a mouse or a dog'. Moreover, he continued, scientists themselves often explained that they could not bring themselves to work on 'higher' animals, including even mice, but that to the contrary they experienced no remorse when working with fish.[17] The benefits of this fish work are not just scientific: they are political and ethical, and consequently also personal, in that users report they find it psychologically easier to conciliate themselves to the reality of what they do.

My own experience confirms Endersby's. For example, a researcher explained to me that working with fish was to them virtually 'guilt free', and that, in their experience, it required less of a 'commitment' than with mice. This difference, they felt, was evidently connected to a lack of 'emotional attachment' (interview, 2018). Not all technicians would agree but an experienced aquarist told me that he'd 'rather work with fish because you don't get the attachment that you would with mice'. And, he explained that, while he takes them 'seriously' and wants them to be 'healthy': 'it wouldn't keep me up at night, if I had to cull some fish at the end of the day' (Frank, interview, 2018).

Such candid speech suggests that whatever expectations exist concerning how technologists should feel and talk about the fish they care for, these are not exactly overwhelming. Indeed, it may be argued that expectations or norms surrounding fish are wholly different to other species commonly used in science, especially mammals and birds. That is, there really are no relevant expectations to wrestle with. Rather, there are genuinely different standards: fish work is guilt free, and society does not expect aquarists to 'relate' to fish at all. Owain Jones' statement, from the year 2000, that fish in general are 'ethically invisible', may thus retain some force today.[18] Despite fish always having possessed much the same legal protections as other vertebrate animals in the UK, there are certainly ways in which fish struggle to maintain substantive parity with other vertebrate species, especially mammals.[19] As Sharp put it (albeit in a North American context), zebrafish are the 'quintessential *other* animal' of animal research.[20]

However, to say that fish are ethically invisible in research aquariums at the present time is an overstatement. Fish, in this context, are not outside the 'moral circle'[21] – though where they are exactly is harder to say, hence the dissonance of the aquarist. Importantly, prevailing opinions about the cognitive and emotional capacities of fish – including their ability to experience pain and suffering – have undergone a sea change since the early 2000s, and this has had a considerable (if not uncontroversial) impact on how informed actors think about the moral status and welfare of this taxa.[22] Fish are no longer officially recommended as a kind of 'relative replacement' for 'higher' animals in research in the UK (though it wasn't long ago that they were).[23] In principle at least, fish from a certain stage of development are no longer assumed to suffer less or be less sentient than other animals (though precisely what stage is debated).

Like all kinds of animal research facilities, research aquariums are entirely saturated with ethical discourses, not to mention their materialised analogues in the form of animal care protocols, technologies, and embodied skills.[24] Some aquariums have a degree of administrative independence (and sometimes physical distance) from other animal facilities on campus. This can result in distinctive local cultures: aquarists do often consider themselves – and are considered by others – a distinctive subgrouping within the animal technology community. Nevertheless, they are not entirely a class apart: they are an element of the same overall regulatory and discursive apparatus.[25] Continuous pursuit of the principles of replace, reduce, and refine (the 3Rs) is the governing principle and ideology, followed, increasingly, by the idea of a 'culture of care'.[26] Aquarists are thus always exposed to and adopt various values, perspectives and expectations derived from within the wider field of animal technology.[27] These changes and institutional realities entail expectations that must then be reckoned with.

Additionally, at least at the rhetorical level, the ethical claims on humanity of an ever wider range of organisms and entities are undergoing re-evaluation: crustaceans, for example, are increasingly thought to feel pain; indeed, the idea that insects and even plants are sentient is seriously discussed.[28] More relevantly, fish welfare in aquaculture, and in recreational and commercial fishing contexts, is increasingly a subject of concern.[29] Philosophers have eviscerated support for the idea that fish represent a legitimate space of ethical exception.[30] Animal activists have begun to take note, while pop-science books exploring the mental and emotional lives of fish are now available.[31] Examples of 'relatable' (and vulnerable) aquatic organisms are now common in popular culture: the 'fish are friends, not food' slogan of *Finding Nemo* is an example – though one, admittedly, set against a very long back-catalogue of less sympathetic representations.[32] In this context, specific communities have been presented as proof that it is in fact possible to 'learn to care' for fish.[33] And if it is seen as possible now, I think, it could prove a short step to being seen as desirable and, perhaps, one day, even obligatory.

While this account of the shifting scientific, ethical, legal, professional, and cultural parameters of human–fish relations is impressionistic, it's important here because it plays a role in constituting the moral landscape within which aquarists participate and must find their moral compass. They are not the general public, who may be forgiven for having only a fleeting interest in fish. My hypothesis then is simply that aquarists, professionals in fish care, are required to navigate between existing, ambiguous, personal moral intuitions, and embodied emotions, and the mixed expectations being constantly forged around them (which they may internalise and come to expect of themselves). The next section looks at one prominent way in which aquarists construct their moral-professional identities: by means of contrasts with others.

Caricatures of the mawkish and the hard-headed

When I asked a leading practitioner about what makes a good aquarist, rather than describing obviously useful attributes like sharp observational skills or a penchant for water chemistry, they focused their comments on what distinguished them from technologists who work with other animals. Indeed, Fae argued that 'normal animal technicians like small cuddly things', and that people who related to 'fluffy [...] mammalian things' made better mouse technologists. Aquarists, she suggested, were by contrast unmotivated by 'relations' with fish. Her staff, she said, were either 'very science minded', meaning they had a background or interest in biology, or were 'very fish orientated', meaning they did things like keep aquariums or go fishing in their leisure time (Fae, interview, 2018). That is, good aquarists (like Fae's staff) are apparently pragmatic and unsentimental.

Indeed, Fae was not alone in implying that it takes a particular kind of person to want to work with fish. 'Normal animal technicians', apparently, do not much care for them, in part because, aquarists often implied, the work was not emotionally satisfying in ways equivalent to working with other animals. Evelyn, also a facility manager, put it succinctly: the difference with fish (or zebrafish anyway), is that you 'get nothing back'. There is no sense of positive 'interaction' leading to 'emotional attachment' when working with these animals (Evelyn, interview, 2018). While another facility manager told me that he occasionally heard of mouse technologists wanting to work with fish (perhaps a shrewd career move), Erica emphasised that most wanted 'nothing to do with them' (Farol, interview, 2018; Erica, interview, 2018). Frank even said he'd seen mouse technicians start crying when told they had to shift to working with fish, something he explained in terms of being scared of the unknown and 'because they're not interested whatsoever' (interview, 2018).

Indeed, I frequently found aquarists ready to depict mouse technologists as more emotional, sometimes in a frivolous sense. Thus, while acknowledging that killing fish just 'doesn't raise the emotions in the same way that culling mice or rats does with animal technicians', Fae, like others, indulged in a degree of caricature (interview, 2018). While interviewing staff at one facility, I noted that the expression 'fluffies', a word connoting the feeling of a cuddly animal and also a sense of schmaltziness, could be used in reference to both animals like mice *and* to the colleagues that specialise in working with them (Eugenie, Gemma, and Fiona, interview, 2018).

Through their (sometimes patronising) opposition to 'normal animal technicians', we can discern the portrayal of a pragmatic, hard-headed, identity for aquarists. Frank agreed with the suggestion that his self-declared inability to 'anthropomorphise' fish makes it easier for him to be 'objective' about them; Harry similarly felt 'that little degree of separation' between technologist and fish 'probably makes your care more professional in a sense, less emotive because you are separated' (Frank, interview, 2018; Harry, interview, 2018). Relatedly, another aquarium manager suggested that the popularity of the word 'care' (as in 'culture of care' or 'Animal Care and Welfare Officer', for example) in animal research circles was a sop for a sentimental public. The word accords well, he implied, with conventional ideas of what relationships between responsible people and mammals ought to involve but it's really misleading: What an aquarist should rather try to do is 'provide good care and make sure that everything is done properly, rather than being caring about them [animals] so that you do it properly' (Felix, interview, 2018). 'Care', for Felix, is thus merely the word we use to designate an outcome (such as achieving high welfare standards), not an emotionally charged input, or some special psychic or embodied property necessary to achieve such an outcome. Indeed, I sensed that talk of emotions and 'relating' generally could be beside the point for some aquarists. Fae, for example, found it 'annoying' that 'what people can relate to and what people believe' has an influence on welfare provision (interview, 2018). Fish welfare, Fae meant, should not be left to caprice, to what animal technologists or anyone else can 'relate' to. Basing one's perspective on this kind of impartial, hard-headed language accords well with scientific norms around objectivity; in the process, however, it provides consolation and may help rationalise personal experiences of emotional distance or disconnect, recoding these relations as positive, even virtuous.

Making a virtue of a necessity

This is where things begin to get more complicated. While it takes a moment to see it, by describing 'normal animal technologists' as unable or unwilling to relate to fish, fish people begin to occupy this space themselves. This is not a matter of consistency (aquarists will also readily declare their own inability to 'relate' to fish) and neither should it be. Yet, somehow, snobbishness aimed at 'fluffies' may at once be transfigured and elevated and come to define the figure of a tender-hearted aquarist. As we noted with Endersby above, aquarists like zebrafish scientists regularly describe themselves as unable to work with 'higher' animals like mammals: they, in other words, may be the sensitive souls who care about animals – and, also, it's the 'fluffies' (and the world at large) who evidently don't care enough about fish.

Connectedly, Erica and I discussed her feelings about the necessity of having to kill animals as a part of the job. Erica admitted to squeamishness at the mere thought of it – nobody likes culling, whatever the species. But it's much worse with mammals than fish, she explained. Referring particularly to a manual method by which mice may be humanely killed (it involves the skilful application of a ruler to the back of the neck) Erica said: 'I'm not really sure if I could do it, I probably could do it but I don't want to. I don't want to get comfortable with that. So yeah, I prefer the fish, there's a level, there's a level there' (interview, 2018).

Zebrafish, by contrast, are always killed by adding a poisonous chemical into their environment. The method is much less proximate, less visceral, and thus less emotionally distressing. But it's clear that the difference does not come down entirely to methodology, which merely reinforces the physical and emotional distance already noted. Fae, after noting a 'distance' to aquarist–fish relations, claimed that she would 'never work with companion animals and I would never work with primates either because I feel that's too close' (interview, 2018). I also recorded a conversation in my fieldnotes with an experienced aquarist named Helga. After noting that she followed the familiar pattern of claiming to care for animals so much she'd be unable to harm a mammal, I wrote that '[she] [thus] takes it that because mouse technicians evidently can [harm mice] because they do, they must be more callous and less caring, or at least more able to remove their personal feelings towards the animals from the situation' (fieldnotes, 2018).

Framed like this, sentimentality is allowed in to consecrate a caring identity, while the caricature of the mawkish mammalian is technologist reversed as the cool pragmatism of those who care for fish is transformed into a kind of moral-emotional positioning of its own. A non-sentimental, disinterested attitude may thus be variously construed as a virtue or a vice, depending on context. For example, Frank, after saying he wouldn't be sleepless after a day's culling in the aquarium, claimed: 'If I had to cull a pig or a dog or a cat, I wouldn't be in the job, I wouldn't do it' (interview, 2018). Now Frank, as it happens, does not work with pigs, dogs, or cats. But he does in fact work with mice, and must necessarily kill them from time to time. Thus he, too, like Erica, must have 'a level': tuned by species, he's found a personally acceptable moral and emotional bandwidth to work in, and this happens to be congruent with what he is required to do according to his job description. Thus, helped by claims of social differentiation, necessity may be turned into a virtue, and comes in turn to anchor personally acceptable positions – though such alignments are not always easily achieved.

On advocacy and speciesism

As we've already seen, aquarists are often very articulate about why people – including 'normal animal technicians' – are unsympathetic towards fish. This is because this is something they can readily sympathise with. Echoing Evelyn's sense of 'getting nothing back' from fish, Erica for instance explained how she felt that fish 'can see you but they don't look at you, so yeah there's like a bit of a disconnect' (interview, 2018). And Harry, an experienced animal technologist who'd recently begun to work with fish for the first time when I spoke to him, explained to me that '[w]ith a mouse you can hold it and stroke it, you can gain comfort, give comfort. You cannot do that with a fish' (interview, 2018). Harry's words are poignant – they contain a sense of loss. Evelyn likewise described feeling 'grumpy', indeed 'traumatised', by this absent connection when she were first forced to switch from working with mammalian species to zebrafish. There may therefore be, in some cases anyway, a process by which aquarists need to adjust their own expectations for emotional fulfilment when they enter into this species of animal care-work.

This makes the effort of many aquarists to position themselves at the vanguard of progressive human–fish relations all the more notable. Consistent also with the unexpected construction of the aquarist as especially sensitive towards the suffering of animals, many will assume the identity of advocate or pioneer of welfare techniques. Aquarists for example often spoke to me about the importance of maintaining equality amongst species. Here's Evelyn:

> They should have the same rights as everything else, and it might be just a fish, but going back a very long time someone told me that it was just a monkey ... So you know, there should be no difference in my [view] ... whether it's a fruit fly or a fish or a monkey or a pig or a mouse, whatever. (Evelyn, aquarist, interview, 2018)

Here, Evelyn places fish on a progressive trajectory of incremental 'rights'. But with the repetition of the word 'should', she offered a hypothetical statement about what is wished for, or desirable in principle. Evelyn's explicit exhortations to remember that zebrafish are equals reveals that this is not how people normally think of them – it requires continuous effort to overcome default state speciesism.

Indeed, in interviews, aquarists saw themselves as advocates for fish, despite the difficulties, and expressed distress at how they get forgotten or neglected within their institutions or society generally. Eugenie, for example, told me how her staff are advocates, arguing with researchers:

> we're for the zebrafish welfare and we stand by that. I mean I've had to say in meetings, you know because they don't really appreciate the welfare of the fish jumping [out] so I've had to say in meetings quite clearly to the PIs [Principal Investigators] I will not stand down on this, this is a welfare issue, you will not allow fish to jump by doing that technique. (Eugenie, aquarist, interview, 2018)

Fae similarly explained how she was spurred into getting 'completely immersed' in the world of fish welfare by what she saw as the 'raw deal' that fish get. Talking of her earlier days in the job, Fae said:

> you know, they were just in these tanks half the time–, at that point you used to pull a tank out and like the fish would be dead or like you would have like a label written in Japanese, and the Home Office were like 'oh they're brackish water species' and they're not, and … so I felt no one really cared and I thought this is ridiculous, like no one actually cares that much. (Fae, aquarist, interview, 2018)

For a number of professionals I spoke with, developing and improving techniques that promote fish welfare and aid science had become a passion, and some have made lasting contributions to the field. As I've discussed elsewhere, aquarists in the UK have made significant technical contributions to the development of fish welfare in the field of husbandry, often exceeding what is expected of their roles and with limited institutional support and economic means.[34] If actions speak louder than words, it would be uncharitable to suggest that these actions are motivated solely by either a desire for professional recognition or to compensate for the discomfort of moral complexity or ambivalence. Nevertheless, advocating equality for fish, or working overtime to develop a new husbandry protocol, certainly helps address the problem of how to be a 'Good Aquarist' in the relative absence of conventional, strong or lasting forms of emotional attachment to these animals – as we'll see below – other means of making these (non)relations more meaningful can and do emerge.

The exception proves the rule

Humans frequently form lasting bonds with individual laboratory animals. The special nature of these largely inter-mammalian relationships is sometimes consecrated in awarding specific animals their own names and giving them pet-like treatment.[35] Such relations, though, are vanishingly rare in zebrafish aquariums. Short lifespans, small size, and the vast quantities in which zebrafish are kept are the obvious ways of accounting for this. Additionally, care for terrestrial lab mammals is often directed at the bodies of individuals, including through physical touch and individually applied veterinary treatments, things which hardly ever occur in zebrafish laboratories.[36] I asked Fae, who in addition to managing a zebrafish aquarium has years of experience working with mammals, whether she'd ever seen a technologist develop a relationship with an individual fish. She replied that she hadn't, and emphasised how different this was to mammalian work: 'I've never seen that, and yet I've never really not seen it with people looking after mammals' (interview, 2018).

Aquarists, however, do aspire to individualise fish in particular ways. They understand that it is individual fish who ultimately suffer the poor welfare they are responsible for helping avoid. One facility publicised their informal motto on Twitter: 'Remember they may shoal but they are all individuals'. In part then, the struggle to remember that fish are individuals is connected to the work of professional fish husbandry: 'seeing skilfully', visually discriminating ill from healthy fish, is a critical part of the job.[37] But endeavouring, however difficult, to see these fish as individuals is a moral activity too: it's seen as the right thing for aquarists to try and do. Fae, despite her scepticism about relations with individual fish, told me that 'I do think about them as individuals as well sometimes, because nobody [else] does' (interview, 2018).

Aquarists, then, may aspire to individualise fish because this seems desirable to them. Contra Fae's generalisation, an experienced aquarist named Gemma confided that she actually did have favourite fish, a 'really pretty' one with 'a really nice pattern'. I asked if the fish had a name. 'No she doesn't have a name, I've not named her', Gemma replied, 'but she's the only female in the tank and she's got a really nice pattern, so much so that I WhatsApped everyone else in the group and said come look at this fish, showed them her pattern on the side, and they did come and see it. So yeah we do have kind of favourites that we look out for' (interview, 2018).

Gemma's attempts to create meaning in her relationships with fish actually go even further than occasionally picking a favourite. After our interview, while I was having lunch with other members of staff, Gemma made contact via text message. She'd forgotten to tell me something she felt was important. Gemma – whose nickname amongst colleagues is the 'Grim Reaper', and whose husband jokingly calls her the 'Butcher of Badlesmere' – said that she wanted to tell me that she refused to cull fish on their birthdays. A colleague laughed, saying they knew all about this idiosyncrasy of Gemma's (fieldnotes, 8 February 2018). With this observance, Gemma responds to the expectation that she is merciless for doing what she does, while imbuing her actions, and fish themselves, with some kind of moral significance.

Indeed, while naming individual fish is clearly rare, I did encounter one case. Like Gemma's 'favourite', Sharkey was easily (and unusually) distinguishable. He was considered gigantic. I never saw Sharkey, but it was estimated to me that he was three times the size of a typical zebrafish. He was housed off the main systems in his own tank, and lived to the age of four (he was allowed to grow so old because he was a part of an ageing study). Sharkey became an aquarium legend. Referring to the 'head fish', one aquarist joked with me that Sharkey was a lifetime inmate: 'if you stripped the walls [of the aquarium] you'd probably find his name carved into one of them – "Sharkey was here"'. Frank also told me, with a wry smile, that people were sad when Sharkey left the place (fieldnotes, 2018).

Such atypical cases do not invalidate the rule about individualisation, though they do seem to prove it. It's hard to know what to make of them. For Sharp, practices like naming and picking favourites generally work to 'subvert the mandated regimes' of impartial laboratory science.[38] These rare events could then be read as subverting the expectation, if not mandate, that aquarists are entirely unsentimental about fish: perhaps they are more like rodent technicians than they like to let on? Elevating select individuals, though, may also be interpreted as a coping device, a way of managing the moral and emotional contradictions of the work. In this case then, they might also be viewed as rituals that respond to the internalised expectation, derived from mammalian experience, about what appropriate human–animal relations in labs should involve – namely, forms of emotional connection that result in moral strain.

Conclusion

This chapter has suggested that, given the reported challenges of 'relating' emotionally to 'the other animal', aquarists may seek other sources of reassurance that they are moral, on the side of the Good. As we saw, one response is to double down on the conventional critique of emotions in science and the adoption of a 'hard-headed', pragmatic persona. This approach on its own, though, seems somewhat out of step with a context in which emotions like empathy towards laboratory animals accrue a progressively positive moral meaning – and indeed, professional significance. Aquarists therefore may engage in other forms of situated cognition, speech and action to construct their identities and present themselves as moral actors who conform to particular social expectations. This chapter therefore followed in a tradition of studying the varied moral and emotional tensions animal technologists live within. But rather than trying to show that aquarists, like other animal technologists, also need to cope with the guilt, fatigue, or anguish caused by complicity with the harming and killing of animals, it explored how aquarists dealt with the relative absence of strong inter-species emotional relations in a context that increasingly seems to expect and approve of them. Technologists, it seems, are increasingly asked not merely to care *for* but to care *about* animals. The case of aquarists is useful for exploring the possible implications of this, particularly in terms of what may be over the horizon as other 'other' species are brought more fully into societies' moral reckoning – though the future is, as ever, uncertain.

This said, and as suggested in the final section, I do not discount the idea that fish, even zebrafish, can exercise a degree of 'sentimental leverage' of their own. I want to agree, in fact, with recent literature on human–fish relations, which, as Allmark-Kent remarks, has emphasised that particular groups of people in specific circumstances 'can and do learn to care for the fishes they encounter' (often in contradistinction to the public in general).[39] There is certainly no doubt that aquarists become exquisitely capable of reading what we take to be signs of suffering in the bodies of fish. Does the tacit knowledge upon which this skill depends ultimately rest on a form 'attunement' to fishy experiences, that is, on a form of inter-species embodied communication that goes beyond anthropomorphic empathy?[40] Future work would profit by examining not only the variety of ways in which aquarists react to, adapt or transform the changing community norms that bear upon them, but also how the limitations of embodied experiences are positively overcome in practice. In this respect, one might liken the quotidian activities of aquarists – including but not reserved to their various claims of moral leadership in this field – to the actions of what Koch and Svendsen called 'pathfinders' who, working their way through changing moral landscape, shape it in turn by their actions.[41]

Notes

1 Julian Groves, *Hearts and Minds: The Controversy Over Laboratory Animals* (Philadelphia, PA: Temple University Press, 1997) on shame and guilt in animal research; Annabella Williams, 'Caring for Those Who Care: Towards a More Expansive Understanding of "Cultures of Care" in Laboratory Animal Facilities', *Social & Cultural Geography* 24.1 (2021), 31–48, DOI: 10.1080/14649365.2021.1939123 on the distinction between 'care' and 'welfare' in this context.

2 On the sociology of moral emotions, see, Jonathan H. Turner and Jan E. Stets, 'Moral Emotions', in *Handbook of the Sociology of Emotions*, ed. by Jan E. Stets and Jonathan H. Turner, *Handbooks of Sociology and Social Research* (Boston, MA: Springer US, 2006), pp. 544–566.

3 Angela Kerton and Jordi L. Tremoleda, 'Emotional Challenges in Our Work with Laboratory Animals: Tools That Support Caring for Others and Yourself', *Animal Technology and Welfare*, 20.1 (2021), 43–60.
4 *The Inevitable Bond: Examining Scientist–Animal Interactions*, ed. by Hank Davis and Dianne Balfour (Cambridge: Cambridge University Press, 1992); Kathryn Bayne, 'Development of the Human–Research Animal Bond and its Impact on Animal Well-Being', *ILAR Journal*, 43.1 (2002), 4–9, DOI: 10.1093/ilar.43.1.4; Jordi L. Tremoleda and Angela Kerton, 'Creating Space to Build Emotional Resilience in the Animal Research Community', *Laboratory Animals*, 49.10 (2020), 275–277, DOI: 10.1038/s41684-020-0637-7.
5 See Max Weber, *The Protestant Ethic and the Spirit of Capitalism*, 3rd Roxbury edn (Los Angeles, CA: Roxbury Publishing Company, 2002).
6 Jamie Lorimer, 'Nonhuman Charisma', *Environment and Planning D: Society and Space*, 25 (2007), 911–932, DOI: 10.1068/d71j; also Mark J. Estren, 'The Neoteny Barrier: Seeking Respect for the Non-Cute', *Journal of Animal Ethics*, 2.1 (2012), 6–11, DOI: 10.5406/janimaleth ics.2.1.0006.
7 Beth Greenhough and Emma Roe, 'Attuning to Laboratory Animals and Telling Stories: Learning Animal Geography Research Skills from Animal Technologists', *Environment and Planning D: Society and Space*, 37.2 (2019), 367–384 (p. 369), DOI: 10.1177/0263775818807720.
8 Lesley A. Sharp, *Animal Ethos: The Morality of Human–Animal Encounters in Experimental Lab Science* (Chicago, IL: University of Chicago Press, 2019); Lesley A. Sharp, 'Animal Research Unbound: The Messiness of the Moral and the Ethnographer's Dilemma', *History and Philosophy of the Life Sciences*, 43.2 (2021), DOI: 10.1007/s40656-021-00426-2. In this chapter I also follow Sharp's understanding of 'morality' as contextual, private, and uncodified, as opposed to 'ethics', which is taken as public and codified – though, like Sharp, I appreciate that the edges are blurry and the words get used interchangeably.
9 Examples include Arnold Arluke, 'Uneasiness among Laboratory Technicians', *Occupational Medicine*, 14.2 (1999), 305–316; Lynda Birke, A. Arluke, and M. Michael, *The Sacrifice: How Scientific Experiments Transform Animals and People* (Lafayette, IN: Purdue University Press, 2007); more recently Wakana Suzuki, 'Improvising Care: Managing Experimental Animals at a Japanese Laboratory', *Social Studies of Science*, 51.5 (2021), 729–749, DOI: 10.1177/030631272 11010223.
10 See, for example, Arnold Arluke, 'Living with Contradiction: Response to Comments', *Anthrozoös*, 3 (1989), 90–99.

11 See Tremoleda and Kerton, 'Creating Space to Build Emotional Resilience'; Joseph T. Newsome et al., 'Compassion Fatigue, Euthanasia Stress, and Their Management in Laboratory Animal Research', *Journal of the American Association for Laboratory Animal Science*, 58.3 (2019), 289–292, DOI: 10.30802/AALAS-JAALAS-18-000092; Keith Davies and Duncan Lewis, 'Can Caring for Laboratory Animals Be Classified as Emotional Labour?' *Animal Technology and Welfare*, 9.1 (2010), 1–6.
12 Sharp, *Animal Ethos*, p. 226.
13 See John Law and Marianne E. Lien, 'The Practices of Fishy Sentience', in *Humans, Animals and Biopolitics: The More-than-Human-Condition*, ed. by Kristin Asdal et al. (Abingdon: Routledge, 2017), pp. 30–47.
14 On anonymisation and allocation of pseudonyms, see Davies et al., Introduction in this volume.
15 The phrase is borrowed from Lene Koch and Mette N. Svendsen, 'Negotiating Moral Value: A Story of Danish Research Monkeys and Their Humans', *Science, Technology, & Human Values*, 40.3 (2015), 368–388, DOI: 10.1177/0162243914553223.
16 Jim Endersby, *A Guinea Pig's History of Biology: The Plants and Animals Who Taught Us the Facts of Life* (London: William Heinemann, 2007), p. 408.
17 Endersby, pp. 407, 408.
18 Owain Jones, '(Un)Ethical Geographies of Human-Non-Human Relations', in *Animal Spaces, Beastly Places: New Geographies of Human–Animal Relations*, ed. by Chris Philo and Chris Wilbert (London; New York: Routledge, 2000), pp. 268–291 (p. 284).
19 Reuben Message and Beth Greenhough, '"But It's Just a Fish": Understanding the Challenges of Applying the 3Rs in Laboratory Aquariums in the UK', *Animals*, 9.12 (2019), 1075, DOI: 10.3390/ani9121075.
20 Sharp, *Animal Ethos*, p. 226, emphasis in original.
21 Vonne Lund et al., 'Expanding the Moral Circle: Farmed Fish as Objects of Moral Concern', *Diseases of Aquatic Organisms*, 75.2 (2007), 109–118, DOI: 10.3354/dao075109.
22 For discussions see, for example, Lynne U. Sneddon and Culum Brown, 'Mental Capacities of Fishes', in *Neuroethics and Nonhuman Animals*, ed. by L. Syd M. Johnson et al., *Advances in Neuroethics* (Cham: Springer International Publishing, 2020), pp. 53–71; Howard I. Browman et al., 'Welfare of Aquatic Animals: Where Things Are, Where They Are Going, and What It Means for Research, Aquaculture,

Recreational Angling, and Commercial Fishing', *ICES Journal of Marine Science*,76.1 (2018), 82–92, DOI: 10.1093/icesjms/fsy067.

23 While it is no longer policy, it is true that many people still think of fish as 'replacements' of some kind. Fish larvae younger than five days old are still considered legal replacements for older fish or other vertebrates, though this is increasingly challenged too. See Lynne U. Sneddon, 'Where to Draw the Line? Should the Age of Protection for Zebrafish Be Lowered?' *ATLA*, 46 (2018), 309–311, DOI: 10.1177/026119291804600605. On the idea of 'relative replacements', see Michael D. Mann, 'Choosing the Right Species in Research', *Laboratory Animals*, 44.7 (2015), 274–278, DOI: 10.1038/laban.742.

24 Robert G. W. Kirk, 'Care in the Cage: Materializing Moral Economies of Animal Care in the Biomedical Sciences, c.1945–', in *Animal Housing and Human–Animal Relations: Politics, Practices and Infrastructures*, ed. by Kristian Bjørkdahl and Tone Druglitrø (Routledge, 2016), pp. 167–184.

25 Message and Greenhough, '"But It's Just a Fish"'.

26 'Culture of care' in animal facilities is a relatively recent and hard to define idea, as well as ethical and bureaucratic objective, see also Kirk, Chapter 5; Roe and Greenhough, Chapter 6, and Tremoleda and Kerton, Chapter 8.

27 Aquarists are often specialists, working only with fish or other aquatic species. This is not universally the case however.

28 For example, Robert W. Elwood, 'Assessing the Potential for Pain in Crustaceans and Other Invertebrates', in *The Welfare of Invertebrate Animals*, ed. by Claudio Carere and Jennifer Mather, *Animal Welfare* (Cham: Springer International Publishing, 2019), pp. 147–177; Paco Calvo, Vaidurya Pratap Sahi, and Anthony Trewavas, 'Are Plants Sentient?' *Plant, Cell & Environment*, 40.11 (2017), 2858–2869, DOI: 10.1111/pce.13065.

29 For example, Maximilian Padden Elder, 'Fishing for Trouble: The Ethics of Recreational Angling', in *The Palgrave Handbook of Practical Animal Ethics*, The Palgrave Macmillan Animal Ethics Series (London: Palgrave Macmillan, 2018), pp. 277–301; P. Sandøe et al., 'Balancing the Needs and Preferences of Humans against Concerns for Fishes: How to Handle the Emerging Ethical Discussions Regarding Capture Fisheries?' *Journal of Fish Biology*, 75.10 (2009), 2868–2871, DOI: 10.1111/j.1095-8649.2009.02461.x; Lund et al., 'Expanding the Moral Circle'.

30 For example, Maximilian Padden Elder, 'The Fish Pain Debate: Broadening Humanity's Moral Horizon', *Journal of Animal Ethics*,

4.2 (2014), 16–29, DOI: 10.5406/janimalethics.4.2.0016; 'The Moral Poverty of Pescetarianism', in *Ethical Vegetarianism and Veganism*, ed. by Andrew Linzey and Clair Linzey (New York: Routledge, 2019), pp. 103–116; Peter Singer, 'Fish: The Forgotten Victims on Our Plate', *Guardian*, 14 September 2010, section Opinion, www.theguardian.com/commentisfree/cif-green/2010/sep/14/fish-forgotten-victims [accessed 12 November 2021].

31 For example, Jonathan Balcombe, *What a Fish Knows: The Inner Lives of Our Underwater Cousins* (London: Oneworld Publications, 2017).

32 The 'consciousness raising' effects of popular culture are admittedly debateable: Driessen found that the film *Finding Nemo* became a popular name for fish and chip shops, as did – I regret to report – the BBC's popular ocean series *Blue Planet*; see Clemens P. G. Driessen, 'In Awe of Fish? Exploring Animal Ethics for Non-Cuddly Species', in *The Ethics of Consumption: The Citizen, the Market and the Law*, ed. by Helena Röcklinsberg and Per Sandin (Wageningen: Wageningen Academic Publishers, 2013), pp. 251–256.

33 Candice Allmark-Kent, 'How to Read Fishes: Science, Empathy, and Salar the Salmon', *Society & Animals*, 30.2 (2021), 170–187 (p. 3), DOI: 10.1163/15685306-12341569.

34 Message and Greenhough, '"But It's Just a Fish"'.

35 See, for example, *The Inevitable Bond*, ed. by Davis and Balfour; Arnold Arluke, 'Sacrificial Symbolism in Animal Experimentation: Object or Pet?' *Anthrozoös*, 2.2 (1988), 98–117, DOI: 10.2752/089279389787058091; Tora Holmberg, 'Mortal Love: Care Practices in Animal Experimentation', *Feminist Theory*, 12.2 (2011), 147–163, DOI: 10.1177/1464700111404206.

36 See, for example, Tora Holmberg, 'A Feeling for the Animal: On Becoming an Experimentalist', *Society & Animals*, 16.4 (2008), 316–335, DOI: 10.1163/156853008X357658.

37 See Tone Druglitrø, '"Skilled Care" and the Making of Good Science', *Science, Technology, & Human Values*, 43.4 (2017), 649–670, DOI: 10.1177/0162243916688093.

38 Sharp, 'Animal Research Unbound'.

39 Allmark-Kent, 'How to Read Fishes, p. 3.

40 See Vinciane Despret, 'The Body We Care for: Figures of Anthropo-Zoo-Genesis', *Body & Society*, 10.2–3 (2004), 111–134, DOI: 10.1177/1357034X04042938; Greenhough and Roe, 'Attuning to Laboratory Animals and Telling Stories'.

41 Koch and Svendsen, 'Negotiating Moral Value', p. 371.

8

Commentaries on culturing and sustaining care

Edited by Beth Greenhough

This book chapter focuses on cultures of care in animal research as an object of both professional and academic interest. The chapter presents three separate commentaries, two from invited respondents to this section of the book, and one from the section editor. The first commentary is from Jordi L. Tremoleda and Angela Kerton, whom both work as professionals in the field of animal research and have a keen interest in nurturing a culture of care for both animals and the humans who work with them. The second commentary is from Eva Giraud, an academic who specialises in animal studies, media, and activism, who offers a more critical take, exploring the incommensurable ethical positions occupied by animal welfarists working within animal research and animal activists operating outside it. The chapter editor's closing commentary then draws these different perspectives into a wider conversation around the tensions and complexities that emerge when different ideas, forms and spaces of care are juxtaposed.

8.1

Balancing the personal and the professional when culturing care in animal research

Jordi L. Tremoleda and Angela Kerton

As laboratory animal veterinarians directly involved in the training and education of animal research professionals, and responsible for animal care and welfare, we would like to raise awareness of the importance of aligning best animal care and best care for the staff working with laboratory animals. The chapters in this section (and this volume) open up questions around: how a culture of care is defined; how it is developed and embedded within our institutional systems and assessed to ensure that it is functioning well; how it impacts on an already empathetic professional community of animal technologists; and how it supports openness and transparency. For us these questions inform the broader question of how both personal and professional attitudes necessarily intersect in a working environment where you are emotionally bound to the care for another being.[1] This is a key topic of discussion in the laboratory animal field, one which invites a diverse range of opinions, shaped by expertise and experience working in varied facilities and systems, and by the specific needs of diverse lab animal species (Message, Chapter 7). In our role the challenge, then, arises of how best to harmonise the development and assessment of cultures of care across institutions, and the role played by innovative training and educational tools in nurturing a good culture of care in our facilities.

Nurturing a good culture of care remains a main ethos for those working in animal research, and it is widely recognised as needing to be well-integrated in our professional training and development. At the same time, addressing an issue that can provoke varied and complex ethical concerns and emotional responses requires guidance and logistical support. It also requires an approach that allows for reflective personal learning and emotional openness, allowing people to express how they feel when working with research animals, and importantly, integrating varied individual perceptions, expertise, duties, institutions, and species. As the chapters in this section showed, how staff approach this will be largely influenced by their individual role within a scientific establishment, their current and past experiences, job responsibilities and satisfaction, and how 'empowered' they feel to speak openly about the subject. For example, someone in a senior governing role, may defensively respond that systems to ensure a 'good culture of care' are well established, and be reassured by an 'open door policy on whistleblowing' and their research and funding success. Young animal technicians or PhD students, with demanding expectations and pressures in their work, are less likely to share such positive perspectives, or openly express any emotions related to their workplace environment. Therefore, it is important that empathy is shown across all professional levels, and importantly, that such sentiment is well-fostered across management and staff, paying a closer attention to the human–human relations in our institutions.[2]

Unfortunately, in the UK (and arguably beyond), current research culture is leading to an unhealthy, competitive environment in which a majority of research staff may not feel comfortable to speak up, and feel constrained and unable to see proactive steps forward.[3] This reality presents a worrying scenario that may directly impact on staff welfare, animal care, and science. Clearly, we need to build up better trust and respect for all the individuals, animals, and personnel that work with them – we need a better culture of care. Yet the lack of clarity as to what defines a good culture of care and how it can be assessed continues to jeopardise further action and commitment from major stakeholders in the animal research field. Indeed, various definitions have been proposed at numerous workshops that have been conducted within the laboratory animal science sector (see Figure 8.1).[4]

Interestingly, most of us would agree that the culture of care relates to high welfare standards, respect for all stakeholders' viewpoints, and excellent communication driven through individual and collective responsibility for good science, animal welfare, and the wellbeing of colleagues. Furthermore, these definitions are aligned with the expectations of society and the regulators. Perhaps where different stakeholders vary is the emphasis we can and would like to place on each of these different elements, and (as noted above) our capacity to take responsibility for actioning these.

To help materialise any plans for improving and assessing our culture of care attitudes in our establishments, it is therefore important to put the concept of 'care' within the animal research environment into context. In this way the chapters in this section serve a useful role in highlighting how cultures of care emerge in practice in different ways for different actors. For some, care may indeed be a 'technological infrastructure';[5] if laboratory animals are not cared for properly the consequences are wide ranging with

Figure 8.1 Responses to 'How would you define a culture of care?' from participants as a laboratory animal science workshop (from Robinson and Kerton, 2021).

critical legal, social, and institutional responsibilities. Care can also be seen as a form of governance, a 'necessary evil' and a 'big-brother' operative with a strong drive for compliance and licensing red-tape, which would translate care into a box-ticking exercise and lead to a lack of engagement from staff (Greenhough and Roe, Chapter 6). This triggers a broader reflection on whether legislative enforcement improves behaviour as a whole or whether a good culture of care could operate and be the norm without it? Working within an existing regulatory framework provides a direct impact through an emphasis on legal compliance, but it is important that a caring attitude builds up organically and naturally within our daily working practice; setting of a more stringent compliance and bureaucratic working schemes may not be the best route to build up an emotional and engaging genuine attitude across staff.

Care can equally be thought of as a set of skills, bound up with a professional identity.[6] Animal technologists or actively involved animal researchers are able to pick up on the smallest changes in animals' wellbeing, and their 'hands-on', attentive attitude will directly impact on the quality of their research interventions and their scientific validity. Attention to detail is critical, as such subtle changes can often indicate the onset of a disease and/or the effects that may confound our scientific findings. Interestingly, such knowledge and 'know-how' expertise seems to be vaguely acknowledged, and the importance of these subtle and difficult-to-assess nuances may be hard to describe or communicate between different stakeholders, and more importantly to those who determine funding and institutional support for research involving the use of animals.

Care can also take the form of a relationship between co-workers[7] and between humans and animals (Greenhough and Roe, Chapter 6). Assessing the nature of this latter relationship is not simple. For example, some people may find it difficult not to get 'over-attached' to animals they work with, especially those animals that display unique characteristics or appear to appreciate greater human contact time. Furthermore for some staff, like animal technologists or welfare researchers, their career path is strongly driven by their commitment to animal care. Thus, working with laboratory animals may create personal challenges for them, particularly when the time comes for them to humanely kill the animals they work with,[8] as has sadly been reflected through the closure of various animal research units during the pandemic. Equally, for others, as Reuben Message (Chapter 7) shows in his chapter, it can be challenging to form the attachments seemingly expected of them. How these relationships are handled by management (e.g., ensuring the provision of emotional support and the sharing or delegation of challenging scenarios such as mass humane culling), and sensitively accounted for by institutional governance (e.g., by assertively acknowledging such emotional challenges and the need for institutional support), will have a great impact on how staff feel supported and valued in their caring role.[9]

But the key question remains: how can we ensure a good standard of care that is well harmonised and valued by research facilities – and, importantly, that is well supported by governance and scientific stakeholders? Despite the development of valuable guidance on the culture of care in biomedical research,[10] there have been some recent reports of failure in care standards within research facilities.[11] Moreover, the risk of emotional dissonance[12] driven by the conflict between emotional responses to being exposed to, and sometimes perpetuating animal stress, and the capacity to express such emotions[13] remains critical to prevent disengagement, frustration, and compassion fatigue, highlighting that more needs to be done to protect the welfare of animals and staff. While such news is totally devastating, we must constructively reflect on our institutional and personal practices, and ensure that our staff continue to be fully supported under often challenging circumstances.[14]

New initiatives are required to facilitate this, providing better training to help researchers promote good cultures of care and new ways for them to raise concerns without fear of reprisals or prejudice. We should seek to offer opportunities for the research community to engage with leadership and management to help institutions and individuals to drive real changes and develop new precedents for assessing the health of our work environment and supporting the emotional satisfaction of staff. It is important to account for the provision of more reflective educational approaches, engaging and well-trained educators and, importantly, institutional support for integrating culture of care training into existing syllabi. It is crucial to integrate the multifaceted nature of the subject, addressing inter-cultural perceptions and ethical values, the value of multi-disciplinary expertise and the importance of the key role of all stakeholders including researchers, animal technologists and management. It is important that initiatives are built on active engagement and participation, bringing a positive uptake by individuals, institutions, and governance. It needs to be inclusive, with opportunities for individuals to talk openly about their experiences to colleagues who may have experienced similar situations. At the same time we need to remain conscious of the challenges of our highly time- and delivery-pressurised work. Therefore we welcome the AnNex initiative 'Care-Full Stories', which uses social science approaches and storytelling to help create new 'culture of care' training resources.[15] Such initiatives will provide a unique opportunity to further assess the impact of emotional openness and attitude explorations on the wellbeing of our staff, its impact on the care of our animals and, importantly, on supporting our scientific community to build up a better working environment.

We hope that we have been able to highlight the constant 'turmoil' that those working with the laboratory animal science field face. From our perspective as lab animal vets and educators, it is critical to recognise the complexity of addressing caring attitudes, but likewise the strength of our multi-disciplinary and multifaceted community, which provides a unique opportunity to openly and safely embrace the emotional dimensions of our professional work. While we should not think of our work as a juxtaposition of care and harm we do agree that there is 'also a form of harm–care analysis within animal research facilities, through which those working there negotiate tensions and pressures in their day-to-day work.' (Greenhough and Roe, Chapter 6, pp. 153–154). Therefore it is important to continue developing strategies to facilitate openness and discussions about how we feel about our work, including its emotional challenges.

8.2

Incommensurable care

Eva Haifa Giraud

I have sat down to write this commentary multiple times. I've tried beginning with theoretical quotations,[16] historical contexts,[17] and provocations that critically engage with the forms of care that are centralised in laboratory ethnographies,[18] but nothing seemed 'right'. In part I struggled because, after being invited by this book's editors to write a commentary from a perspective more explicitly sympathetic to anti-vivisection activism, I was unsure how to do justice to this task.

Critiques of animal research are heterogeneous, consisting of different academic and political threads that form a nexus of their own: from ecofeminism and critical animal studies, to canonical work in feminist science studies.[19] Other interventions range from internal critiques about the necessity and validity of certain forms of research, to welfarist movements who have collaborated with scientific communities in developing legislation, and activism that calls for abolition (which sees anything else as not just compromise but complicity offering legitimacy to vivisection).[20] Indeed, in the view of activists and scholars who hold more critical perspectives – and echoing Kutz's complicity principle where there is 'no participation without implication' – by contributing to a book that does not outrightly condemn animal research I would be seen as complicit in legitimising the practice.[21] The strands and components of anti-vivisectionist critique, therefore, are characterised by different aims, ideologies, and tactics, some of which cannot be brought into dialogue with cultures of care that are oriented toward welfare *within* the laboratory.

In the midst of trying to tie together all of the incommensurable ethical positions articulated across my failed drafts, however, I realised that perhaps there was something important to say about the ethical significance of incommensurability itself.[22] Following this aim, I begin by reflecting on conceptions of a culture of care (as defined by Greenhough and Roe, Chapter 6) and the specific ethical practices that are centralised by this concept.[23] I situate these arguments in relation to broader scholarship that sees care-work within the laboratory as a site of both knowledge generation and ethical obligation, and which has adopted a particular focus on close, bodily relationships between technicians and animals as a site where these ethical responsibilities emerge. This literature, for instance, has engaged with examples including guinea pigs being mourned by caretakers, laboratory rats observed by students, and even acts of killing in the laboratory, to argue that the close proximity afforded by care-work in these settings can foster new forms of attentiveness to non-human animals.[24] As outlined by Greenhough and Roe, in literature focused on the emergence of care within the laboratory, this attentiveness is often seen not only as holding capacity to generate closer attentiveness to the needs of non-human animals but can even create space for resistance and refusal.[25]

During my research I have tried to make sense of culturally specific tensions that surround animal ethics, engaging with social scientific frameworks from critical theory, science and technology studies, and media studies. Here I draw on some of my previous work to discuss the limitations of embodied care practices within the laboratory, referring to this form of care ethics as 'institutionally oriented care', due to its focus on strengthening cultures of care within existing institutional contexts. To conclude, I argue for the ongoing importance of external, activist critique as a lever for social change, even despite (and perhaps because of) its incommensurability with institutional cultures of care.

Institutionally oriented care

The chapters in this section on 'Culturing and sustaining care' articulate a complex understanding of care that speaks to wider literatures on embodied practices of caring. Kirk (Chapter 5), for instance, argues that care in the form of 'subjugated love' has historically held a vital but uneasy position within animal research. Technicians' 'affective bonds', he suggests, 'not least the ability to comprehend and understand the needs of animals in ways that were difficult to reduce to language', were both integral in preparing animals for experiments and something that resisted instrumentalisation. Likewise, in delineating what a contemporary culture of care in the laboratory might mean and how it could be enacted, Greenhough and Roe (Chapter 6) foreground how the practices of contemporary animal technicians are saturated with, and sometimes even motivated by, emotion; everyday care-work is entangled with affect-laden practices that do not always sit easily with rigid laboratory protocols. Message (Chapter 7) both echoes and complicates these arguments. Describing how aquarists negotiate professional expectations that technicians should foster affective bonds with animals, he foregrounds how emotion-work has become increasingly central to securing an identity as a 'good' aquarist.

While the chapters offer different insights about the *role* of care, then, each foregrounds a similar *mode* of care: a care which emerges in the laboratory through relationships between technicians and animals. A focus on this form of care speaks to a wider theoretical paradigm that has made important critiques of the limits of normative rights-based models of animal ethics in responding to the specificity of animal lives. Instead, this body of theory has found ethical potential in situated, affective relationships between humans and non-human animals, as pioneered by influential work from thinkers such as Haraway and Despret. Haraway, for instance, argues that: 'Caring means becoming subject to the unsettling obligation of curiosity, which requires knowing more at the end of the day than at the beginning'.[26] This approach is presented as a radically non-anthropocentric model of care, wherein ethical obligations are generated through humans becoming attuned to non-human animals' needs, learning what matters to them, and responding to these obligations.

For this form of relational care, the critical question is how space can be created for curiosity – and its attendant caring obligations – to flourish in sites (like laboratories) where these potentials are often submerged. Despret, for instance, illustrates her arguments with examples that include Konrad Lorenz's work with goslings and Temple Grandin's slaughterhouse design, arguing that each is an instance of humans learning to respond to the needs of non-human animals through curiosity that fostered partial sensory affinities and new understanding.[27] Lorenz learned how to care for his geese, while slaughterhouses were redesigned to be less distressing for cows. Chapters in this section, likewise, examine the complex potentials of felt responsibility (or to borrow Kirk's use of Haraway, response-ability) experienced by laboratory technicians. As Greenhough and Roe suggest, recognising these relationships is increasingly seen as integral to fostering a healthy culture of care for workers and animals, as well as generative of good science. Against this backdrop, Greenhough and Roe question whether it is fair or accurate to describe the complex role of care within the laboratory purely in terms of instrumentalisation – as Hollin and I suggest in previous work.[28]

In this previous work, we framed our arguments in terms of instrumental – rather than subjugated – love because of concern about the implications of grounding ethics in care that emerges through affective relationships in the laboratory. Effectively, we were emphasising the limitations of care ethics that emerge from within existing institutional settings. We turned to the history of laboratory beagles to foreground these limits. The first large-scale experimental beagle colony was constructed at the University of California Davis (UCD), as part of a wider set of 'lifespan' experiments designed to test the effects of radiation on a living population. At UCD, beagles were either given a short sharp dose of radiation (to replicate the impact of a nuclear blast) or fed radioactive food (to mirror the experience of living in a contaminated environment), but were otherwise allowed to live their lifespan. Importantly, a secondary – but equally influential – set of experiments were also being undertaken at UCD: to develop a model experimental colony, described as a 'utopic' environment by researchers.

When reading laboratory reports, what struck us was that the language utilised by the scientists at UCD to describe the experiments was uncannily close to wording used in contemporary animal studies scholarship to describe the transformative potential of care. In papers, reports, and the core textbook based on the research at UCD, for instance, researchers explicitly emphasised the importance of paying attention to what mattered to the beagles, and transformed their lived environment in response to these observations. Everything, from cage design to the type of gravel used as flooring, to the level of contact between caretakers and beagles, was re-designed to maximise beagle contentedness.

Our arguments in this context were two-fold. Firstly, it is important not to lose sight of the fact that these were experiments that induced radiation poisoning in dogs for human benefit. Ultimately, even though new knowledge and welfare techniques were developed through being closely attuned to beagles' needs, affective engagements were eliminated as soon as they threatened the ends of the experiment. Caretakers who expressed too much love for animals were fired, noisy dogs were de-barked. Our second point was that knowledge gained from these careful encounters fed into future beagle colony design with the aim of minimising the likelihood of beagles expressing disruptive agency in the future. This is not to say that researchers at the colony, such as Solarz (see Kirk, Chapter 5) didn't genuinely care for the animals (indeed, Solarz continued to work with beagles in the context of bed bug control after the experiments ended). We do, however, feel that it is important to recognise that this is an institutionally oriented conception of care, which has the capacity to foreclose – rather than create – ongoing obligations towards non-human animals.

As van Dooren points out in his sympathetic critique of Despret, for instance, through forcing goslings to imprint on him Lorenz might have generated academic insight and personal emotional affinity, but this was also an act of violence that foreclosed any future possibilities for the geese to bond with members of their species.[29] This act of violent-care, in other words, might be born of genuine emotional attachment but still created birds that loved their cages. Likewise, Grandin's insights might be cultivated through careful sensory attunement with cows but the purpose of this knowledge was to generate slaughterhouse designs that make the killing process less 'fricative'.[30] There is a tension here: the social scientist can see a dog that loves its cage and, without taking account of *why* said dog is caged, or *how* it came to love, use that as evidence that an experiment is ethical.

The space opened up by centralising institutionally oriented care has been valuable in disrupting narratives of human mastery, by foregrounding non-human agency in spaces that are conventionally thought of as instrumental. However, this line of argument has also resulted in a situation where practices ranging from slaughterhouse design to pedigree dog breeding are framed as ethical, caring, even loving, while activist work that critiques or contests these practices (from an inevitable distance) on the basis of harm caused to animals, are dismissed as retrograde. This situation is dangerous as it can foreclose discussion of wider contexts that undermine the potentials associated with institutionally oriented care ethics, while at the same time marginalising activist ethics that ask these structural questions. It seems important, therefore, to reflect further on forms of care that emerge beyond institutional settings.

Care outside of a culture of care

One of the reasons I had difficulty writing this commentary was because I struggled to construct a neat narrative about incommensurable ethical positions. Activist care for animals that manifests itself as opposition to animal research, for instance, cannot be brought to the table in dialogue regarding how to strengthen a culture of care within the laboratory. I suggest, however, that there is something critically important about care that is expressed as contestation and opposition, precisely because of its incommensurability with formal institutions.

As Greenhough and Roe note in Chapter 6, animal rights activism has historically played a significant role in articulating opposition that has exerted pressure on mainstream discourse and practice. A paradox, for instance, is that anti-vivisectionism is incompatible with welfarism (thus cannot come to the table as part of the debate), but has also contributed to reform in the wake of practices such as hidden filming (even as these practices have been motivated by opposition to animal research). These dynamics speak to the extensive body of research about the (complex) role of counter-public narratives and movements in fostering social change.[31] Often forced to operate autonomously, in different contexts and discursive spaces from those in which 'mainstream' public opinion is formed, counter-publics have nonetheless been an important realm in which dissenting knowledge is produced that – under the right socio-political conditions – can impact upon mainstream political beliefs and practices.

In contrast, care grounded in affective bonds can be limited precisely because of its compatibility with existing institutional arrangements. Message (Chapter 7), for instance, beautifully elucidates how the researchers' affective bonds with animals are now a normative expectation rather than subjugated, to the point that technicians have to undertake significant emotional labour as part of their professional identity. This form of institutionally oriented care ethics can sometimes even be leveraged for more strategic ends to allay public concern or criticism of animal research. In my book, *What Comes After Entanglement*, for instance, I discussed media depictions of primate research controversies in the early 2000s, in which activist objections were dismissed for being predicated on cartoonlike perceptions of animals.[32] In contrast, newspaper articles, broadcasts, and a high-profile BBC documentary about the research emphasised researchers' care for animals, with practices such as giving rats names framed as testament of emotional bonds. This close link between proximity, expertise, and care depicted the only legitimate forms of care as those expressed by individuals participating in research, and became a common frame through which activist concerns were dismissed.

In sum, while counter-public expressions of care as contestation opens space for critiques of institutions and practices, care ethics emerging from within institutions are – perhaps inevitably – easier to realign with these institutions even if a degree of dissonance exists. This argument might seem obvious, but since the late 2000s, the significance of contentious activism has sometimes been obscured, with normative activist expressions of care contrasted unfavourably with care ethics born of affective encounters. In theoretical contexts, for instance, the repeated refrain about animal activism and veganism is that these movements mark a denial that no way of living is possible without some form of killing.[33] As touched on above, this stands in contrast with depictions of care that emerge from close relationships with non-human animals, which are framed as 'turning things around', overhauling anthropocentric ways of knowing, and giving rise to ongoing ethical response-abilities.[34] It is important to move beyond this framing, and recognise that care which is incommensurable with institutional cultures of care still has a vital political role in asking whether things could be otherwise.

8.3

What constitutes care-in-practice?

Beth Greenhough

As an academic I am aware that care long been a concern for social scientists, with scholars such as Joan Tronto stressing the complexity of the concept of care[35] and the difficulty of defining care 'needs' and 'good care'.[36] Tronto's work has been especially significant in challenging social perceptions of aspects of care-work as 'feminised', which has often led to that work being undervalued. To address this, Tronto suggests focusing on 'the place of caring in concrete daily experience and in our patterns of moral thought' to 'forge a society in which care can flourish'.[37] Feminist care ethics further cautions against moments when the physical labour of providing care (feeding, washing, and so on) can become separated from 'care-as-affection' or 'caring about'. For example, Hochschild[38] marked a distinction between the 'warm' care provided by those who cared about as well as for those they cared for (typically associated with caring for family and friends), and the 'cold', often institutionalised forms of care where the task of caring for the needs of a patient was met, but there was little inclination to care about them.

Echoing these shifts in care theory, a series of scandals in the healthcare sector, such as investigations into malpractice at the NHS mid-Staffordshire Health Trust, drew attention to uncaring workplace cultures and suggested increasing social unease about the forms of cold care that seemingly characterised institutional settings.[39] This led to a demand that patients receive 'effective care from caring, compassionate and committed staff'[40] and resulted in a series of interventions designed to promote a 'culture of care' within institutions, such as the NHS 'culture of care' barometer.[41] This pattern is echoed in the animal research sector, which has also suffered a series of prominent care failings, including the recent British Union for the Abolition of Vivisection allegations concerning the conduct of animal research at Imperial College London.[42]

Amongst those who work in animal research, there is increasing recognition that the established set of guidelines and procedures that shape moral reasoning in animal research, articulated through the 3Rs principles[43] and harm–benefit analysis,[44] may not be sufficient for preventing significant care failings. There is also growing recognition of the largely hidden burden that 'affective' care-work places on staff, seen in both internal and external assessments of the emotional labour and compassion fatigue experienced by those who provide care for laboratory animals.[45] Since I began researching the social and cultural dimensions of animal research in 2013, the phrase 'culture of care' has gone from being a little-known soundbite to becoming a key focus of regulatory concern, training provision, and everyday life in animal facilities across the UK and beyond. Consequently, it was not surprising that when my colleagues and I brought a group of social scientists and humanities scholars together with stakeholders from across the animal research community in 2015 to develop a collaborative agenda for *Social Scientific Research on Laboratory Animal Science and Welfare*, one of the key themes to emerge was around how best to define, develop, and sustain cultures of care.[46]

Questions of care feature strongly in the work of many of those who write about animal research from the perspective of the social sciences and humanities,[47] demonstrating how academic insights can help us work through the complexities of giving and receiving care in the context of animal research. This work is continued by each of the chapter authors in this section. Robert G. W. Kirk (Chapter 5) charts the emergence and co-production of the laboratory animal and the laboratory technician, showing how the norms for this working relationship became established and describing the development of infrastructures for delivering care. Kirk (Chapter 5, p. 128) draws on Daston's 'moral economy of science, to demonstrate that care is both integral and subjected to the production of objective scientific knowledge. Care was expressed through professionalism, a meticulous attention to biosecurity, and the production of Specific Pathogen Free animals, but it was also emergent and excessive. For Kirk (Chapter 5, p. 139) these more emergent forms of care are evidenced in the example of handling; hard to standardise as a set of written instructions, handling must instead be 'attentive and responsive to the feeling of the animal toward the handler'. This contrast – between the more infrastructural, bureaucratic, top-down articulations of care as mandated by regulations and guidelines, and the more bottom-up, everyday forms of care that shape day-to-day encounters between humans, and between humans and animals, in the context of animal research – also emerges in the chapter by Emma Roe and myself (Chapter 6). This illustrates how different aspects and understandings of care do not always sit easily alongside each other, and may even be directly contradictory, such as the Individually Ventilated Cages, which make it safer (at least in terms of biosecurity) for lab animals and technicians to be in close proximity while simultaneously imposing a plastic barrier between them. Reuben Message (Chapter 7) similarly draws out the challenges associated with providing good care, but specifically highlights how this can vary by species. Message shows how tensions may emerge between working in an animal research environment where caring about as well as for animals is increasingly the norm and working with a species to which society seems largely indifferent.

As a researcher, I see these complexities – and more specifically the question of what constitutes care-in-practice – as both object of academic concern and as a remit for ethical praxis: How do the animal technologists I observe, interview, and write about position their role as carers? How do I care, in turn, for my research subjects by protecting their confidence, sharing their stories and working with them to improve staff and animal welfare? I am fascinated by the juxtaposition of the more formal, largely utilitarian ethical values that shape the regulatory architectures of animal research, and the more everyday acts of care I witness and hear about in the lab, such as labouring to make fake waterweed to enrich the lives of zebrafish, or electing to be the one to euthanise a loved animal. I am also troubled by the need for care I perceive as I begin to understand the emotional burdens carried by those who care for research animals and become better acquainted with the stresses and strains of working in animal research. As Jordi L. Tremoleda and Angela Kerton (Chapter 8) note, engaging in discussions about cultures of care with those working within animal research – which are sensitive to the diverse, complex, emotionally difficult, and contradictory forms care can take – remains a key challenge for the field, and one in which I am increasingly personally invested. At the same time, I am conscious of my own positioning within the world of animal research; as Eva Giraud (Chapter 8)[48] reminds us, some forms of care – for example those expressed in some types of animal rights activism – remain incommensurable with those seen from within animal research institutions. In recognising and participating in conversations about culturing care in animal research from within the sector I may exclude myself from other, different conversations. Culturing care in animal research is, then, for me, not so much a matter of rational moral judgement, but a matter of deep and sustained ethnographic engagement, concern, conversation, and collaboration – an academic choice, but also always and unavoidably a social and ethical one.

Notes

1 Keith Davies and Duncan Lewis, 'Can Caring for Laboratory Animals Be Classified as Emotional Labour?' *Animal Technology and Welfare*, 9.1 (2010), 1; Beth Greenhough and Emma Roe, 'Attuning to Laboratory Animals and Telling Stories: Learning Animal Geography Research Skills from Animal Technologists', *Environment and Planning D: Society and Space*, 37.2 (2019), 367–384, DOI: 10.1177/0263775818807720; Beth Greenhough and Emma Roe, 'Exploring the Role of Animal Technologists in Implementing the 3Rs: An Ethnographic Investigation of the UK University Sector', *Science, Technology, & Human Values* 43.4 (2017), DOI: 10.1177/0162243917718066.

2 Annabella Williams, 'Caring for Those Who Care: Towards a More Expansive Understanding of "Cultures of Care" in Laboratory Animal Facilities', *Social & Cultural Geography*, 24.1 (2021), 31–48, DOI: 10.1080/14649365.2021.1939123.

3 Wellcome Trust, *What Researchers Think about the Culture They Work In*, 2020, www.wellcome.org/reports/what-researchers-think-about-research-culture [accessed 25 November 2022].

4 Sally Robinson and Angela Kerton, 'What Does a Culture of Care Look like? Lessons Learnt from a Workshop Survey', *Laboratory Animals*, 50.10 (2021), 263–271, DOI: 10.1038/s41684-021-00852-6.

5 Tone Druglitrø, '"Skilled Care" and the Making of Good Science', *Science, Technology, & Human Values*, 43.3 (2018), 649–670, DOI: 10.1177/0162243916688093, see also Greenhough and Roe (Chapter 6); Giraud (Chapter 8).

6 Druglitrø, '"Skilled Care" and the Making of Good Science'. See also Kirk (Chapter 5); Message (Chapter 7).

7 Williams, 'Caring for Those Who Care'.

8 Greenhough and Roe, 'Attuning to Laboratory Animals and Telling Stories'.

9 American Association for Laboratory Animal Science, *Compassion Fatigue: The Cost of Caring. Human Emotions in the Care of*

Laboratory Animals, www.aalas.org/education/educational-resources/cost-of-caring [accessed 25 November 2022].

10 Norecopa, 'Culture of Care', https://norecopa.no/more-resources/culture-of-care [accessed 25 November 2022]; Thomas Bertelsen and K. Øvilsen, 'Assessment of the Culture of Care Working with Laboratory Animals by Using a Comprehensive Survey Tool', *Laboratory Animals*, 55.5 (2021), 453–462, DOI: 10.1177/00236772211014433; Penny Hawkins and Thomas Bertelsen, '3Rs-Related and Objective Indicators to Help Assess the Culture of Care', *Animals*, 9.11 (2019), 969, DOI: 10.3390/ani9110969; Sally Robinson et al., 'The European Federation of the Pharmaceutical Industry and Associations' Research and Animal Welfare Group: Assessing and Benchmarking "Culture of Care" in the Context of Using Animals for Scientific Purpose', *Laboratory Animals*, 54.5 (2019), 421–432, DOI: 10.1177/0023677219887998 [accessed 1 February 2023].

11 Natalie Grover and Ashifa Kassam, 'Undercover Footage Shows "Gratuitous Cruelty" at Spanish Animal Testing Facility', *Guardian*, 8 April 2021, www.theguardian.com/environment/2021/apr/08/undercover-footage-shows-gratuitous-cruelty-at-spanish-animal-testing-facility-madrid-vivotecnia [accessed 25 November 2022].

12 Megan R. LaFollette et al., 'Laboratory Animal Welfare Meets Human Welfare: A Cross-Sectional Study of Professional Quality of Life, Including Compassion Fatigue in Laboratory Animal Personnel', *Frontiers in Veterinary Science*, 7.114 (2020), DOI: 10.3389/fvets.2020.00114.

13 David Grimm, '"It's Heartbreaking": Labs Are Euthanizing Thousands of Mice in Response to Coronavirus Pandemic', *Science*, 2020, DOI: 10.1126/science.abb8633; J. Preston Van Hooser et al., 'Caring for the Animal Caregiver – Occupational Health, Human–Animal Bond and Compassion Fatigue', *Frontiers in Veterinary Science*, 8 (2021), DOI: 10.3389/fvets.2021.731003.

14 Jordi L. Tremoleda and Angela Kerton, 'Creating Space to Build Emotional Resilience in the Animal Research Community', *Laboratory Animals*, 49.10 (2020), 275–277, DOI: 10.1038/s41684-020-0637-7.

15 Beth Greenhough and Hibba Mazhary, *Care-Full Stories: Innovating a New Resource for Teaching a Culture of Care in Animal Research Facilities*, 2021.

16 My first version, for instance, drew on Vinciane Despret and Donna Haraway – as key theorists who have been engaged with across this collection: Vinciane Despret (trans. Brett Buchanan), *What Would Animals Say if We Asked the Right Questions?* (Minneapolis, MN:

University of Minnesota Press, 2016); Donna J. Haraway, *When Species Meet* (Minneapolis, MN: University of Minnesota Press, 2008).

17 This draft opened with a focus on nascent distinctions between animal welfare and rights, which started to emerge in Victorian England as anti-vivisectionist groups began to split, depending on whether they entirely opposed animal research or were willing to compromise and engage in dialogue to improve welfare. See: Richard D. French, *Antivivisection and Medical Science in Victorian Society* (Princeton, NJ: Princeton University Press, 1975).

18 Initially I made a similar point to my argument in a review forum on Nicole Nelson's nuanced ethnography *Model Behavior*. In my commentary, I foreground challenges associated with lab ethnographies and interviews with researchers, particularly the challenge of maintaining positive relationships, respect, and loyalty to research participants, without feeling compelled to bracket aside critical questions about particular research agendas (here genetic models of alcoholism) or more searching questions about the nature and necessity of animal research in these contexts. Nicole C. Nelson, *Model Behavior: Animal Experiments, Complexity, and the Genetics of Psychiatric Disorders* (Chicago, IL: University of Chicago Press, 2018); Eva H. Giraud, 'Model Behaviour by Nicole Nelson: Complex Ethics', *Studies in History and Philosophy of Science Part C-Studies in History and Philosophy of Biological and Biomedical Sciences*, 82 (2020), DOI: 10.1016/j.shpsc.2020.101267. For a more critical appraisal of this form of animal research, which questions its necessity, see the commentary piece from Wayne Hall, 'Model Behaviour by Nicole Nelson: Is Animal Behaviour Genetics a Degenerating Research Program?' *Studies in History and Philosophy of Science Part C-Studies in History and Philosophy of Biological and Biomedical Sciences*, 82 (2020), DOI: 10.1016/j.shpsc.2020.101268.

19 Rather than include a long list, a helpful piece that encapsulates these tensions is Zipporah Weisberg's 'Broken Promises of Monsters', which is critical of the model of care promoted in Haraway's *When Species Meet* for breaking from her critique of vivisection in earlier work (in the context of Harlow's primate research). Zipporah Weisberg, 'The Broken Promises of Monsters: Haraway, Animals and the Humanist Legacy', *Journal for Critical Animal Studies*, 7.2 (2011), 22–62.

20 Internal debate is captured by Hall's aforementioned criticisms of animal models for alcoholism (see note 3); other internal criticisms include critiques of lack of funding for non-animal models, or the role of commercial interests in preventing information sharing and the disincentive

to share 'failed' results, which leads to repetition between experiments; see, for instance, Judith Hampson, 'Legislation: A Practical Solution to the Vivisection Dilemma?' in *Vivisection in Historical Perspective*, ed. by Nicolaas A. Rupke (London: Routledge), pp. 314–339.

21 Christopher Kutz, Complicity: Ethics and Law for a Collective Age (Cambridge: Cambridge University Press, 2000), p. 122.

22 I use 'incommensurability' in a fairly broad, literal sense in this chapter. For a fully articulated 'ethics of incommensurability', defined as an approach that 'digs into difference and maintains that difference while also trying to stay in good relations' (137) see Max Liboiron, *Pollution is Colonialism* (Durham, NC: Duke University Press, 2021). This ethic foregrounds how those with different commitments (in Liboiron's case, Indigenous Elders and academic peer reviewers) might hold incommensurable understandings of 'what is true and right and good' in research (33). For Liboiron, a recognition of incommensurability is vital in recognising the importance of concrete forms of refusal and contestation (particulary in settler-colonial contexts where dominant land relations, reproduced through certain forms of scientific practice, foreclose Indigenous futures), but can also be a starting point for generating new solidarities. As Liboiron describes: 'Incommensurability means things do not share a common ground for judgement or comparison … Anticolonialism within dominant science. Diversity work in a racist institution. Humility in a tenure application. All are impossible bedfellows that are nonetheless crucial to pursue and indeed happen, yet should never be smoothed over or conflated in that process' (136).

23 See also the articles in Gail Davies et al. (eds), 'Science, Culture, and Care in Laboratory Animal Research', *Science, Technology & Human Values* 43.4 (2018), 603–621, DOI: 10.1177/016224391875703.

24 Haraway, *When Species Meet*; see also Vinciane Despret, 'The Body We Care For: Figures of Anthropo-Zoo-Genesis', *Body & Society* 10.2–3 (2004), 111–134, DOI: 10.1177/1357034X04042938; Tora Holmberg, 'Mortal Love: Care Practices in Animal Experimentation', *Feminist Theory*, 12.2 (2011), 147–163, DOI: 10.1177/14647001114042. For discussion of some of the tensions surrounding these relationships, see: Carrie Friese, 'Realizing Potential in Translational Medicine: The Uncanny Emergence of Care as Science', *Current Anthropology*, 54. S7 (2013), S129–S138, DOI: 10.1086/670805.

25 For an informative summary of literature in which these debates initially emerged, see Beth Greenhough and Emma Roe, 'Ethics, Space, and Somatic Sensibilities: Comparing Relationships between Scientific

Researchers and their Human and Animal Experimental Subjects', *Environment and Planning D: Society and Space*, 29.1 (2011), 47–66, DOI: 10.1068/d17109.

26 Haraway, *When Species Meet*, p. 36.

27 Despret, 'The Body We Care For'.

28 Eva Giraud and Gregory Hollin, 'Care, Laboratory Beagles, and Affective Utopia', *Theory, Culture & Society* 33.4 (2016), 27–49, DOI: 10.1177/0263276415619685.

29 Thom van Dooren, *Flight Ways* (New York: Columbia, 2014), p. 105.

30 For more on how animal resistance is often incorporated into killing to make it less fricative, see Dinesh Wadiwel, *The War Against Animals* (Leiden: Brill, 2015).

31 I am using the term 'counter-public' in line with classic feminist, queer, and anti-racist critiques of Habermas's valorisation of the liberal-bourgeois public sphere, which have foregrounded how certain voices are ordinarily excluded or marginalised from the mainstream public sphere and can only exert pressure via the formation of distinct counter-public identities. For example, Nancy Fraser, 'Rethinking the Public Sphere', *Social Text* 25/26 (1990), 56–80, DOI: 10.2307/466240; Michael Warner, 'Publics and Counterpublics', *Public Culture*, 14.1 (2002), 49–60, DOI: 10.1215/08992363-14-1-49; Sarah J. Jackson et al., *#HashtagActivism: Networks of Race and Gender Justice* (Cambridge, MA: MIT Press, 2020).

32 Eva H. Giraud, *What Comes After Entanglement?* (Durham, NC: Duke University Press, 2019).

33 Haraway, *When Species Meet*, p. 22.

34 Despret, for instance, makes this point in 'K is for Killing' explicitly drawing on Haraway's argument that no way of living is possible without killing; *What Would Animals Say*, p. 85.

35 Berenice Fisher and Joan Tronto, 'Towards a Feminist Theory of Caring', in *Circles of Care: Work and Identiy in Women's Lives*, ed. by Emily K. Abel and Margaret K. Nelson (Albany, NY: State University of New York Press, 1990), pp. 35–62.

36 Daniel Engster, *The Heart of Justice: Care Ethics and Political Theory* (Oxford: Oxford University Press, 2007); Virginia Held, *The Ethics of Care: Personal, Political, and Global* (Oxford: Oxford University Press, 2006); Joan C. Tronto, 'Beyond Gender Difference to a Theory of Care', *Signs: Journal of Women in Culture and Society*, 12.4 (1987), 644–663, DOI: 10.1086/494360.

37 Joan C. Tronto, *Moral Boundaries: A Political Argument for an Ethic of Care* (New York: Routledge, 1993), p. 633.

38 A. R. Hochschild, *The Managed Heart: Commercialization of Human Feeling* (Berkley, CA: University of California Press, 2003).
39 Annabella Williams, 'Caring for Those Who Care: Towards a More Expansive Understanding of "Cultures of Care" in Laboratory Animal Facilities', *Social & Cultural Geography*, 24.1 (2021), 31–48, DOI: 10.1080/14649365.2021.1939123.
40 Robert Francis, *Report of the Mid Staffordshire NHS Foundation Trust Public Inquiry* (London: Department of Health, 2013), p. 67.
41 Anne Marie Rafferty et al., *Culture of Care Barometer: Report to NHS England on the Development and Validation of an Instrument to Measure 'Culture of Care' in NHS Trusts* (London: King's College London, 2015).
42 Steve Brown, *Independent Investigation into Animal Research at Imperial College London*, 2013, http://brownreport.info/ [accessed 26 April 2019].
43 The 3Rs are a commitment to replace, reduce, and refine the use of animals in research. NC3Rs, 'The 3Rs' www.nc3rs.org.uk/the-3rs [accessed 11 October 2021].
44 Harm–benefit analysis is a key principle shaping the evaluation of UK animal research licence applications, see Animals in Science Regulation Unit, *The Harm–Benefit Analysis Process New Project Licence Applications, Advice Note: 05/2015* (London: Home Office, 2015); G. Davies, 'Harm–Benefit Analysis: Opportunities for Enhancing Ethical Review in Animal Research', *Laboratory Animals*, 47.3 (2018), 57–58, DOI: 10.1038/s41684-018-0002-2; Animals in Science Committee, *Review of Harm–Benefit Analysis in the Use of Animals in Research* (London: Home Office, November 2017) www.gov.uk/government/uploads/system/uploads/attachment_data/file/662098/Review_of_harm_benefit_analysis_in_the_use_of_animals_in_research.pdf [accessed 1 February 2023].
45 Keith Davies and Duncan Lewis, 'Can Caring for Laboratory Animals Be Classified as Emotional Labour?' *Animal Technology and Welfare*, 9.1 (2010), 1; Megan R. LaFollette et al., 'Laboratory Animal Welfare Meets Human Welfare: A Cross-Sectional Study of Professional Quality of Life, Including Compassion Fatigue in Laboratory Animal Personnel', *Frontiers in Veterinary Science*, 7.114 (2020), DOI: 10.3389/fvets.2020.00114.
46 Gail Davies et al., 'Developing a Collaborative Agenda for Humanities and Social Scientific Research on Laboratory Animal Science and Welfare', *PLOS ONE*, 11.7 (2016), 1–12, DOI: 10.1371/journal.pone.0158791.

47 See, for example, Tone Druglitrø, '"Skilled Care" and the Making of Good Science', *Science, Technology, & Human Values*, 43.3 (2018), 649–670, DOI: 10.1177/0162243916688093; Lynda Birke et al., *The Sacrifice: How Scientific Experiments Transform Animals and People* (Lafayette, IN: Purdue University Press, 2007); Lesley A. Sharp, 'The Moral Lives of Laboratory Monkeys: Television and the Ethics of Care', *Culture, Medicine, and Psychiatry*, 41.2 (2017), 224–244, DOI: 10.1007/s11013-017-9530-2; Gail Davies, 'Caring for the Multiple and the Multitude: Assembling Animal Welfare and Enabling Ethical Critique', *Environment and Planning D: Society and Space*, 30.4 (2012), 623–638, DOI: 10.1068/d3211.

48 See also Eva Haifa Giraud, *What Comes after Entanglement? Activism, Anthropocentricism and an Ethics of Exclusion* (Durham, NC: Duke University Press, 2019).

Part III

Distributing expertise and accountability

9

(Dis)placing veterinary medicine: veterinary borderlands in laboratory animal research

Alistair Anderson and Pru Hobson-West

Introduction

The veterinary profession is a profession of multiplicities, with broad training and expertise deployed in a wide variety of situations. Each of these situations comes with its own world,[1] ranging from the financial and affective context of human–pet relationships, the capitalist context of agribusiness, and the surveillance and management of public health governance. Veterinarians' varied professional roles entail a variety of technical skills and 'situated expertise'[2] required to treat different species and deal with different non-veterinary stakeholders, such as pet owners, farmers, and animal research scientists. The veterinary profession, broadly defined, has traditionally been understudied in the social sciences, with a nascent literature examining the profession and its challenges, relatings, and mobilities.[3]

This chapter contributes to this emerging social scientific body of work by examining how Named Veterinary Surgeons (NVSs) – a mandated presence in commercial and university UK animal research laboratories – articulate their niche as part of the broader veterinary profession. While other outputs from the Animal Research Nexus Programme (AnNex) focus on the career journey of the individual NVS,[4] and the role of geography in the construction of the laboratory as a positive ethical space,[5] this chapter focuses more specifically on the borderlands[6] that emerge between clinical and laboratory practice when veterinary professionals articulate the practical and personal differences involved in moving between and performing different kinds of veterinary roles. In so doing we point to the complex ways in which veterinarians draw, navigate, and blur boundaries between their professional worlds, despite ostensibly centralised professional regulation.[7]

In what follows we firstly review the limited existing work on the experiences of NVSs. The methods through which interview data were collected and analysed are then briefly described. The findings are presented by analytic theme, covering firstly the way in which the lab/clinic boundary relies on claims by NVSs that they are not 'real' veterinary professionals, secondly the maintenance of the lab/clinic boundary as exemplified in their reflections on engagements with the general public, and thirdly how this boundary is also sometimes blurred, via examples of shared learning across professional spaces and places.

The complexity of the Named Veterinary Surgeon role

In the UK, animal research is regulated via the Animals (Scientific Procedures) Act 1986 (ASPA), which creates a three-way system of licensing. The involvement of veterinary professionals in animal research did not begin with ASPA. However, the Act did make the appointment of NVSs at all licensed establishments mandatory.[8] ASPA created statutory responsibilities for animal welfare for which NVSs, among other named individuals such as Named Animal Care and Welfare Officers, are accountable.[9] These protections, some NVSs have argued, give a degree of consistency to the quality of animal lives within the laboratory as compared with those outside it.[10]

The role of the NVS, however, is not as straightforward as the legislative mandate might initially suggest. Indeed, as reflected on by Dennison's commentary in Chapter 13, all veterinarians have to take an oath when joining the profession, which is to make animal welfare their first priority. As Ashall and Hobson-West summarise, the role of the NVS is 'particularly complex in terms of accountability and professional responsibility, since the NVS is accountable to both the establishment licence holder (under ASPA), whilst *also* having professional responsibilities to the animals under their care, the public, other veterinary surgeons, and the Royal

College of Veterinary Surgeons (under the Veterinary Surgeons Act)'.[11] The implication is that the role is one that requires veterinary professionals to actively navigate the boundary between these two pieces of legislation, exercising professional judgement to reconcile potentially conflicting tensions arising from multiple professional accountabilities within the laboratory. As argued below, this navigation also entails a complex form of boundary drawing, which includes certain kinds of images of those outside the lab, including wider publics.

In addition to the flexibility introduced by several sets of legislation (see Palmer, Chapter 10 in this volume for more on flexibility), the NVS role harbours further complexities in that it is not simply a clinical role but involves a number of other areas of administration and advisory work within the laboratory, and relationship management inside and outside the laboratory setting. As has been neatly summarised by others,[12] vets have a diversity of expertise spanning 'comparative pathology, diagnosis, prognosis, disease prevention and treatment, anaesthesia and surgery, pain recognition and control, breeding control, and euthanasia'. This, it is argued, renders NVSs 'uniquely qualified to provide training, assessment, and supervision on what [are] considered to be veterinary interventions for scientific procedures'.[13] NVSs are consequently not only involved in the direct management of animal health and welfare, but are also involved in training, ethical review, and the implementation and promotion of the principles of replacement, reduction, and refinement (the 3Rs).[14] As with other veterinary professional roles,[15] the NVS position thus requires a specific set of social skills to develop and manage relationships with wider staff involved in animal research. As summarised in a careers section of a major veterinary journal, 'The NVS role requires good communication, good teamwork and good working relationships, with mutual respect for the responsibilities of others'.[16]

In addition to key relationships with others inside the laboratory, it is important to recognise the wider social context in which NVSs operate. For example, previous publications have pointed to the way in which NVSs imagine the wider public or wider audiences. An article published in 2006 reported that NVSs can feel 'caught in the middle', with one noting that 'the anti-vivisectionists don't like you, because you're on the other side, as they perceive it [and] some scientists perceive you as trying to change the way they do their work'.[17] Previous sociological work has also highlighted the way in which NVSs draw on particular images of those outside the lab, in order to navigate and explain their own professional role. In their interview study, Hobson-West and Davies[18] show how NVSs articulate a particular imaginary (dubbed 'societal sentience') of wider public views towards animals and particular species, and that this has an impact both on legislation, and on animal care practices.

More broadly, scholars have sought to articulate the role of publics in the wider animal research debate. For example, Davies et al.[19] have highlighted the positioning of publics as 'stakeholders with opinions that matter', and Hobson-West has previously argued that public opinion is framed as a 'resource in the animal research debate',[20] used by all sides to show themselves as legitimate. Beyond surveys, publics are also enrolled more directly in research governance as 'lay reviewers on funding panels, where their expertise helps align research priorities and practices with public expectations of research'.[21] To return to veterinarians, public opinion is likewise enrolled to frame the veterinary profession as trustworthy, with repeated surveys conducted for the British Veterinary Association (BVA) and the Royal College of Veterinary Surgeons (RCVS) positioning the veterinary profession as one of the most trusted in the UK.[22]

In summary, NVS work exists at the intersection of a number of potentially conflicting sets of perceptions and expectations regarding animal research, for example between 'anti-vivisectionists', as claimed in Smith and Wolfensohn,[23] and the more romanticised view of the profession cultivated by veterinary professional bodies.[24] Within this context, as for other professionals, NVSs have to actively navigate their own place within both animal research and their profession. So how exactly do veterinarians in the laboratory go about this? This chapter focuses on how NVSs draw contrasts between their own role and with clinical practice, and shows how this requires a series of discursive boundaries to be created and blurred.

Methods

One strand of AnNex focuses on publics and professions. As part of this strand of work, NVSs were recruited through snowball sampling via the project team's existing networks and a callout during a specialist conference. The interview agenda and empirical design benefited from the advice of an advisory panel comprising of three NVSs, to ensure that the work was pertinent to veterinary stakeholders. Ethical approval for data collection was granted by the School of Veterinary Medicine and Science at the University of Nottingham (approval number 1800160608), and data collection took place in 2018. All those interviewed were currently employed as an NVS, and many had previously worked in general clinical practice.[25] All vets were associated with a commercial or university research site. This is noted here in order to recognise that while the vast majority of animal research occurs in a laboratory setting, some does occur outside 'in the field', including in the veterinary clinic, where different regulatory boundaries for veterinarians are also particularly important. However, this chapter focuses on the role of the veterinarian in the UK animal research laboratory.

Qualitative interviews were chosen as the research method, firstly due to the lack of existing data on NVS work, and secondly the desire to explore the work of NVSs in detail through their own accounts and in their own language. Interviews were carried out in person at a location identified by the participant. An interview guide was developed and discussed with an expert advisory panel of three NVSs, and was trialled during two pilot interviews with no subsequent alterations. The interview format and order of questions were subject to revision as the data collection progressed. Interviews were transcribed by a third party under a confidentiality agreement. Transcripts were anonymised with all identifiable material regarding names, locations, and organisations removed. Each transcript was assigned a random but gender-specific pseudonym. These transcripts were analysed by the first author using NVivo 12. Codes were also discussed between the authors.

The analysis approach was reflexive Thematic Analysis[26] and involved two cycles of inductive coding. In the first cycle, transcripts were coded line by line in order to prioritise the voices of the participants.[27] The second round of coding aimed to understand the patterns that underpinned these initial categories,[28] and these patterns were used to coalesce this coding into analytic themes. This analysis was a creative process, and the themes did not 'emerge from' the data but were 'active creations' of the analyst.[29] The findings now described are therefore an interpretive story, developed from this data interacting with the biography of the researchers. It is also important to note that what follows does not mirror the format of the interview agenda, nor the phrasing of the interview questions. As is common in qualitative research, we have developed and refined these categories via the analysis. In practice these themes are highly interrelated, but to allow readability we have divided these into three. First, we consider how the boundary between the lab and the clinic is constructed. Second, we consider how this boundary is maintained. And finally, we consider how the boundary is blurred.

Constructing the lab–clinic boundary

The RCVS 2019 Survey of the Veterinary Profession[30] found that the main area of UK veterinary surgeons is small animal practice (52.6%). By contrast, the numbers working in animal research fall into the 'other' category, which comprises 2.8% of the profession and includes consultancy, racing, and government roles. This also helps to explain the lower public profile of the NVS as opposed to other veterinary roles.[31] In the interviews, participants were well aware of their NVS role as niche, and used various discursive strategies to set themselves apart from the mainstream veterinary profession. For example, NVS Peter described laboratory veterinary medicine as the 'poor cousin of the veterinary world'. More specifically, interviewees would often contrast the laboratory role in which they were using their veterinary expertise to the role of a 'normal' or 'typical' veterinarian. Sometimes this contrast was made in passing, as part of wider reflections, or made explicitly, when referring to their own career history.

> I guess initially when I wanted to go to vet school, since I was a little girl, I had wanted to do small animals, so the typical veterinary surgeon. (Natasha, interview, 2018)

> [A]fter my degree I went straight to work, just a normal vet, small animal practice. (Nathalie, interview, 2018)

However, this perception was not always presented in an internalised manner. As shown in the next extract, Melanie recounted the way in which a colleague appeared to distinguish between 'real' veterinary work and 'paper pushing' as the NVS:

> When I worked with [former colleague], two or three years ago, [former colleague] always said to other people always, farmers or whatever, 'Now, I do real veterinary work', but what [former colleague] meant was, 'you don't do any veterinary work here! It's just paper pushing. I don't know what it is, but it's nothing really like what you do in general practice'. (Melanie, interview, 2018)

This portrayal of NVS work as more hands-off than general practice work was recurrent in several narratives. While NVS work was often separated from general practice in these accounts as being part of a different 'world' (multiple interviewees), experience developed in general practice was still constructed as professionally invaluable in the laboratory:

> A few years ago, twice actually, we had new graduates come straight into NVS and it was disastrous, it was absolutely disastrous, because we can't provide enough procedural work to get them slick. That's different in different institutions of course but here, suddenly you need those skills, they have to come from somewhere and if you haven't been in practice before, where do you learn it? We don't have the frequency to teach people enough. (Maddison, interview, 2018)

This was not solely an issue for new graduates however, as Olivette contended that as an NVS, one can lose practical skills and become less effective despite having practice experience:

> I personally feel just because I'm a vet, if I haven't touched or done a certain procedure for months or even years, why am I suddenly magically going to be able to do it? I've got a slight shake, I used to be able to shake my way into any vein, but I'm not sure if I could shake my way into any mouse vein if I haven't had any recent training. [...]

> It doesn't give you a gold pass just because you've got the degree if you haven't used your hands for a while. (Olivette, interview, 2018)

Despite these examples highlighting the value of previous clinical experience, the particular language used by NVSs to contrast general and laboratory practice was still striking. For example, Maddison described their previous role in general practice when they were 'a real vet' as their 'previous life', and implied the NVS was not seen as a 'proper job':

> I was an equine vet in my previous life and the practice was being sold and I was simply looking for a job in the area and I thought I'll do this for a short time before I find a proper job ... I'm very welfare driven in the whole approach to my post and I feel that's really rewarding and I must say, as a real vet, my understanding of ethical things was minimal compared to what the NVS role provides you with. I regret that now when I look back on my time in practice, there's a lot of things I would've done very differently now. (Maddison, interview, 2018)

However, it is important to stress that using such language did not mean the NVS role was presented negatively in terms of job satisfaction. Indeed by contrast,[32] some veterinarians did note the advantages of the NVS role as compared to general practice, both for themselves and the animals. For example, some veterinarians claimed that they could care more effectively for more animals in the lab, and more effectively influence others. This can be understood as a form of resistance to social movement campaigns, which have criticised veterinary involvement in animal research.[33]

Other participants also reflected on the personal advantages of the NVS role. In the following example, Paul appreciates that the NVS role allows him to 'have a life' before nevertheless going on to repeat the equation of general practice with 'real vet' work:

> You have a life [as an NVS], that's what I would say. Although I don't feel like I'm a real vet anymore. [...] I used to do a lot of routine surgery, routine stuff, three, four hours a morning in a busy, small practice, and I like working with my hands. And also, you're problem solving the whole time. The medical challenges, the diagnosis, that's the art and science of veterinary medicine, it's diagnosis and then of course treatment and hopefully seeing a happy result. Also, the instant cure of surgery, I used to really enjoy that. Probably getting

> back to why I did veterinary in the first place, I think the real vet is the practitioner. (Paul, interview, 2018)

Overall, this section has demonstrated the way in which, during interviews, NVSs drew discursive boundaries between general clinical veterinary practice and the NVS role. That these roles are described as different is not in itself surprising. Indeed, the RCVS Code of Professional Conduct[34] frames the NVS role as more *advisory and managerial* than *clinical and technical*, given that the NVS 'provides advice on the health, welfare and treatment of animals', is 'entrusted with the necessary management authorities' and 'should advise licence holders and others on implementing the 3Rs'. Rather, what struck us during the analysis was the particular language used to draw these distinctions. By constructing general practice as 'real' veterinary practice, these discourses effectively situate the NVS role as on the margins of the broader profession. What is also striking is the way in which language use could be seen to assume notions of power. Despite the rapid feminisation of the profession, Clarke and Knights[35] argue that vets in practice who they interviewed reproduced a masculine narrative of mastery and orderliness, as opposed to the 'skillful performer' of practice who effectively mobilises communication and relationship management skills.[36] One could interpret the data in this section as a further reproduction of this gendered narrative, in that clinical practice is presented as hierarchically above the more managerial and advisory NVS work. We return to why this might matter in the conclusion.

Traversing the lab–clinic boundary

The previous section focused on how NVSs in our study drew discursive boundaries between general veterinary practice and the specific NVS role. This section changes tack to explore how NVSs negotiate their relationship to wider publics outside the laboratory environment. This, we found, was an important part of the way the lab–clinic boundary was maintained.

During interviews, NVSs recounted personal fears of being attacked by organisations or individuals committed to ending animal research. Such fears have been reported in previous interview studies with others, including senior laboratory scientists.[37] Some veterinarians described how this fear still drove personal behaviour. For example, Mia, who works both in general practice and in the animal research laboratory, described holding back on sharing some information with her children for security reasons:

> We did have an incident where we were targeted as a veterinary practice by [group] and so I was really concerned about them turning up outside the school gates and by inadvertently one of my children saying something in a school debate or [...] actually revealing that [they] knew rather more about the topic than [they] should have done, that it might have led someone to put two and two together and put them in some sort of danger. (Mia, interview, 2018)

However, in the present study, the most common narrative was that these safety fears were predominately in the past. Nevertheless, what we found striking was the way in which NVSs still recounted complex interaction with wider publics, not only or directly due to their association with potentially controversial science, but to do with their *professional status as a veterinarian*. In short, the analysis points to a careful line being walked, between on the one hand a perceived public image of a typical or ideal veterinarian in clinical practice, and on the other their actual lived reality of laboratory work. What matters here is that some interviewees saw their role as potentially transgressive in the public imagination. The space of animal research, for example, was not considered to be an instinctive fit with the romanticised view of the veterinary profession personified in the UK by individuals like James Herriot. Indeed, the semi-fictional image of Herriot was specifically invoked by multiple interviewees regarding their everyday conversations with members of the public, for example in the hair salon:

> When you say 'I'm a vet' they picture James Herriot or the small animal vet in a white coat. That's it. So, you're safe saying 'I'm a vet'. (Maddison, interview, 2018)

> It becomes complicated and it's affected me in that I'm now hesitant to tell people I'm a vet because the next question is always, 'Oh where?', and I got around it the other day for the person cutting my hair and it seemed to work, I may go this route, in that I just said,

> 'Oh, I used to be a cattle vet but now I'm a specialist vet and I study diseases' and it's at that point they change the conversation. So, it's difficult because I always used to have a good conversation with people outside, 'Oh, you're a cattle vet, I've seen James Herriot' or whatever. (Nicole, interview, 2018)

As Nicole illustrates, the atypicality of the NVS role from the general perception of what a veterinary professional does – or where they belong – precipitates a requirement to navigate the presentation of their professional image in conversation with people external to animal research. This is confirmed by Nathalie, Martin, and Maeve, who all recalled a reaction of surprise from people they spoke to upon discovering that their veterinary role was in an animal research laboratory:

> The classic is, 'How can you do that? You are a vet, you love animals?' And that's when you explain that I love animals, that's why I do it, because I'm the only party who's there for them. (Nathalie, interview, 2018).

> I've had nobody ever saying, 'that's dreadful', and lots of people were surprised that people [in animal research] employ vets. If they talk to me a bit longer [...] they say, 'I'm so glad that people like you exist'. (Maeve, interview, 2018)

> To begin with, some of them are ... surprised maybe. Because obviously you're a vet, it's assumed that because you're a vet you care about animals and you care about their wellbeing, and then that sounds opposed to doing research with them but then you explain what the framework you work under is, what is the legislation, what are the benefits of doing what we do, what would be the downsides of not doing it and I haven't had any issues with that. (Martin, interview, 2018)

As with Nicole, Martin's account thus demonstrates that discussion with the public about their work requires some labour of translation to find a way to communicate the positive value they see in their work in the face of the contradiction that the NVS role is assumed to foster in the public imagination. In the extracts above, such personal labour is presented as ultimately successful. However, this is not always the case. In the following detailed example, Melody recalls her difficult experiences of talking about her work. First, she recounts a conversation with a pharmaceutical representative who was sponsoring a continuing professional development meeting she was at, but ironically failed to see the full link between animal research and pharmaceuticals. Second, she recalls a disappointing dinner interaction, where she divulged her role to the person sitting next to her:

> I had one rep … and I said what I was doing. She pulled a right face and I thought, 'you're selling this and I know your company has just bought a number of animals from us to do … ', so clearly she had no idea either. She was a vet selling medicines who thought that it was not really very nice that I was involved in animal research from veterinary medicine because in [organisation] we did veterinary as well as human medicines. Generally, that's not been a pleasant experience. Even more recently I was at a presentation dinner sitting next to [a business person] and I thought 'Let's try this with a non-vet person'. He seemed like a very posh, well-educated kind of person. He was horrified that I even told him and I said, 'you seem like a trustworthy person', but I thought, 'we're kidding ourselves to think that people are ready for this' … He said, 'you shouldn't be so open with that'. (Melody, interview, 2018)

In Melody's account, she stresses that neither proximity to animal research nor being 'well-educated' were barriers to having a negative attitude towards animal research and the veterinarians involved in it. According to Melody, then, we are 'kidding ourselves' that wider audiences are 'ready' to appreciate the role of the vet in the lab. As analysts, our contention is that we need to appreciate the wider social context of veterinary practice (and not just the wider social context of animal research) in order to make full sense of these data.

As already highlighted, professional bodies such as the BVA and RCVS are keen to stress that the veterinary profession is regarded by publics as highly trustworthy.[38] Previous publications have also highlighted 'love for animals' as a major pull factor for veterinary occupations,[39] and have pointed to the way in which, unlike in human medicine, the UK veterinary profession has largely been left to regulate itself.[40] Irvine and Vermilya[41] also stress that among veterinarians, women are often stereotyped as being attracted to the veterinary profession for nurturing and maternal reasons.

Such work helps to paint a picture of the wider social context in which NVSs operate, and helps to explain why such careful walking of the lab–clinic boundary is deemed necessary. Working in animal research as a veterinary professional was perceived by interviewees as sometimes being anathema in the mind of the general public, contradicting a perception of a more feminised veterinary professional as animal-lover and healer. While the previous section illustrated how NVSs themselves presented their work as atypical for the veterinary profession, their reported experiences in communicating with publics imply that there is not only a public imagination of what a veterinary professional *is*, but also of where veterinary professionals (do not) belong.

Blurring the lab–clinic boundary

In the section above, on 'constructing the lab–clinic boundary', the analysis illustrated the way in which NVSs themselves drew a clear boundary between the 'real' veterinarian in general practice, and the NVS role on the margins of the profession. The second section argued that NVSs have to engage in specific labour to walk this boundary between the lab and the clinic during interactions with those outside the laboratory. This final section considers the ways in which interviewees' detailed examples of their work to blur the apparent boundary between the lab and the clinic.

Some interviewees were keen to discuss the way in which skills or expertise they had developed in laboratory animal research could benefit their or others' work in general practice. For example, several NVSs noted positively that laboratory animal veterinary medicine had a more developed level of knowledge and technique regarding the treatment of small animals such as rats than could be routinely found in general practice. While this was described by Mia, an NVS who also works in practice, as a 'side-benefit of research', others argued that this transfer between worlds was not that common. One reason for this, Oliver argues, is that there are only specific areas of laboratory animal veterinary medicine that are usefully transferable between fields:

> When I first started work for [organisation] the friend who recruited me said 'Oh, it'll make you a better rat vet', and he was absolutely right. But there is some dichotomy in the knowledge because it doesn't matter how much I know about Sendai Virus or Mouse Norovirus, that's never really going to be very much use in my work as a first opinion general practitioner. Whereas, for example, knowing how to anaesthetise them and how to stitch them up so they don't undo the stitches is very useful. (Oliver, interview, 2018)

Interestingly, given the previous discussion about not sharing information about their role in the lab, Oliver stresses that his city centre clients remain unaware of how he has developed these higher level skills.

> They don't know why I'm a rat expert. So it's now got to the stage where I'm comfortable, I've spayed quite a few rats and hamsters for people who wanted ... we've got quite a lot of rat-owning clients, being a city centre practice ... So as I say, we get quite a few requests now to spay the rats, for example, to try and reduce the risk of mammary tumours. (Oliver, interview, 2018)

Both Oliver and Mia noted that, while they had developed greater knowledge and skill in rat medicine, they were rarely asked how these had developed and they felt that neither general practice clients nor other veterinarians realised that it was as a result of working in laboratories. Mia also recounted an example of telephoning the author of a textbook to ask for advice in treating a rat, but still maintains that more widely her colleagues would be unaware that such knowledge would have been developed via animal research:

> I'm not certain that vets in practice realise that it's come from research though. If I asked my assistant where they thought it came from I'm not sure that they would realise that most of it had been gleaned over the years from people trying things [...] in NVS work. (Mia, interview, 2018)

Later in his interview, Oliver also noted that, given concerns about making things public, he would personally be cautious about advertising to his clients where his enhanced knowledge had come from:

> I wouldn't promote it in the practice [...] I think within my own practice I would worry whether that might affect my client base, whether

> it might, I suspect if it was widely known I suspect my clients would split into two camps, one who supported me and one who said, well, we don't like that and we'll go somewhere else. (Oliver, interview, 2018)

The NVSs interviewed also provided further examples of how techniques, skills, or knowledge gained in the laboratory can benefit veterinary care in general practice. For example, Mia highlighted benefits from diagnostic tools like the rabbit grimace scale,[42] which she had pushed to introduce in their clinic:

> I've been going on for ages and I wanted to introduce pain scoring in practice for a long time and no one had really listened because, obviously, as an NVS it's what you do all the time, […] and I was able to say, 'But do you know that comes from research?' […] That's something that was probably pushed in our practice because I was so familiar with it in research. (Mia, interview, 2018)

Euthanasia was also raised as an example of where experience in the lab could potentially benefit clinical practice. Maddison gave an example of how this 'influence' could potentially happen at scale using social media. He mentioned a veterinary Facebook group, with thousands of registered vets as members, which included requests for advice that he felt well placed to provide:

> It's really quite good. I was reading some of the posts and they talked about euthanasia with small animals and I thought 'my god!', and 'where can you learn this, that and the other?' If they looked up a document written for the research on euthanasia, we have so much material that they don't seem to know about. I find that quite strange because somebody said where can you find this, that, and the other, and I could give you about 10 different references straight away. (Maddison, interview, 2018)

Crucially, however, in some examples such as euthanasia, NVSs also reported that the influence could go the other way: that experience in practice can and should impact on laboratory care. To return to Mia again, she noted a requirement introduced through ASPA that mirrors a standard practice in the clinic:

> Coming from practice the other way though, I think one of the biggest things that's been really important is the change with ASPA

> where you have to ensure that the animal is definitely dead by a second method and that's something that's always been done in practice. So, [in practice] you would never contemplate euthanising an animal and not listening for a heartbeat. You just wouldn't, would you? So, actually having a secondary method, that's really reassuring that the animal is definitely dead before anything else happens to it. (Mia, interview, 2018)

Overall, then, this section has focused on the way in which NVSs' accounts seem to blur the boundaries between the worlds of clinical and laboratory practice. In Maddison's example above this could take place through the use of wide-reaching social media platforms, while in Mia's example this took the form of legislative change. Indeed, there is evidence that professional organisations are keen to celebrate or encourage shared learning; for example, in April 2021 a webinar was organised entitled 'Ethical Challenges – How Can Laboratory and Clinical Vets Support Each Other in Decision Making?'[43] Despite such efforts, one potential contribution of our analysis is to highlight the challenges that remain for individual NVSs to talk openly about this shared learning. What their accounts imply is that it is not just levels or depths of expertise that are made to matter, but rather where legitimate expertise comes from.

Conclusion

The veterinary profession is one of multiplicities, and the broad training that veterinarians receive makes them suited for a wide range of occupations – in small animal clinics, on farms, in government, and, as with the qualitative interviewees whose accounts are drawn upon in this chapter, in laboratories. In summary, our reflexive thematic analysis suggests that NVS work exists at the intersection of a wide and conflicting set of expectations and assumptions, with NVSs engaged in the simultaneous drawing, navigation, and blurring of boundaries between professional worlds and spaces.

Despite the broad technical expertise that all vets are deemed to possess by virtue of their training, there are some clear images in these NVS interview accounts of what a 'real' veterinary professional is, what the public's imagined veterinary professional does, and where that professional belongs. These accounts are complex and contradictory, with the NVSs arguably partly complicit in reproducing a hierarchy within the profession, by tying 'real' veterinary work to the technical and scientific clinical work of general practice, contrasted with the administrative and managerial work of an NVS. Indeed, it is this supposed atypical nature of their role that creates a specific imperative to walk the border between general and lab practice in their interaction with others.

More specifically, the accounts assume a particular vision of the 'general public'. Previous sociological work[44] argued that NVSs imagine the public as having a particular attitude or sensitivity towards *animals*. This chapter extends the literature by focusing instead on the imagined public's attitude or sensitivity towards *veterinarians*. This imaginary sees laboratory veterinary work as anathema to the caring and animal-loving veterinary professional encapsulated by a popular-imagination figure like James Herriot. We cannot here confirm whether this image of the veterinary profession is indeed held by wider publics – that would require a different empirical research project. However, what we can conclude is that this public imaginary is made to matter in multiple ways, for example by changing what NVSs tell their children about their work, or how they withhold the provenance of their uncommon expertise in specific species from clients.

Taken together, these complex strands point to both conceptual and material spillovers between the boundaries of general veterinary practice and laboratory veterinary work. This creates complex borderlands that veterinary professionals navigate as they move between different spaces marked by inherent regulatory tensions,[45] and also carry different personal and professional identities and imaginaries of veterinary work into and out of these spaces. NVSs grappled with their identity as veterinary professionals, reproducing masculine narratives of mastery and orderliness around the conception of a 'real' veterinary professional as someone involved in technically challenging clinical work rather than the 'skillful performer' of practice who effectively mobilises communication and relationship management skills.[46] Socially, conversations with the public were described as challenging engagements, as veterinary professionals were imagined as trustworthy and maternal animal lovers – consistent with the image presented by veterinary professional bodies[47] and the genuine occupational pull factor of love for animals[48] – at odds with the controversial space of the laboratory. Moving beyond the refuge of identifying simply as a veterinary surgeon consequently required the development of strategies for engagement or avoidance of conversation about laboratory veterinary work. We would argue that such conclusions were only possible by appreciating the interview accounts of NVSs as multiply-displaced veterinary professionals, rather than a narrower analysis which could have analysed the accounts of NVSs as animal research professionals. For example, only by appreciating the prior career trajectories of NVSs and the shared learning as they have travelled between spaces of veterinary expertise,[49] is it clear that the boundaries between the worlds of the lab and the clinic are experienced by these professionals as porous.

Indeed, we hope that the reflections presented in this chapter may interest scholars with an interest in the complex ethical boundary drawing of other types of actors, for example health professionals, whose work sometimes requires movement between the lab and the clinic.[50] In terms of veterinarians, however, we also hope that this work will be situated not just within social scientific work on the animal research laboratory, but also within the oft-contested and wider history of veterinary expertise. The veterinary profession in the UK has succeeded in carving out a professional monopoly over animal healthcare regulated under the Veterinary Surgeons Act 1966, which defines the art and science of veterinary surgery and medicine as covering diagnosis, and the medical and surgical treatment of animals.[51] However, while expertise remains 'precarious' and impermanent for individuals, it is also historically and spatially contingent.[52] For example, this chapter has shown the way in which NVSs present the 'real vet' as involving technical work in practice rather than the administrative and advisory role of the NVS. However, it is possible that this clinical ideal type may already be fading from prominence, in an increasingly commercialised veterinary industry that challenges veterinary professionals to adopt and adapt the skills of salespeople.[53]

That the veterinary profession as a whole stands at somewhat of a crossroads is argued in a recent article in the *Veterinary Record* (the flagship journal of the BVA). Gardiner[54] argues that the profession faces several contemporary challenges that should prompt reflexivity in considering questions such as '"What should the veterinary profession look like?", "How many vets do we need?", "What areas will they be working in?", and "What role should veterinary schools play?"' The complexity of the borderlands described in this chapter may thus exemplify the respective challenges to, and lack of consensus on, the place, value, and role of veterinary professionals in contemporary society. As long as veterinarians are in the laboratory, grappling with these big issues is, we would argue, necessary in order to fully appreciate the workings and nuances of the animal research nexus.

Notes

1 María Puig de la Bellacasa, '"Nothing Comes Without its World": Thinking with Care', *The Sociological Review*, 60 (2012), 197–216, DOI: 10.1111/j.1467-954X.2012.02070.x.

2 Gareth Enticott, 'The Local Universality of Veterinary Expertise and the Geography of Animal Disease', *Transactions of the Institute of British Geographers*, 37 (2012), 75–88, p. 79, DOI: 10.1111/j.1475-5661.2011.00452.x.

3 Laure Bonnaud and Nicolas Fortané, 'Being a Vet: The Veterinary Profession in Social Science Research', *Review of Agricultural, Food and Environmental Studies*, 102 (2021), 125–148, DOI: 10.1007/s41130-020-00103-1. Caroline Clarke and David Knights, 'Practice Makes Perfect? Skillful Performances in Veterinary Work', *Human Relations*, 71 (2018), 1395–1421, DOI: 10.1177/0018726717745605. Caroline Clarke and David Knights, 'Who's a Good Boy Then? Anthropocentric Masculinities in Veterinary Practice', *Gender, Work & Organization*, 26 (2018), 267–287, DOI: 10.1111/gwao.12244. Gareth Enticott, 'Mobile Work, Veterinary Subjectivity and Brexit: Veterinary Surgeons' Migration to the UK', *Sociologia Ruralis*, 59 (2019), 718–738, DOI: 10.1111/soru.12239. Pru Hobson-West and Annemarie Jutel, 'Animals, Veterinarians and the Sociology of Diagnosis', *Sociology of Health & Illness*, 42 (2020), 393–406, DOI: 10.1111/1467-9566.13017. Leslie Irvine and Jenny Vermilya, 'Gender Work in a Feminized Profession: The Case of Veterinary Medicine', *Gender & Society*, 24 (2010), 56–82, DOI: 10.1177/0891243209355978. Patricia Morris, *Blue Juice: Euthanasia in Veterinary Medicine* (Philadelphia, PA: Temple University Press, 2012).

4 Alistair Anderson and Pru Hobson-West, '"Refugees from Practice"? Exploring Why Some Vets Move from the Clinic to The laboratory', *Vet Record*, 190 (2022), e773, DOI: 10.1002/vetr.773.
5 Alistair Anderson and Pru Hobson-West, 'Animal Research, Ethical Boundary-Work, and the Geographies of Veterinary Expertise', *Transactions of the Institute of British Geographers*, 48 (2023), 491–505, DOI: 10.1111/tran.12594.
6 See, for example, Gareth Enticott, 'Navigating Veterinary Borderlands: "Heiferlumps", Epidemiological Boundaries and the Control of Animal Disease in New Zealand', *Transactions of the Institute of British Geographers*, 42 (2017), 153–165, DOI: 10.1111/tran.12155.
7 Pru Hobson-West and Stephen Timmons, 'Animals and Anomalies: An Analysis of the UK Veterinary Profession and the Relative Lack of State Reform', *The Sociological Review*, 64 (2016), 47–63, DOI: 10.1111/1467-954X.12254.
8 Gerard Brouwer-Ince, 'The Work of the Named Veterinary Surgeon', *Vet Record* (2013), 173, p. i, DOI: 10.1136/vr.f5886.
9 Pru Hobson-West and Ashley Davies, 'Societal Sentience: Constructions of the Public in Animal Research Policy and Practice', *Science, Technology, & Human Values*, 43 (2018), 671–693, DOI: 10.1177/0162243917736138.
10 Sarah Wolfensohn and Paul Honess, 'Laboratory Animal, Pet Animal, Farm Animal, Wild Animal: Which Gets the Best Deal?' *Animal Welfare*, 16 (2007), 117–123. Anderson and Hobson-West, '"Refugees from Practice"?'; Palmer et al., 'When Research Animals Become Pets and Pets Become Research Animals: Care and Death in Animal Borderlands', *Social and Cultural Geography* (2022), 1–19, DOI: 10.1080/14649365.2022.2073465.
11 Vanessa Ashall and Pru Hobson-West, 'The Vet in the Lab: Exploring the Position of Animal Professionals in Non-Therapeutic Roles', in *Professionals in Food Chains*, ed. by Svenja Springer and Hervig Grimm (Wageningen: Wageningen Academic Publishers, 2018), 291–295, p. 292, DOI: 10.3920/978-90-8686-869-8_45.
12 Ghislaine Poirier et al., 'ESLAV/ECLAM/LAVA/EVERI Recommendations for the Roles, Responsibilities and Training of the Laboratory Animal Veterinarian and the Designated Veterinarian under Directive 2010/63/EU', *Laboratory Animals*, 49 (2015), 89–99, p. 93, DOI: 10.1177/0023677214557717.
13 Poirier et al., 'ESLAV/ECLAM/LAVA/EVERI Recommendations for the Roles, Responsibilities and Training of the Laboratory Animal Veterinarian and the Designated Veterinarian under Directive 2010/63/EU'.

14 William Russell and Rex Burch, *The Principles of Humane Experimental Technique* (London: Methuen & Co Ltd, 1959). Ngaire Dennison and Anja Petrie, 'Legislative Framework for Animal Research in the UK', *Practice*, 42 (2020), 488–496, DOI: 10.1136/inp.m3920. Maggie Lloyd et al., 'Refinement: Promoting the Three Rs in Practice', *Laboratory Animals*, 42 (2008), 284–293, DOI: 10.1258/la.2007.007045.
15 Clarke and Knights, 'Practice Makes Perfect?'; Morris, *Blue Juice*.
16 Ngaire Dennison and Anja Petrie, *Working as a Named Veterinary Surgeon*, www.vetrecordjobs.com/myvetfuture/article/working-as-a-named-veterinary-surgeon-dennison-petrie/ [accessed 28 September 2021].
17 Kerri Smith and Sarah Wolfensohn, 'Caught in the Middle', *Nature*, 444 (2006), 811, DOI: 10.1038/444811a.
18 Hobson-West and Davies, 'Societal Sentience'.
19 Gail Davies et al., 'The Social Aspects of Genome Editing: Publics as Stakeholders, Populations and Participants in Animal Research', *Laboratory Animals*, 56.1 (2021), 88–96, 88, DOI: 10.1177/00236772 21993157.
20 Pru Hobson-West, 'The Role of "Public Opinion" in the UK Animal Research Debate', *Journal of Medical Ethics*, 36 (2010), 46–49, p. 46, DOI: 10.1136/jme.2009.030817.
21 Davies et al., 'The Social Aspects of Genome Editing'.
22 Royal College of Veterinary Surgeons. *Vets Amongst the Most Trusted Professionals, According to Survey*, www.rcvs.org.uk/news-and-views/news/vets-amongst-the-most-trusted-professionals-according-to-rcvs/ [accessed 30 April 2021]. Vet Futures Project Board (2015) *Public Trust in the Veterinary Profession*, www.vetfutures.org.uk/resource/public-trust-in-the-professions-may-2015/ [accessed 30 April 2021].
23 Smith and Wolfensohn, 'Caught in the Middle', *Nature*, 444 (2006), 811, DOI: 10.1038/444811a.
24 Royal College of Veterinary Surgeons, *Vets Amongst the Most Trusted Professionals*. Vet Futures Project Board (2015) *Public Trust in the Veterinary Profession*.
25 See Alistair Anderson and Pru Hobson-West, '"Refugees from Practice"?'
26 Virginia Braun and Victoria Clarke, 'Reflecting on Reflexive Thematic Analysis', *Qualitative Research in Sport, Exercise and Health*, 11 (2019), 589–597, DOI: 10.1080/2159676X.2019.1628806. Virginia Braun et al., 'Thematic Analysis', in *Handbook of Research Methods in Health Social Sciences*, ed. by Pranee Liamputtong (Singapore: Springer Singapore, 2019), pp. 843–860, DOI: 10.1007/978-981-10-5251-4.

27 Johnny Saldaña, *The Coding Manual for Qualitative Researchers* (Los Angeles, CA: Sage, 2016), DOI: 10.1038/444811a.
28 Victoria Clarke and Virginia Braun, 'Using Thematic Analysis in Counselling and Psychotherapy Research: A Critical Reflection', *Counselling and Psychotherapy Research*, 18 (2018), 107–110, DOI: 10.1002/capr.12165.
29 Clarke and Braun, 'Using Thematic Analysis in Counselling and Psychotherapy Research'.
30 Royal College of Veterinary Surgeons, *Vets Amongst the Most Trusted Professionals*.
31 Anderson and Hobson-West, '"Refugees from Practice"?'
32 See for more detail Anderson and Hobson-West, '"Refugees from Practice"?'
33 Animal Aid, *Say 'NO' to Animal Experiments at the Royal Veterinary College*, www.animalaid.org.uk/rvc/index.html [accessed 28 September 2021]. Anderson and Hobson-West, 'Animal Research, Ethical Boundary-Work, and the Geographies of Veterinary Expertise'.
34 Royal College of Veterinary Surgeons, *Named Veterinary Surgeons*, 2023, www.rcvs.org.uk/setting-standards/advice-and-guidance/code-of-professional-conduct-for-veterinary-surgeons/supporting-guidance/named-veterinary-surgeons/ [accessed 28 May 2021].
35 Clarke and Knights, 'Practice Makes Perfect?'
36 Clarke and Knights, 'Practice Makes Perfect?'; Clarke and Knights, 'Who's a Good Boy Then?'
37 See, for example, Pru Hobson-West, 'Ethical Boundary-Work in the Animal Research Laboratory', *Sociology*, 46 (2012), 649–663, DOI: 10.1177/0038038511435058.
38 Royal College of Veterinary Surgeons, *Vets Amongst the Most Trusted Professionals*. Vet Futures Project Board (2015), *Public Trust in the Veterinary Profession*.
39 Clinton Sanders, 'Annoying Owners: Routine Interactions with Problematic Clients in a General Veterinary Practice', *Qualitative Sociology*, 17 (1994), 159–170, p. 164, DOI: 10.1007/BF02393499.
40 Hobson-West and Timmons, 'Animals and Anomalies'.
41 Leslie Irvine and Jenny Vermilya, 'Gender Work in a Feminized Profession: The Case of Veterinary Medicine', *Gender & Society*, 24 (2010), 56–82, DOI: 10.1177/0891243209355978.
42 Victoria Hampshire and Sheilah Robertson, 'Using the Facial Grimace Scale to Evaluate Rabbit Wellness in Post-Procedural Monitoring', *Lab animal*, 44 (2015), 259–260, DOI: 10.1038/laban.806.

43 European Society of Laboratory Animal Veterinarians. *ESLAV-LAVA-UFAW Ethical Challenges Webinar, 27/04/2021*, www.eslav.org/events/eslav-lava-ufaw-ethical-challenges-webinar-27-4-2021/ [accessed 7 June 2021].
44 Hobson-West and Davies, 'Societal Sentience'.
45 Ashall and Hobson-West, 'The Vet in the Lab'.
46 Clarke and Knights, 'Practice Makes Perfect?' Clarke and Knights, 'Who's a Good Boy Then?'
47 Royal College of Veterinary Surgeons, *Vets Amongst the Most Trusted Professionals*. Vet Futures Project Board (2015), *Public Trust in the Veterinary Profession*.
48 Sanders, 'Annoying Owners'.
49 Anderson and Hobson-West, '"Refugees from Practice"?'
50 Steven Wainwright et al., 'Ethical Boundary-Work in the Embryonic Stem Cell Laboratory', *Sociology of Health & Illness*, 28 (2006), 732–748, DOI: 10.1111/j.1467–9566.2006.00539.x.
51 Andrew Gardiner, 'The "Dangerous" Women of Animal Welfare: How British Veterinary Medicine Went to the Dogs', *Social History of Medicine*, 27 (2014), 466–487, DOI: 10.1093/shm/hkt101. Hobson-West and Timmons, 'Animals and Anomalies'. Abigail Woods, 'From Practical Men to Scientific Experts: British Veterinary Surgeons and the Development of Government Scientific Expertise, C. 1878–1919', *History of Science*, 51 (2013), 457–480, DOI: 10.1177/007327531305100404. Abigail Woods, 'Between Human and Veterinary Medicine: The History of Animals and Surgery', in *The Palgrave Handbook of the History of Surgery*, ed. by Thomas Schlich (London: Palgrave Macmillan, 2018), pp. 115–131.
52 Angela Cassidy, *Vermin, Victims and Disease: British Debates over Bovine Tuberculosis and Badgers* (Springer Nature, 2019). Gareth Enticott, 'The Local Universality of Veterinary Expertise and the Geography of Animal Disease', *Transactions of the Institute of British Geographers*, 37 (2012), 75–88, DOI: 10.1111/j.1475-5661.2011.00452.x. Gareth Enticott, 'Mobile Work, Veterinary Subjectivity and Brexit: Veterinary Surgeons' Migration to the UK', *Sociologia Ruralis*, 59 (2019), 718–738, DOI: 10.1111/soru.12239. Gardiner, 'The "Dangerous" Women of Animal Welfare'. Abigail Woods, 'The Farm as Clinic: Veterinary Expertise and the Transformation of Dairy Farming, 1930–1950', *Studies in History and Philosophy of Science Part C: Studies in History and Philosophy of Biological and Biomedical Sciences*, 38 (2007), 462–487, DOI: 10.1016/j.shpsc.2007.03.009.

53 Zoe Belshaw et al., 'Motivators and Barriers for Dog and Cat Owners and Veterinary Surgeons in the United Kingdom to Using Preventative Medicines', *Preventive Veterinary Medicine*, 154 (2018), 95–101, DOI: 10.1016/j.prevetmed.2018.03.020. Clarke and Knights, 'Practice Makes Perfect?' Jason Coe et al., 'A Focus Group Study of Veterinarians' and Pet Owners' Perceptions of the Monetary Aspects of Veterinary Care', *Journal of the American Veterinary Medical Association*, 231 (2007), 1510–1518, DOI: 10.2460/javma.231.10.1510. Morris, *Blue Juice*.

54 Andrew Gardiner, 'It Shouldn't Happen to a Veterinary Profession: The Evolving Challenges of Recruitment and Retention in the UK', *Vet Record*, 187 (2020), 351–353, p. 351, DOI: 10.1136/vr.m4096.

10

'Field folk': citizen scientists and the Animals (Scientific Procedures) Act

Alexandra Palmer

Introduction

Ethan is a university-based ornithologist, whose research regularly involves working with non-professional naturalists, specifically bird ringers. In an interview (2018), he describes such collaborations with non-professionals as 'at the very core of field ornithology', referencing the discipline's long tradition of using data collected by non-professional birders, and the finer line between professionals and amateurs in ornithology compared with most scientific fields.[1] Such non-professional involvement in research is today commonly referred to as 'citizen science'. This is a term developed in the 1990s, which ornithologist Rick Bonney[2] defined as involving public contributions of observations to scientific research, while sociologist Alan Irwin's[3] definition emphasised opening up science processes and policy to the public. Bonney's definition has become the more popular of the two, although the term has increasingly incorporated elements of both ideas. Alongside astronomy, wildlife research of the kind that Ethan and his ringer collaborators undertake tends to dominate in the public profile of citizen science.[4]

Ethan's research involves activities like collecting blood samples from free-living birds in the UK. Such activities are deemed to exceed a certain threshold of invasiveness,[5] which means that conducting them requires licensing under the Animals (Scientific Procedures) Act 1986 (ASPA; for further background, see Myelnikov, Chapter 1). As the personal licence holders, Ethan and his university-based research associates (e.g., PhD students and postdocs) are the only people involved in the project authorised to conduct blood sampling and other licensed procedures under ASPA. However, volunteer bird ringers help with other elements of the research, like setting, monitoring, and extracting birds from the net, and attaching rings to help identify the birds next time they are caught.

Many of the ringers Ethan works with are highly experienced – indeed, their training to become licensed ringers took longer than that required for securing an ASPA licence. Nonetheless, Ethan described being asked by the Home Office – the authority charged with regulating animal research under ASPA – to oversee the training and techniques of his volunteers. This, to Ethan, is problematic, since 'I can't turn round to somebody who's been ringing for 40 years and tell them that they're not handling a bird correctly and they're not ringing it properly', especially since he personally holds fewer years of ringing experience. Ethan viewed some of his volunteer citizen scientists as having greater expertise than himself at handling and ringing birds, and felt that he should therefore not be required to oversee them – doing so would make little sense and disrespect the ringers.

The reflections of 'Ethan' (a pseudonym) serve as the starting point of this chapter, which is concerned with how volunteer citizen scientists engage with ASPA. In considering this, it connects with a longstanding interest among historians and science and technology scholars in the relationships between professional and non-professional scientists, including amateur naturalists. Among other subjects, scholars have explored how non-professionals acquire their knowledge, and how this process of knowledge accumulation may reflect different values between the two (potentially quite socially separate) groups.[6] Furthermore, scholars have considered how professionals have perceived non-professionals' expertise, and their (not always successful) efforts to establish and maintain common understandings and consistent methods when working with non-professionals.[7] Common across much of this literature is an ethical concern with making science more accountable to the public, with citizen science frequently viewed as a way of fostering a mode of science that is responsive to public concerns, and makes use of non-scientists' knowledge.[8] This literature also interrogates what it means to be an 'expert', and how non-scientific expertise – of farmers[9] and Indigenous peoples,[10] for example – is (or isn't) incorporated into scientific research and policy-making.

These themes are extended in this chapter by demonstrating that in ASPA-regulated wildlife research, citizen scientists are simultaneously excluded from positions of authority, yet also able to directly contribute to animal care and research practice. To make this case, I use the concept of 'knowledge-control regimes' developed by Hilgartner. A knowledge-control regime is a socio-technical arrangement – guided by legal and quasi-legal mechanisms, and informal shared understandings – that produces categories of agents and allocates those agents certain 'entitlements and burdens'.[11] A knowledge-control regime may serve many goals, such as allocating 'epistemic authority' (in this case, who has authority for managing protocols and protecting animal welfare during research),[12] maintaining quality, and constructing professional jurisdictions. Exploring how ASPA as a knowledge-control regime simultaneously includes and excludes citizen scientists highlights the value of flexibility and long-term, trust-based relationships for ensuring alignments between expertise and authority, and the inherent (and arguably intentional) exclusivity of ASPA and other licensing-based systems of animal research regulation.

The chapter examines two themes that emerged from research relating to citizen scientists and their involvement in ASPA. First, it examines how citizen scientists engage with ASPA, showing that there are regulatory, institutional, and social barriers to their involvement. Second, it explores how misalignments between expertise and authority are negotiated under ASPA via flexibility and trust-based relationships between inspectors and researchers. Finally, it concludes by discussing what the case of citizen scientists can tell us about knowledge-control regimes in animal research, and the challenges of engaging publics in science.

Methods

This chapter presents qualitative research undertaken as part of the wider Animal Research Nexus Programme (AnNex). The particular strand of AnNex presented here focused on non-laboratory animal research in the UK in veterinary clinics, farms, zoos, fisheries, and wildlife research field sites. Under ASPA, such sites are commonly classified as Places Other than Licensed Establishments (POLEs). A key theme that emerged from the POLEs research was the complex interface between ASPA-regulated wildlife research and citizen science. This subject became sufficiently important that we collaborated with stakeholders to run a panel discussion event on the subject of citizen science regulation.[13] Research therefore primarily focused on ASPA-regulated research outside of the laboratory, but also looked at citizen science.

Interviews (sometimes in collaboration with other AnNex researchers) with 30 people, and 24 lengthy informal conversations with others, were conducted between 2018 and 2020. All but five interviews were conducted in person, with the remainder taking place over phone or video call due to logistical challenges, participants' preferences, and COVID-19-related restrictions. Potential participants were identified via online searches, snowballing, and existing contacts within the AnNex team. Together, 22 of these conversations focused primarily on wildlife research, with two interviews and three informal discussions focusing almost exclusively on wildlife citizen science. A further 10 involved discussions with Named Veterinary Surgeons (NVSs) and Home Office inspectors and covered a broad range of topics. Participant observations were also conducted during visits of one to two days to five non-laboratory research projects, and during shorter site visits, a wildlife research training course, and relevant conferences (including one focused on wildlife citizen science).

Interview transcripts, fieldnotes, and relevant documents were analysed using qualitative data analysis software NVivo. Coding was inductive, with themes emerging from the research rather than being pre-determined ahead of data collection. All interviews were conducted with written consent from participants. This research was granted ethical approval by the Central University Research Ethics Committee of the University of Oxford (Reference Number: SOGE 18A-7). Due to the sensitive nature of the topic, a policy of using pseudonyms was adopted (using letters C, and E through H, as per the AnNex policy; see Introduction) and de-contextualisation, when necessary.

The POLEs research project offers a good example of what can be achieved by using insights from the social sciences to encourage new thinking in animal research. In addition to the professional/citizen scientist interface, another key theme that emerged was a sense that non-laboratory research is neglected compared with laboratory research. For instance, there are no published statistics on the number of projects or animals used at POLEs, few dedicated networks intended to share information and provide support for researchers working at POLEs, and arguably less attention to non-laboratory work in animal research guidelines compared with the laboratory. In light of this lack of attention and support, some stakeholder collaborators elected to write a grant proposal aimed at creating a network for wildlife researchers, following from a stakeholder-focused workshop we held on the subject of non-laboratory research.[14]

Citizen scientists and ASPA licensing

Let us begin by considering how citizen scientists can become involved in ASPA-licensed research. To do so, it is important to briefly explain who citizen scientists are. The categories of 'professional' and 'citizen' scientist are not completely separate. Professional wildlife researchers sometimes simultaneously identify as 'birders' or similar, and may watch and catch wildlife in their spare time. Indeed, I was once invited out on a weekend recreational birding expedition by a professional ornithologist (though the trip was cancelled due to bad weather). Several researchers I encountered in interviews and fieldwork also narrated their career trajectories in zoology and field biology as first inspired by citizen science activities in their formative years.

However, a large contingent of citizen scientists never go on to become professionals and are self-conscious about their non-professional identity. For example, one presenter at a citizen science conference I attended in 2019 began his presentation by saying, 'I'm not a scientist, not a researcher, but I am a birder'. As other scholars have demonstrated, these non-professional science groups may develop unique shared identities, terminologies, values, and methods of sharing knowledge.[15] For example, birders were once jokingly described to me as 'men over 60 in checked shirts' (fieldnotes from 2019 conference). This summary may have some factual basis given that bird-related activities sometimes described as 'competitive',[16] such as ringing and 'twitching',[17] in the UK and US are strongly male-biased,[18] although the anorak is more typically referenced in descriptions of the birder's attire.[19]

The conference also made clear that citizen science communities may be tightly knit social groups with hierarchies and an 'in' crowd. At previous iterations of the conference, names had been assigned to specific tables at the main conference dinner. Because this practice had caused tension, it was formally stopped in the year that I attended; however, attendees informally re-created this system themselves, which meant that outsiders like myself struggled to find seats given our lack of community contacts. My research also reinforced that certain individuals within wildlife citizen science communities hold a higher status than others, and that citizen science communities such as birding are characterised by both cooperation and competition.[20] In particular, several participants suggested that the highest ranks are held by those in possession of high levels of skill in trapping and marking animals, and can demonstrate this skill through the possession of difficult-to-obtain licences for riskier trapping and marking techniques (e.g., cannon netting). In an interview (2019), researcher Hugh cynically summarised the situation as a 'hierarchy of elitisms among citizen scientists', citing the examples of mist netting bats, and the use of high-tech equipment like radio transmitters, as 'more elite activities' within the 'batter' community. Hierarchies and networking are therefore not only important features of professional science, but also in some areas of non-professional citizen science.

Despite this sense of community, citizen science is by definition characterised by a lack of institutional affiliation.[21] As we have discussed in more detail elsewhere, this can pose a problem for citizen scientists seeking to secure an ASPA licence, given the substantial amount of funding and institutional knowledge that may be required to successfully complete ASPA paperwork, and even to understand what the law covers.[22] In the words of interviewee and citizen scientist Calum, requiring ASPA licences for attaching tracking devices to birds 'would scupper most of it' due to citizen scientists' perception that they cannot secure licences (2020). In this sense, institutional knowledge – such as that possessed by local Animal Welfare Ethical Review Bodies – is a fundamental part of how ASPA's knowledge-control regime functions, as securing a licence requires (or, at least, is believed to require) familiarity with the 'knowledge format' of the ASPA licence application.[23] Those without this institutional knowledge feel as if they are excluded, even if this is not technically the case.

Citizen scientists' feeling of exclusion from ASPA is not entirely surprising, since ASPA was not historically intended to regulate citizen science. As we have discussed in more detail elsewhere,[24] the Home Office has never had an interest in regulating bird ringers and other citizen scientists, in part because doing so would dramatically increase inspectors' workloads, and citizen science activities are viewed as low-risk in terms of animal welfare and public acceptance. Indeed, the undesirability of straying into regulation of this large community was raised in evidence given by the British Ecological Society to the 1980 House of Lords Select Committee on the Laboratory Animals Protection Bill (a precursor to ASPA).[25] One might also interpret certain criteria determining which activities require ASPA licensing as being drafted with the exclusion of citizen science in mind, such as the 'identification threshold' whereby ringing, tagging, and marking animals for science do not require ASPA licensing provided they 'cause only momentary pain or distress (or none at all) and no lasting harm'.[26] Furthermore, as Valverde has observed, the licence as a technology of governance aims to restrict when, where, how, and by whom certain activities can be done, as a way of reducing opportunities for violations of the law.[27] Licensing is therefore inherently, and to some extent intentionally, exclusionary.

Given citizen scientists' feeling of being unable to secure ASPA licences, my research suggested that there is push-back from citizen science communities about the encroachment of ASPA into favoured citizen science activities, and satisfaction when favoured activities are moved out of ASPA's remit. For example, birders reported that plucking birds' feathers was until recently regulated by the Home Office under ASPA, but has now shifted to the authority of the British Trust for Ornithology, which issues licences for bird ringing (fieldnotes from conversation with birders, 2019). From various conversations, I gathered that citizen scientists were pleased to have this activity regulated by an authority with which they are already familiar. Informal conversations also suggested that in the past feather plucking may have been done by some birders without ASPA licences, even when this was legally required.

In short, the Home Office has little interest in regulating citizen scientists, and citizen scientists in turn don't want to be regulated under ASPA.[28] The main people calling for ASPA to cover more citizen science activities are therefore the occasional animal welfare advocate (though wildlife, and specifically citizen science, tend not to be key campaign subjects),[29] and professional wildlife researchers concerned about risks to animal welfare or perceived unfairness if they (unlike citizen scientists) are expected to secure licences for borderline activities that arguably don't fall under ASPA.[30] The end result of this situation is that citizen scientists do not hold ASPA licences themselves. If they are involved in ASPA-regulated research it is typically as skilled volunteers rather than licence holders.

However, there may be additional reasons why citizen scientists do not tend to lead ASPA projects, which are less related to ASPA and its associated regulatory mechanisms and more a reflection of how professional and citizen scientists regard their own expertise and relationships with one another. Citizen scientists tended to show deference towards professionals for their analytical and project leadership abilities. For example, in an interview, citizen scientist Clive described reaching out to a professional researcher after it became clear that:

> I can data collect for two weeks while I'm on leave, but then I'm rather stuck to make sense of it. So we'd ringed 30 thousand birds [in a particular location], you know, which is a lot of birds. And we realised that we were unable to deal with the data. We were the citizen scientists – we needed the scientist side to help us here. (Clive, citizen scientist, interview, 2019)

Clive added that having a professional researcher on board was important for helping the citizen scientists understand 'the regulations' as they began to undertake more sophisticated techniques, which posed greater risks to birds and therefore required further regulatory oversight. In short, citizen scientists are sometimes sceptical about their own abilities,[31] at least when it comes to data analysis and understanding regulations. This means that it is not just professionals who engage in 'demarcation work' by maintaining and reinforcing separation between professionals and volunteers,[32] but also non-professionals.

At the same time, citizen scientists commonly expressed a sense of pride in their greater field-based expertise compared with professional researchers. For example, during a question and answer session at the conference, one citizen scientist observed that the amateur–professional relationship can be mutually beneficial, since the amateur can't do statistics but the professional researcher 'probably hasn't been out in the field for 50 years' (a comment which received much appreciative laughter). Thus, non-professional scientists may understand their possession of a highly developed skill as conferring dignity and respect, and as restoring their status in a context where they are likely to be viewed as the inferior party.[33] The self-identification of citizen scientists as 'field folk', as one birder put it in an informal discussion (2018), therefore implies both insecurity and pride.

Negotiating expertise under ASPA

Citizen scientists tend to only participate in ASPA-regulated projects as skilled volunteers working for licence-holding professional researchers. However, this subordinate role may be at odds with citizen scientists' technical expertise, which might be mutually recognised by professional and citizen scientists – as demonstrated by Ethan's comment at the beginning of this chapter. Another example came during the conference I attended, in which a researcher acknowledged one volunteer in particular, who goes out nearly every day for the project and keeps comprehensive genealogies of the animals under study. The researcher observed that really the citizen scientist should be the one giving the talk. Researchers and citizen scientists alike may therefore be well aware that expertise comes in multiple forms, and is not exclusively possessed by professional scientists.[34] This example also illustrates how professionals may be willing to undertake 'welcoming work' whereby they (at least partially) deconstruct professional–volunteer boundaries by welcoming specific volunteers into their domain.[35] At the same time, a common theme throughout the conference was the claim that researchers tend to 'parasitise' the hard work of volunteers (as one speaker put it), such as by not giving them proper acknowledgement in publications, thereby advancing their own careers while failing to build up the status of volunteers.[36]

These complex negotiations of expertise, respect, and responsibility are dealt with under ASPA through a range of mechanisms. First, the law itself features a series of named roles, such as the NVS, Named Animal Care and Welfare Officer, and holders of a project licence and personal licence. Each of these parties is granted certain responsibilities (e.g., offering independent advice on animal welfare in the case of the NVS)[37] based on their role in the research and their expertise; they are therefore required to have completed certain training requirements and demonstrate certain skills. For example, personal licence holders have to undertake an authorised training course, in addition to being signed off as competent in specific techniques, before undertaking licensed procedures under ASPA.

However, in practice there may be a mismatch between expertise and legal responsibilities assigned by ASPA. This may occur if highly skilled individuals are not legally permitted to undertake tasks that they're good at, or if people responsible for oversight have less expertise than those they are overseeing. This situation does not just occur with professional and citizen scientists (as in Ethan's case), but also in other contexts where researchers are supported by colleagues or volunteers with a high degree of skill. For example, wildlife researcher Geoff explained that one of his colleagues with no ASPA licence is highly skilled at fitting collars onto their mammalian research subjects (fieldnotes, 2018). Because the research animals are under anaesthesia when their collars are fitted, this step is meant to be done by a licence holder, though as we will see later Geoff was able to change this in practice. Similarly, I visited a veterinary study that involved orally administering an experimental medical treatment to animals at people's homes, and in one case at a charity where an experienced veterinarian was on site. In this case, the on-site veterinarian expressed incredulity that she was not permitted to administer the experimental treatment as she did not have a project licence. Allowing this to occur would have made the research process smoother, as the researcher otherwise had to repeatedly drive considerable distances between his research locations (fieldnotes, 2019). In another example, an agricultural researcher described it as 'ridiculous' that he had to 'train [farmers] how to look at their own sheep', since the sheep under study remained under the care of farmers while being technically under the researcher's ASPA licence (fieldnotes, 2019).

However, as we have also discussed elsewhere,[38] these issues are often successfully resolved via flexibility and discretion, which in turn rely on long-term, trust-based relationships between actors. As explained by wildlife researcher Geoff, he was able to reach a compromise with his Home Office inspector whereby the colleague who is an expert at attaching collars could do so, so long as Geoff (the licence holder) is present in the room and regularly looks over to check. Geoff's narration of this story implied that he, like Ethan, felt that it was not right for him to be required to supervise someone who is even more highly skilled than himself. Still, this flexible resolution did at least mean that the person with the greatest technical expertise could do the activity they're skilled at, even if they are still under the supervision of the lesser expert. Similarly, the veterinary researcher indicated that he planned to ask his Home Office inspector if they could work something out to allow the on-site charity veterinarian to administer and monitor treatments, implying that informal conversations between researchers and inspectors often lead to productive compromises.

This flexibility might be viewed as not only a feature of relationships between actors (namely regulators and researchers), but as an inherent property of ASPA itself and its associated guidance. For example, ASPA guidance indicates that the assessment of wild animal health after capture can be assessed by either a vet or by an 'other competent person': wording which allows regulators, vets, and researchers to negotiate who qualifies as competent in specific cases.[39] ASPA is therefore to some extent a deliberately flexible piece of policy, and this particularly shines through in guidance documents indicating how the law might be adapted to work well on the ground in different contexts. While many participants complained about inconsistencies between inspectors (a risk with any flexible system of governance),[40] several participants viewed flexibility as one of ASPA's key strengths. As former Home Office inspector Heather argued in an interview (2019), although she has heard people complain that ASPA is 'open to interpretation', to her, 'part of its strength was that … there was the ability to overlay on top of the legislation an element of common sense'. In other words, Heather believes that ASPA was designed to deliberately allow various (perhaps inevitable) inconsistencies or problems – such as mismatches between expertise and responsibility – to be dealt with on a case-by-case basis via common sense and negotiations between researchers, regulators, and others involved in animal research.

Thus, while ASPA on the surface appears to assign responsibilities to specified roles based on assumed expertise, in practice it is deliberately flexible to allow 'entitlements and burdens'[41] to be rearranged among actors as needed. Some might argue that this flexibility is less a feature of ASPA specifically and more of laws in general, and particularly of licensing systems, which in practice tend to involve a great deal of discretion and even 'epistemological creativity'.[42] It is true that flexibility, regulator–regulatee relationships, and the 'enforcement style' of regulators (e.g., how flexible they are willing to be)[43] are key to how other laws work, such as in food safety.[44] At the same time, ASPA specifically was sometimes talked about in this study as more flexible relative to other laws with which researchers interacted. For example, Geoff identified the wildlife health law regulated by the Department for Environment, Food and Rural Affairs as more prescriptive (and therefore, in his view, worse) than ASPA around wildlife vaccination protocols (interview, 2018), and another researcher made a similar argument about trapping licences issued under the Wildlife and Countryside Act by Natural England and its equivalents.

Furthermore, ASPA can be viewed as more flexible than other systems of animal research regulation, notably that of the US. As Davies argues, ASPA distributes epistemic authority more widely amongst key actors than the US system.[45] While the US system is less centralised, with licensing taking place at the institutional rather than national level, it also gives greater authority to vets and particular scientists compared with the UK. Davies proposes that this is a product of a greater reliance in the US on *performance* standards (where desired animal welfare standards are described, but methods flexible) rather than *engineered* standards (which focus on specifying animal care methodologies and technologies, e.g., cage sizes), and a more formalised role of scientific experts in shaping policy than in the UK.

Yet for several interviewees, the flexibility that lies at the heart of ASPA is being lost. Flexibility was often spoken of as a product not only of ASPA itself, but also of trust and understanding between inspectors and researchers built up over years of working together. But several researchers complained that they are experiencing more frequent changes of inspectors, perhaps due to inspectors' increasingly demanding workloads.[46] Furthermore, researcher Hugh felt that increasing workloads contribute to inspectors becoming 'more narrow and less flexible'. For example, inspectors might say:

> 'If you can't show me X, Y, and Z then you're not doing it.' Now in the past the inspector may have said, 'Oh that's interesting I can see why you want to do it a different way. Okay, start off like this but if you see something come back to me and then we'll work our way through this to try and get to an end point.' And I feel that inspectors don't have the time to do that now. (Hugh, researcher, interview, 2019)

Recent shifts in regulatory practice might be argued to make these concerns more acute. Since July 2021, inspectors have no longer been allocated to individual establishments. Rather, regulation has been split into three teams covering 1) regulatory advice, 2) compliance assurance, and 3) licensing.[47] This means that researchers receive advice from multiple regulators rather than primarily engaging with a single inspector. While this approach is expected to have several advantages, such as the reduction of wait times,[48] it may also further impede the development of productive inspector–researcher relationships that were so often highlighted by participants as crucial for ASPA to work well, and flexibly, in practice.

Conclusion

What, then, can we learn about the knowledge-control regime of ASPA from the case of citizen scientists, or 'field folk'? First,

I demonstrated that citizen scientists almost always engage in ASPA as skilled volunteers rather than licence holders. This is important because it suggests that there are regulatory and institutional barriers to entry for non-professionals into science that impede making science accessible to the general public. This move to encourage 'open science' and citizen science has been adopted (albeit, arguably, in limited ways) by many funders and government bodies.[49] That said, barriers to entry also appear to be linked to how citizen scientists perceive their own abilities and those of researchers, and how they envision the professional–citizen scientist relationship. Thus, while the exclusion of citizen scientists from positions of authority is partly a product of ASPA and its implementation, it is also connected with longstanding and complex relationships between professional scientists and amateur naturalists.

While they are rarely in positions of authority in ASPA-regulated research, citizen scientists may still be regarded – including by the professional researchers in charge – as the greatest experts when it comes to specific tasks, such as catching and marking animals. Even in the lab such misalignments between authority and expertise can occur; for example, animal technologists may develop considerable expertise in activities that they are not strictly responsible for, such as how to design breeding protocols to get the desired number of research animals with minimum surplus (see Peres and Roe, Chapter 12). I have argued that while ASPA sets out default arrangements whereby responsibility is allocated to certain actors based on their presumed expertise, it is also deliberately flexible: a feature which for many participants was the Act's greatest strength, but which is also at risk of being lost due to increasing inspector workloads and frequent changes of inspectors.

The case of 'field folk' therefore offers several key lessons about ASPA as a knowledge-control regime. First, it illustrates the centrality of flexibility to ASPA's implementation, and the risks posed by changes that could undermine the development of long-term, trust-based relationships between researchers and inspectors. Secondly, it highlights how ASPA, like all licensing-based systems of regulation, limits who can conduct animal research, with the result that citizen scientists feel excluded from securing licences. This in turn perpetuates a situation in which citizen scientists are subordinate to professionals. While the advantages of this approach include ensuring that animal researchers have extensive institutional support (which citizen scientists lack), it also impedes any efforts to directly involve publics in conducting scientific research.

Notes

1 Thomas R. Dunlap, *In the Field, Among the Feathered: A History of Birders and Their Guides* (Oxford: Oxford University Press, 2012); Daniel Lewis, *The Feathery Tribe: Robert Ridgway and the Modern Study of Birds* (New Haven, CT: Yale University Press, 2012).

2 Rick Bonney, 'Citizen Science: A Lab Tradition', *Living Bird*, 15.4 (1996), 7–15.

3 Alan Irwin, *Citizen Science: A Study of People, Expertise and Sustainable Development* (London: Routledge, 1995).

4 Alexandra Palmer et al., 'Getting to Grips with Wildlife Research by Citizen Scientists: What Role for Regulation?' *People and Nature* (2020), DOI: 10.1002/pan3.10151; Melissa Eitzel et al., 'Citizen Science Terminology Matters: Exploring Key Terms', *Citizen Science: Theory and Practice*, 2017, 1–20, DOI: 10.5334/cstp.96; Caren B. Cooper and Bruce V. Lewenstein, 'Two Meanings of Citizen Science', in *The Rightful Place of Science: Citizen Science*, ed. by Darlene Cavalier and Eric B. Kennedy (Tempe, AZ: Consortium for Science, Policy & Outcomes, Arizona State University, 2016), pp. 51–62.

5 This is the 'lower threshold', defined as causing the animal 'a level of pain, suffering, distress or lasting harm equivalent to, or higher than, that caused by inserting a hypodermic needle according to good veterinary practice'. Home Office, *Guidance on the Operation of the Animals (Scientific Procedures) Act 1986* (London: Home Office, 2014).

6 Anne Secord, 'Science in the Pub: Artisan Botanists in Early Nineteenth-Century Lancashire', *History of Science*, 32.3 (1994), 269–315, DOI: 10.1177/007327539403200302.

7 Susan Leigh Star and James R. Griesemer, 'Institutional Ecology, "Translations" and Boundary Objects: Amateurs and Professionals in Berkeley's Museum of Vertebrate Zoology, 1907–39', *Social Studies of Science*, 19.3 (1989), 387–420, DOI: 10.1177/030631289019003001; Sally Eden, 'Counting Fish: Performative Data, Anglers' Knowledge-Practices and Environmental Measurement', *Geoforum*, 43.5 (2012), 1014–1023, DOI: 10.1016/j.geoforum.2012.05.004; Rebecca

Ellis and Claire Waterton, 'Caught between the Cartographic and the Ethnographic Imagination: The Whereabouts of Amateurs, Professionals, and Nature in Knowing Biodiversity', *Environment and Planning D: Society and Space*, 23.5 (2005), 673–693, DOI: 10.1068/d353t; Kevin C. Elliott and David B. Resnik, 'Scientific Reproducibility, Human Error, and Public Policy', *BioScience*, 65.1 (2015), 5–6, DOI: 10.1093/biosci/biu197; Hauke Riesch and Clive Potter, 'Citizen Science as Seen by Scientists: Methodological, Epistemological and Ethical Dimensions', *Public Understanding of Science*, 23.1 (2014), 107–120, DOI: 10.1177/0963662513497324; Lewis; Raf de Bont, *Stations in the Field: A History of Place-Based Animal Research, 1870–1930* (Chicago, IL: University of Chicago Press, 2015).

8 Georgina Born and Andrew Barry, 'Art-Science: From Public Understanding to Public Experiment', *Journal of Cultural Economy*, 3.1 (2010), 103–119, DOI: 10.1080/17530351003617610; Philip Mirowski, 'The Future(s) of Open Science', *Social Studies of Science*, 48.2 (2018), 171–203, DOI: 10.1177/0306312718772086.

9 Brian Wynne, 'May the Sheep Safely Graze? A Reflexive View of the Expert–Lay Knowledge Divide', in *Risk, Environment and Modernity: Towards a New Ecology*, ed. by Scott Lash et al. (London: Sage, 1996), pp. 27–83.

10 *Ethnobiology*, ed. by E. N. Anderson et al. (Hoboken, NJ: John Wiley & Sons, 2012).

11 Stephen Hilgartner, *Knowledge and Control in the Genomics Revolution* (Cambridge, MA: MIT Press, 2017), p. 9.

12 Gail Davies, 'Locating the "Culture Wars" in Laboratory Animal Research: National Constitutions and Global Competition', *Studies in the History and Philosophy of Science*, 89 (2021), 177–187, DOI: 10.1016/j.shpsa.2021.08.010.

13 Palmer et al., 'Getting to Grips with Wildlife Research by Citizen Scientists'.

14 Alexandra Palmer et al., 'Animal Research beyond the Laboratory: Report from a Workshop on Places Other than Licensed Establishments (POLEs) in the UK', *Animals*, 10 (2020), 1868, DOI: 10.3390/ani10101868.

15 Secord, 'Science in the Pub'; John Connell, 'Birdwatching, Twitching and Tourism: Towards an Australian Perspective', *Australian Geographer*, 40.2 (2009), 203–217, DOI: 10.1080/00049180902964942.

16 Kenneth Sheard, 'A Twitch in Time Saves Nine: Birdwatching, Sport, and Civilizing Processes', *Sociology of Sport Journal*, 16.3 (1999), 181–205, DOI: 10.1123/ssj.16.3.181.

17 'Twitching' is the practice of travelling long distances in order to see a new bird species.

18 Caren Cooper and Jennifer Smith, 'Gender Patterns in Bird-Related Recreation in the USA and UK', *Ecology and Society*, 15.4 (2010), DOI: 10.5751/ES-03603-150404.

19 Connell, 'Birdwatching, Twitching and Tourism'; Mark Cocker, *Birders* (London: Random House, 2012).

20 Roni Berger, 'Conducting an Unplanned Participant Observation: The Case of a Non-Birder in Bird Watchers' Land', *Forum: Qualitative Social Research*, 18.1 (2016), DOI: 10.17169/fqs-18.1.2730; Connell, 'Birdwatching, Twitching and Tourism'; Sheard, 'A Twitch in Time Saves Nine'; Dunlap, *In the Field, Among the Feathered.*

21 Lisa M. Rasmussen, 'Confronting Research Misconduct in Citizen Science', *Citizen Science: Theory and Practice*, 4.1 (2019), 10, DOI: 10.5334/cstp. 207; Palmer et al., 'Getting to Grips with Wildlife Research by Citizen Scientists'.

22 Alexandra Palmer and Beth Greenhough, 'Out of the Laboratory, into the Field: Perspectives on Social, Ethical and Regulatory Challenges in UK Wildlife Research', *Philosophical Transactions of the Royal Society B: Biological Sciences*, 376.1831 (2021), DOI: 10.1098/rstb.2020.0226.

23 Mariana Valverde, *Law's Dream of a Common Knowledge* (Princeton, NJ: Princeton University Press, 2003).

24 Alexandra Palmer et al., 'Edge Cases at the Boundaries of Animal Research Law: Constituting the Regulatory Borderlands of the UK's Animals (Scientific Procedures) Act', *Studies in the History and Philosophy of Science*, 90 (2021), 122–130, DOI: 10.1016/j.shpsa.2021.09.012.

25 House of Lords, *Report of the Select Committee on the Laboratory Animals Protection Bill* (London: HMSO, 1980), 2: Minutes of Evidence and Appendices, p. 301.

26 Animals in Science Regulation Unit, *Animals (Scientific Procedures) Act 1986: Working with Animals Taken from the Wild, Advice Note: 02/2016* (London: Home Office, 2016), p. 7.

27 Valverde, *Law's Dream of a Common Knowledge.*

28 Palmer et al., 'Getting to Grips with Wildlife Research by Citizen Scientists'; Palmer et al., 'Edge Cases at the Boundaries of Animal Research Law'.

29 Palmer and Greenhough, 'Out of the Laboratory, into the Field'.

30 Palmer et al., 'Edge Cases at the Boundaries of Animal Research Law'.

31 Elizabeth Cherry, 'Birding, Citizen Science, and Wildlife Conservation in Sociological Perspective', *Society & Animals*, 26.2 (2018), 130–147, DOI: 10.1163/15685306-12341500.

32 Marianne van Bochove et al., 'Reconstructing the Professional Domain: Boundary Work of Professionals and Volunteers in the Context of Social Service Reform', *Current Sociology*, 66.3 (2018), 392–411, DOI: 10.1177/0011392116677300.
33 Secord, 'Science in the Pub'.
34 Wynne, 'May the Sheep Safely Graze?'; Secord, 'Science in the Pub'; *Ethnobiology*, ed. by Anderson et al.
35 van Bochove et al., 'Reconstructing the Professional Domain'.
36 David B. Resnik et al., 'A Framework for Addressing Ethical Issues in Citizen Science', *Environmental Science & Policy*, 54 (2015), 475–481, DOI: 10.1016/j.envsci.2015.05.008; Riesch and Potter, 'Citizen Science as Seen by Scientists'.
37 Palmer et al., 'When Research Animals Become Pets and Pets Become Research Animals'; Alistair Anderson and Pru Hobson-West, '"Refugees from Practice"? Exploring Why Some Vets Move from the Clinic to the Laboratory', *Veterinary Record*, 191.1 (2021), e773, DOI: 10.1002/vetr.773.
38 Palmer and Greenhough, 'Out of the Laboratory, into the Field'.
39 ASRU, *Working with Animals Taken from the Wild.*
40 Davies, 'Locating the "Culture Wars" in Laboratory Animal Research'.
41 Hilgartner, *Knowledge and Control in the Genomics Revolution*, p. 9.
42 Valverde, *Law's Dream of a Common Knowledge*, p. 26; Davies, 'Locating the "Culture Wars" in Laboratory Animal Research'.
43 T. William and G. R. Gormley, 'Regulatory Enforcement Styles', *Political Research Quarterly*, 51.2 (1998), 363–383, DOI: 10.1177/106591299805100204.
44 Jenifer A. Buckley, 'Food Safety Regulation and Small Processing: A Case Study of Interactions between Processors and Inspectors', *Food Policy*, 51 (2015), 74–82, DOI: 10.1016/j.foodpol.2014.12.009.
45 Davies, 'Locating the "Culture Wars" in Laboratory Animal Research'.
46 Palmer and Greenhough, 'Out of the Laboratory, into the Field'.
47 Home Office, 'ASRU Operational Newsletter, 29 June 2021', *GOV.UK*, 2021, www.gov.uk/government/publications/asru-operational-newsletter-29-june-2021/asru-operational-newsletter-29-june-2021-accessible-version [accessed 19 July 2021].
48 ASRU, *ASRU Change Programme Stakeholder Meeting* (London: Animals in Science Research Unit, Home Office, 2021) www.lasa.co.uk/asru-change-programme-documents/ [accessed 1 February 2023].
49 Born and Barry, 'Art-Science'; Mirowski, 'The Future(s) of Open Science'.

11

'Knowledge is power, and I do want to know more': exploring assumptions around patient involvement in animal research

Gail Davies, Richard Gorman, and Gabrielle King

Introduction

We have titled this chapter using words from Tina[1] who became involved in research on dementia as a patient representative after retiring from a health-related career and caring for a family member. Her words weave together the complex relations between knowledge, power, and possibility that are emerging as patients[2] are increasingly involved in biomedical research.

This chapter reflects upon issues of knowledge and power from the perspective of different stakeholders, including patient representatives like Tina, who are involved in conversations around animal research. We examine a common but contradictory set of assumptions about patient involvement in animal research, which were expressed by some of the patient representatives, scientific researchers, and involvement professionals in our research, and contested by others. These assumptions may be held by individuals or institutions, and they may reflect aspects of past and present experience. However, they often remain unexamined, leading to missed opportunities for conversations across different perspectives or even miscommunications. These assumptions include that: 1) 'patients don't want to know about animal research'; 2) 'patients will always support animal research'; 3) 'patients don't have relevant expertise to contribute to the policy and practice of animal research'; and 4) 'patients won't make a difference to research using animals'.

In what follows, we locate these assumptions through the recent rise of research involvement, which seeks to include patient perspectives in biomedical research, and earlier periods of animal research activism, which aligned patient voices with research controversies. We then introduce our research before exploring each assumption through qualitative interviews with patients, scientists, and involvement professionals. In conclusion, we suggest that examining these assumptions can help foster two-way learning and create space for patient voices and different perspectives in biomedical research.

The rise of patient involvement in research

Patient involvement refers to research carried out in active partnership with people affected by health conditions. The National Institute for Health Research (NIHR)[3] defines involvement as doing research *with* people, rather than *to*, *about*, or *for* them. There are ongoing debates around the terminology and use of involvement,[4] however, involving patients in research is now embedded in research policy and practice across universities, hospitals, the pharmaceutical industry, and the third sector.

Research involvement has grown in response to demands to empower patient voices in decision-making about the research that affects them, and wider efforts to improve the relevance, quality, and accountability of health research and funding.[5] Involvement often has a range of objectives from satisfying moral and ethical imperatives around patient rights,[6] seeking legitimacy through fostering public acceptance and accountability,[7] to incorporating substantive contributions from patients that enhance the relevance and quality of research using the experiential knowledge they develop through living with a health condition.[8] Research involvement started with patient activism around clinical trials, where direct links were made between involving people in research and improving the relevance of research. Sociologist Steven Epstein tracked the work of AIDS activists and organisations in the 1980s 'who challenged researchers' approaches to conducting trials, which had overlooked patients' preferred outcomes'.[9] Collaborations between researchers and patients in clinical trials have now moved to the medical mainstream and are seen as 'vital if the uncertainties that matter most to patients are to be reduced'.[10]

This institutionalisation of involvement has seen patients move from 'advocacy groups' working outside to becoming members of 'working groups' within research policy.[11] The UK Department of Health set up the 'Consumers in NHS Research' to support patient involvement in 1996.[12] This body has subsequently changed names,[13] but the role of patients has been strengthened, with the 2001 Research Governance Framework for Health and Social Care stipulating that patients 'should be involved at various stages of the development and execution of research projects where appropriate'.[14] Patient involvement is now required in most aspects of health research, with many funders only supporting research that demonstrates the active involvement of patients.[15] However, aspirations for 'patients and the public to be involved in all stages of research'[16] are ambitious. The Shared Learning Group on Involvement in Research, a working group comprised of national voluntary sector organisations in the UK, noted in 2019 that 'much has been written about involving people in clinical, public health and social care research. Much less has been written about involving people in laboratory-based research'.[17]

While involvement in laboratory research remains less common, there is growing evidence of the different ways patients can make positive contributions to reimagining practices and modes of 'upstream' research. Caron-Flinterman et al. proposed several ways in which the experiential knowledge of patients could improve the relevance and quality of biomedical research in 2005.[18] The subsequent literature from social scientists, involvement practitioners, and biomedical researchers now includes many claims made for the value of patient involvement including: more collaborative relationships with patient groups,[19] better understandings of the links between disease and illness,[20] alternative models of translational research,[21] more effective organisation of clinical trials,[22] and even reducing the use of animals by increasing value and reducing waste in research.[23] However, Caron-Flinterman et al., and other health researchers building on this work, also recognise the complexities of involving patients in ways that are meaningful to them in biomedical research.[24] Meaningful involvement depends on creating mutually respectful relationships and recognising the value of patient expertise.[25] Many of the barriers to involvement are exacerbated when research involves the use of animals, due to contemporary sensitivities and past experiences.

The history of patient voices in animal research activism

Creating meaningful involvement around animal research is additionally challenging because of recent histories of public activism and institutional secrecy. Patient stories were increasingly incorporated into public debates around animal research in the 1980s. Pro-research groups used patient testimonies as part of public relations strategies, when they were inserted into the 'battle of moral images' as they were understood to have an 'emotional appeal to match that of animal activists'.[26] Patient voices were interjected into increasingly polarised public debates around animal research, which were strongly structured around pro- and anti-animal research positions. This use of patient testimonies was controversial at the time, with some scientists suggesting that patient stories were 'too emotional' and had the potential to undermine their claims based around scientific rationality.[27]

The initial assumption by many establishing these groups was that patients would always advocate for animal research. In the US, Incurably Ill for Animal Research was set up to bring patient perspectives into political debates to counter anti-animal research activism in the late 1980s.[28] In the UK, the Research for Health Charities Group, launched in 1991, aiming to improve 'understanding of the role played by animals in medical research as well as encouraging good research practice among its members',[29] tended to construe patients primarily as the powerless beneficiaries of medical research, with researchers 'helping victims to live a better life'.[30] Other groups did have more active patient inclusion from the start. The organisation Seriously Ill for Medical Research, also founded in 1991 by patient activist Andrew Blake, was a key mechanism through which 'the voice of the patient' in support of animal research began to be raised.[31]

However, other patient groups began to organise around opposing perspectives. Groups such as Disabled and Incurably Ill for Alternatives to Animal Research (DIIAAR) and Incurably Ill Because of Animal Research advocated for funding to be reallocated from animal research to care services or human clinical research in the US.[32] Groups like DIIAAR drew on disability rights activism, and were critical of the way patient narratives around disability had been co-opted to present illness as 'pitiable, always in need of a cure, and as a barrier to a full life'.[33] The UK group Seriously Ill Against Vivisection (see Figure 11.1) was set up in direct opposition to Seriously Ill for Medical Research, and challenged the way they felt patient voices were being co-opted by pharmaceutical companies, specifically stating 'we are appalled these companies claim to speak for us'.[34]

These particular groups are no longer active, but this history resonates as both moves towards greater openness and involvement

Figure 11.1 Hilary Walsh, centre left, and Gail Record, centre right, founding members of Seriously Ill Against Vivisection, are joined by others to demonstrate about the practice of testing drugs on animals outside the Research Defence Society in central London (Source: PA Images/Alamy Stock Photo).

in animal research bring renewed attention to the dynamics of conversations with patients. Many organisations in the UK have now signed up to the Concordat on Openness on Animal Research, meaning there are more resources to support researchers and institutions to have more open conversations about their use of animals in research.[35] However, the openness agenda is mainly focused on fostering discussions in public rather than processes for involving patients. This difference is important. Involved patients do not stand in as proxies for the public interest.[36] Those studying patient involvement argue that patients are not lay in medical contexts but hold valuable 'experiential knowledge' from having and coping with a condition.[37]

These earlier encounters have left a complex legacy of experiences and assumptions that can be difficult to navigate. There is no single way to manage the difficult mix of emotion, advocacy, and evidence in conversations between researchers and patients. A nexus approach to animal research offers an opportunity to hold together the present and past of patient involvement around animal research, understand these changing imaginaries,[38] listen across different perspectives, and support the diversity of conversations that patients want to have today.[39]

Researching stakeholder perspectives

The conversations presented in this chapter took place during research completed at the University of Exeter as part of the Animal Research Nexus Programme (AnNex) from 2017 to 2022. The research explored the increasing interfaces, and enduring gaps, between practices of patient involvement and animal research in the UK. It focused on learning from those stakeholders who were already doing patient involvement in biomedical research and who were able to reflect on both its possibilities and difficulties. We attended a range of patient engagement and involvement activities as observers and carried out two recruitment surveys with biomedical researchers and patient groups. We interviewed 59 preclinical scientists, patients and their representatives, and involvement and engagement professionals to understand their different perspectives, and organised two workshops to develop conversations between them.

The patients we spoke to were involved in setting research agendas, developing research proposals, reviewing grant applications, sitting on grant panels, monitoring research projects, talking at events, and helping put research into practice. We spoke to people of different genders, from diverse socioeconomic and ethnic backgrounds, and with occupations ranging from students to retired people. We included open questions on identity in the recruitment questionnaire to monitor inclusivity,[40] and our sample records a larger number of white women from professional backgrounds, middle-aged and upwards, which broadly reflects current patterns of research involvement in the UK. These people were making contributions to biomedical research across a range of areas, including cancer research and neurodegenerative and genetic conditions. Their experience of involvement ranged from one to over ten years. They were mainly involved in basic and preclinical research that sought to understand biological processes and test the safety and efficacy of potential treatments.

The research received ethical approval from the University of Exeter. All the names in this chapter are pseudonyms. We allocated names using specific letters (R, S, T, and W) distributed across the AnNex Programme to ensure that the people who took part in our research were given unique and consistent pseudonyms (see Introduction). Our analysis involved thematically mapping the distribution of assumptions, concerns, and hopes about patient involvement in animal research across different stakeholder groups, with a focus on understanding what involvement might mean, and achieve, for different people. We have written about the complex and situated dynamics of these encounters in a number of academic articles[41] and reported our research for the stakeholders who took part in it.[42]

In this chapter, we explore some of the assumptions that may act as barriers to moving from talking about research *to* patients, or speaking *for* patients in animal research, to having conversations about research *with* patients. Our aim is not to suggest that all patients want to have conversations about animal research, but to ensure that these conversations can be meaningful for those, like

Tina, who 'do want to know more'. We hope that by foregrounding the perspectives of those already involved in biomedical research, and listening to how they navigate the complexities of animal research, we can contribute to moving this conversation forward in productive new ways.

Assumption #1: Patients don't want to know about animal research

The first assumption we address is that patients don't want to know about animal research. This was reported by some involvement professionals and patients themselves. If patients do not want to know, it is important that they can continue to be involved in other kinds of clinical and preclinical research without this additional difficult aspect. Being involved in research can be rewarding and purposeful, but it also requires emotional work.[43] However, the people we spoke to wanted to have the opportunity to choose the right information for them, highlighting the importance of not assuming what patients want to know about animal research.

In many ways, patients told us similar things to earlier research on public engagement with animal research. These studies indicate that, for many people, acceptance of animal research is conditional on health benefits being realised and suffering minimised.[44] Prior research also indicates that many people do not want to encounter images of animal research, even if they are supportive.[45] As patient representative Teresa puts it:

> I didn't want to see it, I've got a friend who's very, very into animal rights, she always puts stuff on Facebook, I can't see things like that. I'd rather just hand the unpleasantness over to [the scientists] and let them get on with it. I don't really want to know. I know they use mice and I know they probably do some horrible things to them. But I actually don't need to know this. I wouldn't want to stop it. (Teresa, interview, 2018)

People could support animal research and want to know more about how animals were being treated, even if they still did not want to see them. Rhoda wanted to know how animals were treated but did not want to meet them face-to face because of her connection through research involvement.

> I did look at one application recently who said, 'We're going to have a meet the animal session' and I thought 'No, I really don't want to meet the animals.' I like to know that they're happy and content, but I don't want to look at them and think, 'what's going to happen to you tomorrow?' That went too far the other way. (Rhoda, interview, 2018)

Many images of animal research circulating in the media are criticised as out of date representations, and practices around both openness and involvement have included efforts to update these images.[46] However, for patients, this sensitivity was not only about the physical recoil of upsetting pictures. The patients we spoke to saw themselves as having a distinctive perspective and responsibility that came with their proximity to research.

This closeness has two dimensions. First there is the shared 'susceptibility to injury and illness' that Greenhough and Roe suggest patients can feel in common with animals used in research.[47] This leads to a sense of mutual vulnerability and enhanced sense of corporeal responsibility.

> I think that's the real issue, because as patients we've all been, I hate to use the word, but we have all been in some way, a guinea pig, you know, it's a cliché but we have been analogous to the mouse in a specific situation. (Tristan, interview, 2018)

> I have a much better understanding now of what it's like to be vulnerable. With this diagnosis, as things get harder you become much more aware of how vulnerable you're becoming, you're going to become more and more vulnerable, and be at the whim of other people and I just feel that's how animals are. (Tabitha, interview, 2018)

The second aspect of connection is around what we might call procedural responsibility. Being involved in research meant patients wanted to know more to inform their contributions to decision-making, part of what Druglitrø calls the 'procedural care' practices that now surround animal research.[48] Rhoda reported 'worrying about it in the middle of the night' if she did not know what the animals would experience. Tabitha acknowledged that conversations about animal research were difficult but important to her research evaluation and sense of responsibility.

> I do like to know the detail. I really want to know that they're being treated humanely and that any suffering is minimised, and I do want to know that. I don't want to have my brain working overtime thinking I've just given something a high score for research, and I don't understand what the animals have to go through. [...] I think it would be good if we had more of these conversations [...] reducing that anxiety would be a huge help to me personally and reducing the guilt. (Tabitha, interview, 2018)

For Rachel, it was important to know how animal suffering was going to be minimised. This helped her make a meaningful contribution.

> If I knew an animal was going to suffer [...] I'd want to know how it was being minimised. I want to know that whatever is happening to those animals, any suffering or pain is minimised and it's definitely not prolonged and that if they are killed afterwards, that's humane and quick. I would want to know. I don't think there's anything I wouldn't want to know. If it's relevant to that research and it helps me to provide some kind of informed contribution, then I would want to know. (Rachel, interview, 2018)

Patients who want to know more about animal research talk about their involvement in decision-making through an informal harm–benefit analysis.[49] They are interested in the potential benefits of animal research for their condition, but they also want to know about the animals' experience and what is being done to reduce harms. Anxiety can come from not knowing and being left to make judgements without information on these crucial points. Many institutional communications about animal research, fostered by the Concordat on Openness, have made the links between research practices and potential outcomes more publicly visible. But Tina wanted to 'learn a lot more about how many animals would be used, where the animals were coming from, how they'd been treated', and suggested that for her this learning 'hasn't really seemed possible' (interview, 2018).

Patients are not a homogenous group and rather than assuming patients do not want to talk about animal research, our interviews suggest it is important to provide opportunities for people to be in control of the information about animal research they want to know more about, including around suffering and reducing harms. The potential for these conversations to increase feelings of personal responsibility for animal suffering also indicates the need for caution and careful support for patients who become involved in animal research. This leads directly to our next point, about not assuming that patients will always support animal research.

Assumption #2: Patients will always support animal research

The second assumption we want to unpack is that patients will always support animal research. This assumption might be traced back to the 1980s and 1990s, when patient testimonies were mobilised to foster public support for animal research. It was also something that some people we spoke to reflected on in private. Tim, a patient representative explained that 'your perspective changes when you get an illness like this. I don't know but I imagine people who get chronic illnesses would be far more supportive of animal research in relation to research on that illness' (interview, 2018). Like many of the points in this chapter, this assumption contains important elements of truth,[50] but it is not a simple, or binary position.

Some of the complexities that come from assuming patients will always support animal research are evident in the processes of putting together engagement activities and involvement panels. Conversations between patients and researchers need to be safe and supportive, but this can lead to them being selective. Historical experiences of activism against animal researchers in the UK meant that many involvement professionals we spoke to recounted stories of people being screened out of attending events or joining review panels on the basis of their background and beliefs. Sue explained:

> When I first came to the charity, I've been here six years, the medical director, when he interviewed what was the user committee, if they said they were opposed to animal research, they were chucked off, well not chucked off but they were not recruited, so obviously that is a self-selecting group. (Sue, interview, 2018)

The atmospheres around animal research have changed in some ways, but institutional screening often continues when patient review panels are put together. The membership of a group that reviews a research proposal, or conducts an ethical review of animal research,[51] constructs a particular version of that patient or public voice.[52]

Shirley, an involvement professional for a large medical charity, explains that patient recruitment processes can include questions about people's views in this area: 'Depending on the activity we're interviewing for, we'll often ask people "How do you feel about animal research because it is something that we as an organisation do?"' (interview, 2018). Other organisations reported a similar stage in recruitment, explaining they would not exclude people with different views, but like Shirley acknowledged it did likely lead to a consensus around a 'middle ground of not loving animal research but understanding that it's probably necessary' (interview, 2018).

This compromise will be reassuring to many and help facilitate some conversations, but it also shapes expectation of what a 'patient voice' is and can exclude other perspectives. For example, Tabitha suggests she doesn't have a voice about the use of animals in this context.

> We don't really have a voice in what goes on as far as research using animals, I think there's obviously the anti-vivisectionists, you can do stuff that way, but not as a role in research networks. (Tabitha, interview, 2018)

Those involved in research are asked to bring their own lived experiences, and sometimes the views of others, to their roles. However, some aspects of personal identity and experience, including hesitancy around animal research, are difficult to bring into these conversations. People have many beliefs and experiences that are important to them, which they have to negotiate in involvement. If these challenge animal research, the assumption that patients will always support animal research means important views may be difficult to bring into this space. People risk being left needing to seek support elsewhere, reducing the diversity of perspectives that are included in conversations.

> With the transgenic mice, I was a little bit apprehensive about it because I am a [member of a particular religion]. One of our tenets is not to kill any living being, so I in fact, consulted with my meditation teacher about whether I might be committing murder or committing the killing of the drosophila or the rats by participating in the monitoring – even though I'm not directly involved in the scientific part of it. (Win, interview, 2018)

Questions that might be seen as undermining the consensus about animal research can also be difficult to ask. Here our interviews echo work by Michael and Brown, which indicates that patients offered contingent support for animal research and also wanted to back multiple strands of research, any one of which might yield something useful.[53] The patients we spoke to often wanted to ask questions about alternatives to animal research. Some found these questions hard to ask as they didn't want to seem critical of the research or felt these conversations would be shut down. Others did feel able to ask these questions and wanted to talk about the range of research options that might deliver benefits but recognised this line of questioning could be seen as difficult or disruptive.

> I tend to question quite avidly: why would a mouse model be useful in this particular situation, why can't the researchers be looking for an alternative model, maybe a stem cell? That might be some of the feedback I give quite often actually because as a general principle, I would like to see less animal research. That's my ethical standpoint. [...] but there just doesn't seem to be an interest in doing alternative types, looking at prevention or lifestyle or looking at environment factors that might be of interest and so on. (Tina, interview, 2018)

People come to research involvement with a range of personal experiences and gain further expertise through their involvement. Being open to learning from their diversity of experiences requires attending to conversations about research that move beyond the assumptions that patients will always advocate for animal research. It also requires careful consideration of the nature of patient expertise.

Assumption #3: Patients don't have relevant expertise to contribute

The rise of patient involvement offers those with the experiential knowledge that comes from living with a health condition the opportunity to inform research priorities and practices.[54] Experiential knowledge is not fixed;[55] it is constantly renegotiated as experiences and lives change.[56] The diversity of this expertise is reflected in the range of terms used to describe people involved in research, from patient panels, research volunteers, consumers, experts by experience, and lay faculty, to people affected by health conditions. Some may reject the term patient altogether. As Rebecca, who works for a medium sized medical charity suggests 'I think our patients are trying very hard not to be patients. But of course, they have this wealth of knowledge and understanding of what living with [this condition] is about and I think we have always tried to ensure that's taken into account' (interview, 2018). However, this can be a challenge if there are assumptions that patients don't have relevant expertise to contribute to animal research.

The experiential knowledge that patients bring to involvement can struggle to find a place within laboratory research, as opposed to clinical research, as it takes place at an early stage where there is still a perceived distance between researchers and other stakeholders.[57] Direct links to patient priorities can also be harder to identify in animal research, or when regulatory processes for safety and efficacy testing in animals appear to offer less scope for experiential knowledge to affect research practices. Ruby reflects on her experience of living with a health condition but says: 'I just don't quite know how it translates to actual patient involvement because it seems like researchers have their own opinions about what they would like to do' (interview, 2018). Many of the patients we interviewed valued meeting people affected by similar health conditions and talking to researchers, but they were given little guidance on how their experience and expertise were used to inform research practices.

There were three main ways in which people stressed the relevance of their expertise, which could be used to inform future involvement. Firstly, those involved in research that we spoke to felt their lived experiences of health could and should be taken into account when deciding the priority issues for research. Understanding themselves to have relevant knowledge, people valued the opportunity to link this to the next steps for research and issues around experimental design. Rachel explained that 'whilst there might be a lot of technical terms, actually we can add value by looking at it through the eyes of people living with dementia. Do we see it as a priority? How would that impact us?' (interview 2018).

Secondly, many people combined their lived experiences of health with the expertise gained through their involvement in biomedical research. As Rachel says of one study she reviewed: 'I think I've got enough experience now to be able to go and make a meaningful contribution but also, this is something that really resonates' (interview 2018). Drawing on experiences of involvement over a period of time and becoming more familiar with how processes work was particularly valuable to patients when discussions involved animal research.

Thirdly, while we use the word patient, people bring diverse expertise beyond their embodied experience of health. The patient label can even be problematic in this context as it risks reducing people to their health experiences or their bodies. As Raymond puts it:

> One of the most important things is that volunteers are rarely recognised as having skills and experience that could be of consequence in their new environment. [...] I tend to put it down to this lack of understanding, which I've come across in many contexts, that volunteers are not just bodies to be used but people with all sorts of skills who can actually help the organisation. (Raymond, interview, 2018)

Many people talked about how their careers had given them skills they used in their involvement role, including project planning and management, or training and development, which could help realise benefits from research.

> Part of my role in my civil service career was about drafting specifications and commenting on them, not that I could comment necessarily on the scientific aspect but just to see whether it made any sort of sense. Because some people might find it surprising, you can actually

> quite a lot of the time see whether there is some logic to the proposal, whether you understand the science or not. And I found I could bring that to the table, which is why I decided to apply to the biomedical panel. (Toby, interview, 2018)

Ensuring the relevance of people's varied expertise requires understanding the diversity of what people bring to involvement, and careful planning to ensure that it is valued. This then links to our last assumption, which is that patient voices won't make a difference to animal research.

Assumption #4: Patients won't make a difference to research using animals

The final challenge we want to explore is how patients are understood to make a difference to animal research, or more often the question of *whether* they make a difference. Evaluating the outcomes of patient involvement is an increasingly important issue.[58] Patient involvement requires considerable work from everyone, and there are significant problems around legitimacy and efficacy if involvement is merely directed to researchers ticking the box on the funding proposal. This can again be seen as an instrumentalisation of patient experiences, albeit now moved from media stories to grant applications.

At this point, many of the issues we have explored above come together. The effects of involvement on clinical research are often quite direct, with tangible outcomes defined through patient benefit and researcher gains.[59] However, these are more difficult to trace in laboratory research and the involvement professionals we spoke to recognised they struggled to make the case in this context. Rosalyn, an involvement professional in a small charity, says:

> I think it's far more challenging than clinical research, I think the role of the patient, carer, consumer, whatever term you want to use, I think it's quite well laid out now with regards to clinical research. I think it's a little less clear when it comes to animal research, what role consumers can play, the value of the contribution they can make and what that contribution would look like. So yes, I do think it's more difficult. (Rosalyn, interview, 2018)

These difficulties are compounded through the encounters and conversations around animal research. We found that many people organising patient involvement remain uncertain about the outcomes of involvement on animal research, reflecting many of the assumptions we have explored above. Some felt ill-equipped to start these conversations. Some assumed that patients would contribute only to public communications or ethical discussions about animal research. Those who did try to stage conversations around the use of things like animal models found it hard to make direct links to patient expertise and experience. Sian, a senior university researcher, suggests:

> They didn't really feel comfortable talking or giving any kind of feedback on the animal research at all. When we asked them about how they felt about the animal model, the conversation was very restricted and very limited. We've actually got them saying, 'We just don't feel able to comment on this'. They felt uncomfortable, felt they didn't have enough knowledge to talk about it. They didn't think they had a valid opinion. And they didn't really want to have an opinion on it, to be honest. (Sian, interview, 2018)

We want to be clear about these difficulties as the issues are too important to gloss over in this chapter. However, several of these issues can be addressed through careful planning so that involvement tasks are clear, everyone taking part is briefed sensitively, able to access the information that they want, and feels confident that they know how their input will be used. There is still a long way to go; more experience is needed in staging these sorts of conversations around animal research, with more people sharing honest assessments of the outcomes. It is also important to review the assumptions around these conversations so that patient involvement works for the people it is meant to benefit. This means ensuring that patients are in control of the information they want, able to bring a diversity of perspectives into discussions, and are guided in understanding how their expertise could contribute to the focus and organisation of animal research.

When we talked to people who did feel that they were making a difference, this most often came from organisations who had ensured that patient involvement was not located at the end of the research development process when decisions had already been made. Placing patient perspectives front and centre also enabled people to build relationships, knowledge, and experience together, whether it involved animals or not. As Rachel reflects, being brought into the conversation matters:

> And maybe that is one of the ways that bringing people in with the condition actually does help. Because quite often, I will be putting on one of the comments on some of the research material [that what] I see is there seems to be flavours, suddenly it's all to do with inflammatory conditions or whatever, there is something that ripples throughout the scientific world and you see lots of bids coming in, from lots of different organisations and one of my favourite comments is about shouldn't there be a collaborative bid here? Why are we funding several strands of what looks to me to be the same line of research? Why are they doing this in isolation? (Rachel, interview, 2018)

When involvement worked well, people taking part in research talked about the benefits of knowing their condition was being taken seriously by researchers; they identified the links they had been able to make between researchers, drug companies, and clinical contexts; and they noted anecdotally that researcher grant hit rates had gone up. Few saw themselves as advocating for animal research, but many did report more confidence in the regulation and oversight of animal research in the UK. Tabitha became involved in research following her diagnosis nearly ten years ago. She reflects on what involvement means to her and the people she has worked with:

> It works for everybody because it makes it a better experience for the people with dementia when they're actually being studied, and the researchers are obviously going to get better results because they're seeing the real picture. [...] I think another upside of patient involvement is that it gives you a purpose, it gives you some feeling of some value, that you do still have a value. (Tabitha, interview, 2018)

Conclusion

This chapter has sought to explore some of the assumptions circulating amongst scientists, involvement professionals, and even patients themselves, that can function as barriers to developing more meaningful conversations with patients around animal research in involvement processes. We have mapped out four different assumptions, each of which embodies important truths but presents only a partial perspective on the conversations taking place between patient involvement and animal research. In exploring these assumptions, we have shown how they create gaps and frictions in the complex systems and interactions that make up patient involvement in biomedical research in the UK. We have also followed the opening provocation from Tina to indicate how moving beyond these assumptions requires being reflective and critical of standpoints and power.

We suggest it is important to continue to develop these conversations in more complete and careful ways. In this chapter, we have indicated that opportunities exist to organise patient involvement that enables everyone to have more open, honest, and meaningful conversations around animal research, through better planning, training, relationship building, and organisational investments. We have written these up in a practical guide for those starting out doing patient involvement around animal research.[60] We also suggest patient involvement can contribute to helping realise the benefits from animal research[61] by ensuring that research is more relevant to the experiential knowledge of patients and through making use of the wider experiences and expertise that they hold. But more than this, we suggest that meaningful patient involvement in this area requires acknowledging the complex relations that patients have with laboratory animal research, including feelings of efficacy and anxiety, shared vulnerability, or moral responsibility.

We conclude by suggesting this is a further area where a nexus approach to animal research is important for understanding how the history of different perspectives influences current encounters.[62] If involvement in animal research is about doing research *with* patients, then this also means doing involvement *with* an attentiveness to and recognition of this context. The record of anti-vivisectionist movements and the use of patient 'voices' to promote research cannot be decoupled from the way that different experiences and knowledges come together around animal research today. It is important to reflect on this history but recognising the complexity of these encounters need not be a barrier to starting conversations. Here our research connects to other work across the wider AnNex programme, whether identifying changing responsibilities for the cultures of care in which animal research is located (see Kirk, Chapter 5, and Greenhough and Roe, Chapter 6) or recognising that openness needs to start from the conversations that people want to have (see Roe et al., Chapter 14). There is much that everyone can learn if patient involvement around animal research is fully understood as a 'conversation that supports two-way learning'.[63] And this particular conversation has only just begun.

Notes

1 Tina and all other names used in this chapter are pseudonyms.
2 We use the term patient in this chapter to refer to people who speak from the position of someone affected by a health condition in patient involvement events. This includes those diagnosed with a health condition, family members, friends, carers, and others. There are many different terms used by people organising and taking part in these events and many different views on the preferred terms. We have used patient here as the term most likely to be recognised by those coming new to patient involvement from conversations around animal research. However, as we discuss later in the text, we want to acknowledge that many people are uncomfortable with this term.
3 'National Institute for Health Research', *People in Research*, www.peopleinresearch.org/ [accessed 31 October 2022].
4 Josephine Ocloo and Rachel Matthews, 'From Tokenism to Empowerment: Progressing Patient and Public Involvement in Healthcare Improvement', *BMJ Quality & Safety*, 25.8 (2016), 626–632, DOI: 10.1136/bmjqs-2015004839.
5 Andy Gibson et al., 'Theoretical Directions for an Emancipatory Concept of Patient and Public Involvement', *Health*, 16.5 (2012), 531–547.
6 Jennie Popay and Gareth Williams, 'Public Health Research and Lay Knowledge', *Social Science & Medicine*, 42.5 (1996), 759–768, DOI: 10.1016/0277-9536(95)00341-X; R. Telford et al., 'Consumer Involvement in Health Research: Fact or Fiction?' *British Journal of Clinical Governance*, 7.2 (2002), 92–103, DOI: 10.1108/146641002 10427606.

7 Janneke E. Elberse et al., 'Patient–Expert Partnerships in Research: How to Stimulate Inclusion of Patient Perspectives', *Health Expectations*, 14.3 (2011), 225–239, DOI: 10.1111/j.1369-7625.2010.00647.x.
8 Popay and Williams, 'Public Health Research and Lay Knowledge'; J. Francisca Caron-Flinterman et al., 'The Experiential Knowledge of Patients: A New Resource for Biomedical Research?' *Social Science & Medicine*, 60.11 (2005), 2575–2584; Iain Chalmers, 'What Do I Want from Health Research and Researchers When I Am a Patient?' *BMJ*, 310.6990 (1995), 1315–1318, DOI: 10.1136/bmj.310.6990.1315.
9 Hazel Thornton, 'Patient and Public Involvement in Clinical Trials', *BMJ*, 336.7650 (2008), p. 904. See also Steven Epstein, *Impure Science: AIDS, Activism, and the Politics of Knowledge* (Berkley, CA: University of California Press, 1998).
10 Thornton, 'Patient and Public Involvement in Clinical Trials', p. 903.
11 Hazel Thornton, 'Patients and Health Professionals Working Together to Improve Clinical Research: Where Are We Going?' *European Journal of Cancer*, 42.15 (2006), 2454–2458.
12 Patricia Wilson et al., 'ReseArch with Patient and Public InvOlvement: A RealisT Evaluation – the RAPPORT Study', Southampton (UK): NIHR Journals Library; 2015, https://pubmed.ncbi.nlm.nih.gov/26378332/ [accessed 1 February 2023].
13 This body has been through several different transformations, as INVOLVE, and as part of the NIHR Centre of Engagement and Dissemination from April 2020. See www.nihr.ac.uk/explore-nihr/campaigns/supporting-patient-and-public-involvement-in-research.htm [accessed 31 October 2022].
14 Jonathan Boote et al., 'Consumer Involvement in Health Research: A Review and Research Agenda', *Health Policy*, 61.2 (2002), 213–236, DOI: 10.1016/S0168-8510(01)00214-7.
15 Boote et al., 'Consumer Involvement in Health Research'; Thornton, 'Patients and Health Professionals Working Together to Improve Clinical Research'; Wilson et al., 'ReseArch with Patient and Public InvOlvement: A RealisT Evaluation'.
16 Thornton, 'Patient and Public Involvement in Clinical Trials', p. 904.
17 Shared Learning Group on Involvement, *Involving People in Laboratory Based Research: A Discussion Paper*, 2016, http://slginvolvement.org.uk/wp-content/uploads/2016/09/Lab-based-research-FINAL-10-8-16-public.pdf [accessed 31 October 2022].
18 Caron-Flinterman et al., 'The Experiential Knowledge of Patients'.
19 J. Francisca Caron-Flinterman et al., 'Patient Partnership in Decision-Making on Biomedical Research: Changing the Network', *Science,*

Technology, & Human Values, 32.3 (2007), 339–368; Elberse et al., 'Patient–Expert Partnerships in Research'.

20 Felicity Callard et al., 'Close to the Bench as Well as at the Bedside: Involving Service Users in All Phases of Translational Research', *Health Expectations*, 15.4 (2012), 389–400; Marjaana Jones and Ilkka Pietilä, 'Personal Perspectives on Patient and Public Involvement – Stories about Becoming and Being an Expert by Experience', *Sociology of Health & Illness*, 42.4 (2020), 809–824, DOI: 10.1111/1467-9566.13064.

21 Callard et al., 'Close to the Bench as Well as at the Bedside'; Marianne Boenink et al., 'Giving Voice to Patients: Developing a Discussion Method to Involve Patients in Translational Research', *NanoEthics*, 2018, 1–17, DOI: 10.1007/s11569-018-0319-8.

22 Amy Price et al., 'Patient and Public Involvement in the Design of Clinical Trials: An Overview of Systematic Reviews', *Journal of Evaluation in Clinical Practice*, 24.1 (2018), 240–253, DOI: 10.1111/jep. 12805.

23 Rustam Al-Shahi Salman et al., 'Increasing Value and Reducing Waste in Biomedical Research Regulation and Management', *The Lancet*, 383.9912 (2014), 176–185, DOI: 10.1016/S0140-6736(13)62297-7; Iain Chalmers et al., 'How to Increase Value and Reduce Waste When Research Priorities Are Set', *The Lancet*, 383.9912 (2014), 156–165, DOI: 10.1016/S0140-6736(13)62229-1.

24 Michel Callon and Vololona Rabeharisoa, 'Gino's Lesson on Humanity: Genetics, Mutual Entanglements and the Sociologist's Role', *Economy and Society*, 33.1 (2004), 1–27; Carlos Novas, 'The Political Economy of Hope: Patients' Organizations, Science and Biovalue', *BioSocieties*, 1.3 (2006), 289–305, DOI: 10.1017/S1745855206003024.

25 James Maccarthy et al., 'Facilitating Public and Patient Involvement in Basic and Preclinical Health Research', *PLOS ONE*, 14.5 (2019), 1–16; Anthony G. Wilson et al., 'Patient and Public Involvement in Biomedical Research: Training is Not a Substitute for Relationship Building', *Annals of the Rheumatic Diseases*, 78.11 (2019), 1607–1608, DOI: 10.1136/annrheumdis-2019-215595; Richard Gorman and Gail Davies, 'When "Cultures of Care" Meet: Entanglements and Accountabilities at the Intersection of Animal Research and Patient Involvement in the UK', *Social & Cultural Geography* (2020), 1–19, DOI: 10.1080/14649365.2020.1814850.

26 James M. Jasper and Dorothy Nelkin, *The Animal Rights Crusade: The Growth of a Moral Protest* (New York: Free Press, 1992); F. Barbara Orlans, *In the Name of Science: Issues in Responsible Animal Experimentation* (New York: Oxford University Press, 1993).

27 Anon, 'Double-Talk on Animals: NIH Seems More Ready to Risk its Reputation than to Meet Serious Critics on Animal Care', *Nature*, 4.5963 (1984), 2.
28 Dennis M. Feeney, 'Human Rights and Animal Welfare', *American Psychologist* (1987), 7.
29 Mary Ann Elston, 'Attacking the Foundations of Modern Medicine? Anti-Vivisection Protest and Medical Science', in *Challenging Medicine*, ed. by Jonathan Gabe et al. (London: Routledge, 1994), pp. 196–219.
30 Research for Health Charities Group, 'Meeting the Challenge of Disabling Disease: Animal Research Brings Hope', 1997, available via the Wellcome Collection, www.wellcomecollection.org/works/vtfhtw4a [accessed 31 October 2022].
31 Elston, 'Attacking the Foundations of Modern Medicine?'
32 Orlans, *In the Name of Science.*
33 Sunaura Taylor, *Beasts of Burden: Animal and Disability Liberation* (New York: The New Press, 2017).
34 Seriously Ill Against Vivisection, 'About Us', 2002, website archived at https://www.web.archive.org/web/20021122025734/http:/www.siav.org/ [accessed 31 October 2022].
35 Wendy Jarrett, 'The Concordat on Openness and its Benefits to Animal Research', *Laboratory Animals*, 45.6 (2016), 201–202, DOI: 10.1038/laban.1026; Larry Carbone, 'Open Transparent Communication about Animals in Laboratories: Dialog for Multiple Voices and Multiple Audiences', *Animals*, 11.2 (2021), 368, DOI: 10.3390/ani11020368.
36 Gail Davies et al., 'Which Patient Takes Centre Stage? Placing Patient Voices in Animal Research', in *GeoHumanities and Health*, ed. by Sarah Atkinson and R. Hunt (Cham: Springer, 2020), pp. 141–55; Mio Fredriksson and Jonathan Q. Tritter, 'Disentangling Patient and Public Involvement in Healthcare Decisions: Why the Difference Matters', *Sociology of Health & Illness*, 39.1 (2017), 95–111.
37 Caron-Flinterman et al., 'The Experiential Knowledge of Patients'.
38 Renelle McGlacken and Pru Hobson-West, 'Critiquing Imaginaries of "the Public" in UK Dialogue around Animal Research: Insights from the Mass Observation Project', *Studies in History and Philosophy of Science*, 91 (2022), 280–287, DOI: 10.1016/j.shpsa.2021.12.009.
39 Gail Davies et al., 'Animal Research Nexus: A New Approach to the Connections between Science, Health and Animal Welfare', *Medical Humanities*, 46.4 (2020), 499–511, DOI: 10.1136/medhum-2019-011778.

40 We use pronouns he and she when discussing participant perspectives in line with what we understand to be the preferred pronouns of our respondents from our recruitment survey.
41 Davies et al., 'Which Patient Takes Centre Stage?'; Gorman and Davies, 'When "Cultures of Care" Meet'.
42 Richard Gorman and Gail Davies, *Patient and Public Involvement and Engagement (PPIE) with Animal Research*, 2019, AnNex website, www.animalresearchnexus.org/sites/default/files/patient_and_public_involvement_and_engagement_ppie_with_animal_research.pdf [accessed 1 February 2023]; Gail Davies et al., *Informing Involvement around Animal Research: Report and Resources from the Animal Research Nexus Project*, 2022, AnNex website, www.animalresearchnexus.org/publications/informing-involvement-around-animal-research [accessed 1 February 2023].
43 Alison Faulkner and Rose Thompson, 'Uncovering the Emotional Labour of Involvement and Co-Production in Mental Health Research', *Disability & Society*, 38.4 (2023), 537–560, DOI: 10.1080/09687599.2021.1930519; Gorman and Davies, 'When "Cultures of Care" Meet'.
44 Catherine A. Schuppli et al., 'Understanding Attitudes towards the Use of Animals in Research Using an Online Public Engagement Tool', *Public Understanding of Science*, 24.3 (2015), 358–374; Gail Davies et al., 'The Social Aspects of Genome Editing: Publics as Stakeholders, Populations and Participants in Animal Research', *Laboratory Animals*, 56.1 (2021), 88–96, DOI: 10.1177/0023677221993157.
45 J. Z. Turner, 'I Don't Want to See the Pictures: Science Writing and the Visibility of Animal Experiments', *Public Understanding of Science*, 7.1 (1998), 27–40, DOI: 10.1177/096366259800700103.
46 Wendy Jarrett, 'A Picture Paints a Thousand Words: The Value of Images and Video for Understanding Animal Research', *Laboratory Animals*, 45.10 (2016), 402–403, DOI: 10.1038/laban.1121.
47 Beth Greenhough and Emma Roe, 'Ethics, Space, and Somatic Sensibilities: Comparing Relationships between Scientific Researchers and Their Human and Animal Experimental Subjects', *Environment and Planning D: Society and Space*, 29.1 (2011), 47–66, DOI: 10.1068/d17109.
48 Tone Druglitrø, 'Procedural Care: Licensing Practices in Animal Research', *Science as Culture* (2022), 1–21, DOI: 10.1080/09505431.2021.2025215.
49 Gail Davies, 'Harm–Benefit Analysis: Opportunities for Enhancing Ethical Review in Animal Research', *Laboratory Animals*, 47.3 (2018), 57–58, DOI: 10.1038/s41684-018-0002-2.

50 McGlacken and Hobson-West, 'Critiquing Imaginaries of "the Public" in UK Dialogue around Animal Research'.
51 Carbone, 'Open Transparent Communication about Animals in Laboratories'.
52 Davies et al., 'The Social Aspects of Genome Editing'.
53 Mike Michael, 'Publics Performing Publics: Of PiGs, PiPs and Politics', *Public Understanding of Science*, 18.5 (2009), 617–631, DOI: 10.1177/0963662508098581; Mike Michael and Nik Brown, 'Scientific Citizenships: Self-Representations of Xenotransplantation's Publics', *Science as Culture*, 14.1 (2005), 39–57, DOI: 10.1080/09505430500041769.
54 Caron-Flinterman et al., 'The Experiential Knowledge of Patients'.
55 Jones and Pietilä, 'Personal Perspectives on Patient and Public Involvement'.
56 Categories of researcher and patient can overlap. Some people we spoke to had come to involvement from former careers as health professionals or scientific researchers. And some talked about going on to train for a research career themselves. While it is important not to assume a shared level of scientific understanding in involvement, it also is important not to go into conversations assuming a lack of technical expertise.
57 Caron-Flinterman et al., 'The Experiential Knowledge of Patients'.
58 Susan Jill Stocks et al., 'Application of a Tool for the Evaluation of Public and Patient Involvement in Research', *BMJ Open*, 5.3 (2015), e006390, DOI: 10.1136/bmjopen-2014-006390.
59 S. Staniszewska et al., 'GRIPP2 Reporting Checklists: Tools to Improve Reporting of Patient and Public Involvement in Research', *BMJ*, 358 (2017), j3453, DOI: 10.1136/bmj.j3453.
60 Davies et al., *Informing Involvement around Animal Research.*
61 Davies, 'Harm–Benefit Analysis'.
62 Davies et al., 'Animal Research Nexus'.
63 Kristina Staley and Duncan Barron, 'Learning as an Outcome of Involvement in Research: What Are the Implications for Practice, Reporting and Evaluation?' *Research Involvement and Engagement*, 5.1 (2019), 14, DOI: 10.1186/s40900-019-0147-1.

12

'Bred, but not used': understandings of avoidable and unavoidable waste in animal research

Sara Peres and Emma Roe

Introduction

In 2018, reporting on UK animal research statistics included for the first time the number of animals 'bred, but not used'.[1] The category includes animals that are by-products from the breeding of a specific genetically altered (GA) animal, are bred to maintain a live 'tick-over' animal colony, or are research-ready but do not get used in experiments. This new statistic adds to the annual publication of the number of licensed 'procedures' carried out on animals, across species, as defined by the Animals (Scientific Procedures) Act (ASPA). At a stakeholder meeting held in 2018 at the start of the Animal Research Nexus Programme (AnNex) to help scope out our research, we sensed apprehension from two attendees, both highly engaged experts in laboratory animal research and welfare, about the first public release of these UK figures. The question that troubled them was how these numbers would be understood and received by the public. The statistics are already hard for many people to interpret, while also being the focus of debate by different interest groups. There had been a steady reduction in the number of animal procedures in UK research from the 1990s until the early 2000s.[2] At this point numbers started to rise due to increased breeding of GA animals, which involves an experimental procedure. Since 2015 procedure numbers have been falling again.[3] This chapter explores the apprehension about the release of this new statistic by discussing findings about the industry's language and meanings attached to animals 'bred, but not used' in a regulated scientific procedure.

We were curious to understand why our two meeting attendees conveyed concern that the descriptive category 'bred, but not used' conjured up ideas that these research animals would be considered surplus, disposable, or waste, language that might denote a lack of respect for animals' lives, or inefficiencies within the supply chain. Our research interviews with people involved in the breeding and supply of research animals confirmed that the language of surplus and waste is commonly used for certain groups of animals. This concern connects to ongoing discussion about how far the regulatory protection of animals in science should be extended to breeding and supply animals. The UK government discussed this during the initial development of ASPA[4] and conversations have continued. The breeding of vertebrate animals used in scientific procedures has to take place in a licensed establishment, which sets standards for care and accommodation, but in the majority of cases the process of breeding itself is not counted as a 'licensed procedure' under ASPA. The current guidance on ASPA does now include guidance on reducing waste, yet uneasiness persists about the number of surplus or 'bred, but not used' animals. For example, surplus animals were framed as a public concern in the late 1990s following parliamentary questions,[5] and professional bodies within the animal research sector also confronted their own internal concerns during this period.[6]

The Additional Statistics of 2018 showed that 1.81 million animals (of which 1.45 million were mice)[7] were 'bred, but not used'. The report was welcomed as a step towards greater transparency,[8] but, contrary to our meeting attendees' fears, received little wider public interest. Instead, the publication of these new statistics has been of greater significance for those inside the industry, prompting further reflection on the complex array of practices that can lead to animals being 'bred, but not used'. These reflections within the industry are the focus in this chapter. To understand thinking and practice around surplus animals, we undertook qualitative research in 2018–2019, immediately following the publication of the new figures. After discussing our theoretical framing and research methods, we introduce our research participants' distinction between animals deemed avoidable and unavoidable waste, which is not captured in the 2018 'bred, but not used' statistics.

We then explore proposals for reducing waste from regulatory guidance, and from our own research, with a particular focus on the outsourcing of breeding facilities. We discuss tensions between researchers' expectation of immediate availability of research animals and the distress experienced by those responsible for killing surplus animals. We then conclude with recommendations for regulatory attention.

Theoretical framing

This section discusses our theoretical framing and contribution to the literature for our social science readers. We use a cultural economic analytic framework, which closely follows the thinking of post-Marxist feminist economic geographers such as Gibson-Graham,[9] alongside socio-material approaches to things, in this case live animals, becoming waste.[10] This approach moves away from understanding and describing economic relations through a lens focused solely on the operation of capital. Instead, our approach pays greater attention to the role of emotions as part of the distributed practices involved in breeding a point-of-sale animal commodity to consider the materiality of aliveness as waste and thereby to 'decentre the object of commodity fetishism'.[11] A capital-ocentric approach in this case would look at how animals are commodified as research tools, with the repeated breeding of litters of mice producing surplus-value and enabling capital accumulation.[12] The young mice as surplus-value are positively valued when they are exchanged for capital and become used in one or more animal procedures. However, we also are aware that the social forms and institutions around animal research produce a second group of surplus animals that do not achieve a use-value for experimental research; to use the Home Office nomenclature these are the 'bred, but not used' animals. In defence and explanation of why surplus is normalised, Smith argues that the production of surplus combats a social crisis from scarcity and with that assists the 'social emancipation of human society as a whole from nature';[13] in other words, mouse lives are a unit of production and it benefits human society to have a ready supply of this product. However, an explanation based only on the operation of capital does *not* explain how the industry has different reasoned sentiments, feelings, and ethical practices attached to this second group of surplus, and sub-groups of surplus animals that lie within it.

To study the differently reasoned sentiments, feelings, and practices around animals 'bred, but not used', we turn to the diverse economies approach pioneered by Gibson-Graham.[14] This approach involves paying greater attention to how emotional work shapes markets and economic practices. Our study identified how ethical and social factors lead to the subdivision of the group of animals 'bred, but not used' into those labelled as 'avoidable waste' and 'unavoidable waste'. For example, we examine where, when, and how live animal resources become 'avoidable waste', which is a category of greater ethical concern than those viewed as 'unavoidable waste'. 'Unavoidable waste' often references those animals necessarily bred as part of the process of making specialised, often GA, animals as scientific tools.

Our diverse economies approach also involves responding to the experiences and expertise of those who look after animals that become 'waste'; who undertake not only practical labour but also affective and emotional labour, for example when they cull these animals. Acknowledgement of their affective labour is not visible through current regulatory guidelines or facility budgeting practice, although it may shape happiness in the workplace and staff turn-over. We find that this affective labour is a finite and limited embodied resource. We suggest that it is not only practical labour that shapes how animals are valued, but also affective labour of scientists or animal technicians. In other words, animal values are shaped not just by how people care *for* them, but also how they care *about* them,[15] and this in turn diversifies animals' value. Following a diverse economies approach, we show that animals have value well beyond their use-value as resources for scientific research. In doing so, we illustrate the importance of extending a culture of care into research animal breeding and market activities (for more on the culture of care see Greenhough and Roe, Chapter 6).

Finally, our research highlights the importance of being alert to the changing practices and meanings associated with reduction as people carry out their work in the laboratory. This ethical principle is, in practice, relational, that is, a distributed effect of the connections between various agents, places, and events, and not solely a principle or right ascribed to an individual.[16] Conversations about avoidable and unavoidable waste in animal breeding do not take place through application of the 3Rs at any one point, but permeate the practices and infrastructures that work to actually reduce or increase animal numbers, revealing how these come to matter for people and animals across the breeding system.

Research methods

Between 2018 and 2019, we carried out interviews with 27 participants (some of whom were interviewed in pairs) working across nine UK facilities involved in the breeding, supply, and procurement of research animals. Participants included animal care technicians, facility managers, Named Veterinary Surgeons (NVSs), and researchers, and those with experience both within and outside of the commercial contract research industry. Interviews sought to understand participants' working-life experiences related to the breeding and supply of research animals. In addition, there were two short periods (five days in each location) of ethnographic participant observation with staff involved in animal care, administration, and research. All participants were given pseudonyms, using letters I to L in line with AnNex's policy (see Introduction). Research was approved by the University of Southampton's ethics committee. The transcripts were inductively coded using NVivo software by Peres, using a coding schema devised by Peres and Roe. We also draw on some documentary sources to assist in our analysis of regulation around animal breeding and supply. As we move through the sections of our analysis, we use a diverse economies approach to explore the emotional and affective resources of those tasked with handling the caring and killing of wasted animal lives. We start with a review of how feelings are attached to the terms 'avoidable' and 'unavoidable waste' by those who breed, care for, and supply research animals, before exploring how different economic and ethical values align around efficiencies in science with implications for those involved in the production of animals seen as waste.

How are avoidable and unavoidable waste defined?

The numbers included in the 2018 'bred, not used' statistics conceal an important distinction between avoidable and unavoidable surplus made by our research participants. Laura, who is in a leadership position at a research institution, makes the point that, as far as the Home Office returns go, 'there's no definition between bred and [can] not [be] used, and bred and not needed'. Thus, the category 'bred, but not used' is equated uneasily with the inference that all these animals are surplus to requirements, or waste. For Laura, defining what, in her words, she calls 'surplus' animals should mean the subtraction of those animals it was necessary to breed, but are not used:

> Surplus isn't what animals we breed and don't [include in the reporting] return, it's what we didn't need to breed. So, there's quite a big difference between what you breed and you can't use, because [they are] a consequence of a genetic cross or a consequence of a husbandry practice, and what you breed and you shouldn't have bred. That's quite different. So, in terms of what we breed and we can't use, that's unavoidable [breeding]. (Laura, research leadership, research institution, interview, 2018)

It was therefore important for our research participants to distinguish between those practices and rationales that led to the breeding of unnecessary animals, compared with breeding that is difficult to avoid. Gretchen, an NVS,[17] argues for these distinctions:

> you need to distinguish between bad planning, where people don't think it through, and they don't want to waste time, and therefore they always want to have animals on the ground ready to use, and … genetic altered strains where you … will always have a percentage of mice that are unsuitable for the research. And the only thing you can do is literally then see if the animals can be used for something else, but you cannot avoid having this wastage or surplus because to produce the suitable animals you will automatically produce the non-suitable animals. (Gretchen, NVS, university sector, interview, 2019)

It is evident how different kinds of waste are met with different situated moral judgements and actions. Breeding GA strains may involve the production of many 'unusable' animal genotypes, leaving care staff carrying an emotional burden when culling them. Yet this can be seen as *unavoidable* waste: a consequence of the biological process of strain-making that does not pose a challenge to the underlying principles of the 3Rs. There is often no obvious option for reduction here. The alternative term offered by Gretchen of 'surplus' carries less moral hazard compared with the more negative term 'wastage', yet she uses both; this perhaps conveys her own moral conundrum. The more ethically manageable term of 'surplus' follows the efforts to create an animal that is an exchangeable commodity, yet with acceptance that there will not be a use or a buyer for every genetically altered mouse life made in the process. Avoidable waste, on the other hand, was described by Gretchen as a consequence of prioritising the ready availability of animals, or the outcome of 'bad planning', leading to avoidable overproduction of animal lives as 'wastage'.

This binary framing carries all the moral and economic meanings associated with and conveyed by the term 'waste' discussed in the waste literature.[18] For example, while live animal waste can be regarded as socially unacceptable, in reality its social distinction is contingent on changing spatio-temporal relations of the animal research nexus. Animal life as waste, as an outcome of avoidable human practices, weighs heaviest on those who see how such waste could be reduced, or who deal with the consequential culls. The lack of wider public outcry at the 'bred, but not used' statistic supports this reading. Thinking with Gay Hawkins,[19] some live animals becoming known as waste is in fact the stuff of politics, conveying socio-technical changes of our time around the politics of animal research and shifting societal relations to the sentient animal. The live animal as waste is not an inert object but is driving conversations and concerns that have the potential to reshape the operations of animal research – something we are contributing to in this chapter by highlighting the distinction between avoidable and unavoidable waste. In much the same way, it matters what is in surplus to understand the ethical response to its existence. Animals as surplus and animals as waste both sit uncomfortably with those acknowledging animal lives.

We learnt from our participants what practices lead to animal 'waste'. Researchers may be primarily concerned with the availability of animals to ensure that no *time* is wasted, especially given the competitive nature of contemporary biomedical research and the perceived pressure to produce results and publish them as quickly as possible.[20] Hence, it is important to consider that, although powerful, 'waste' may be a relative term – or, at least, one applied not only to animals, but also to other resources (such as time and money) that are required to produce scientific outputs. A research culture that prioritises speed may place higher value on researchers not wasting time waiting for animals to be bred for their work, as opposed to the consequential waste of animal lives if they are treated as a resource that is always ready to hand.

Where is responsibility for reducing waste located?

Current regulatory guidance in the UK conveys efficiency as a central goal for breeding. For example, it is in the title of the 2016 guidance document published by the Home Office to share best practice on breeding GA animals.[21] Efficiency as a term speaks to prudent, careful allocation of resources to minimise wastage – not just of animals, but also of financial and labour resources. In other words, using efficiency arguments can seemingly align ethical and economic factors. For example, we heard of efforts to demonstrate that outsourcing animals from a separate breeding facility saves the user on animal husbandry costs (sometimes known as 'hotel charges'), along with a reduction in in-house surplus animals. Equally, there is a clear acknowledgement in the regulatory guidance (both at EU and UK level) that the dynamics of supply and demand for research mice can be unpredictable, and that matching the two is complex. Consequently, the regulatory guidance leaves open considerable flexibility, allowing that breeding practices are contingent on the local context and the needs of the project.[22] Responsibility for breeding, according to the guidance, falls to establishment and personal licence holders, who lead research experiments using animals. They are not only accountable for the animals that they procure, through breeding or buying, but also must ensure that the production of surplus animals is minimised.[23] It is notable that no specific statement is made in the guidance to show care for those tasked with culling surplus animals, although they could broadly be referred to in the 'local context' category.

In its guidance on APSA's operation, the Home Office provides different strategies for avoiding wastage of animals. Specifically, they recommend:

1. planning experiments with enough time to breed to requirements, and applying the principle of Reduction by designing experiments with an accurate number of animals;
2. justifying any special characteristics in the experimental population (e.g., sex and age) that may make animals unusable;
3. sharing animals and tissues with local users 'wherever feasible';
4. 'Question[ing] the need for small, often in-house breeding colonies of common strains' where they are available from larger colonies;
5. using cryopreservation;
6. keeping a record of surplus animals and reviewing reasons for overbreeding; and
7. a role for the AWERB in awareness-raising, policy-making, coordination, and rationalisation of breeding *vis a vis* users' needs.[24]

We note in these strategies how there is nothing specific about the implications of promoting a 'culture of care' in facilities as a strategy to guide policy on this issue.

The rest of this chapter addresses practices discussed in our empirical data that implement recommendations (1) and (4). We chose these because they are commonly discussed in our data, and because they most clearly speak to how breeding practices extend beyond a facility's walls to encompass other agents through which research animals are procured. These strategies involve outsourcing breeding to places where demand and supply can be better matched and where breeding expertise has been developed, such as commercial suppliers or university breeding 'cores'. Outsourcing is a widely used strategy to reduce surplus waste, though not universally applied. Moreover, advance planning and breeding on-demand are framed as ways to make breeding efficient through enabling easier management of supply and demand. Therefore, we find that outsourcing is a particularly interesting aspect to examine, as it illustrates how the ethics of making life are deeply embedded in social relations between different parts of the supply chain, such as between breeders and the researchers who are their customers (hence, we refer to 'researcher-customers'). Outsourcing can therefore point to how a culture of care might be extendable beyond the institution.

What are the implications of outsourcing breeding for ethics, economics, and expertise?

The work of commercial suppliers or large breeding facilities is increasingly at the forefront of innovations in colony management. We consider how their use is a form of outsourcing a service, and is an example of how positive affective and exchange-value may align behind an economic practice. These facilities have been managing supply and demand at greater scales and for far longer than the academic sector. Therefore, they can act as centres of expertise for breeding. For instance, the Jackson Laboratory in the US operates both as a (commercial, yet non-for-profit) supplier of mice and a provider of resources and training in colony management.[25] Indeed, as economic geographer Bronwyn Parry notes, this institution's commercial strategy uses its 'reputation for the fidelity and stability of their mouse strains' as a selling point, with intellectual property protected through trademarks rather than patenting. In other words, the Jackson Laboratory emphasises its craft in colony management, and the ability to produce a genetically 'true to form' mouse of a given Jackson strain.[26] However, underlying the strategy of outsourcing breeding is the hopeful expectation that the larger commercial or institutional breeder is responsible for, and able to, better manage supply in relation to demand in order to minimise surplus. By centralising demand in this way, it is hoped, suppliers can allocate small numbers of mice to many different buyers, and so reduce waste. The emotion of hope is important to recognise here; it is not known how well commercial supplies meet these expectations.

We found that many UK institutions were avoiding in-house small breeding colonies and instead sourcing animals from either a commercial supplier or, where available, an institutional breeding 'core'. Yet there is still notable variation in how such outsourcing is implemented between institutions and for different colonies. For instance, Leonard reports that his institution (see quote below, which we have not independently verified) increased their sourcing of animals from commercial operations some decades ago. We understand this shift as inspired by ongoing arguments that to ensure the genetic integrity of the animal model being used, it is better to purchase a specific sub-strain from suppliers. This enables users to have a degree of quality assurance that the colony of origin has been carefully managed to minimise 'genetic drift', a phenomenon whereby isolated colonies of mice can become increasingly genetically different over generations.[27] Despite these services being on offer, an interview with a Named Animal Care and Welfare Officer from the same institution as Leonard revealed how the scientific work at the institution meant that they still held and bred from GA breeding colonies for work requiring timed pregnancies. Therefore, we interpret Leonard's words as referring to the move towards buying increasingly standard, off-the-shelf models for the reasons outlined below:

> decades ago, [the University] said we can't make this work. Therefore, commercial operations make it work and I can't answer for them with regards to the amount of wastage, but any wastage for them is uneconomical so they will be very clever in the way they design their production schedules, and of course their prices and their catalogue. So we've said it's far easier to just buy these animals in, rather than take up valuable space in one's institution and then unfortunately have to kill more animals than you actually sell, if that's the right word, to your local scientific community. And so we said let's stop, so we did. (Leonard, facilities leadership, university sector, interview, 2018)

Notable here is how commercial companies' production schedules, prices, and catalogue are admired as 'very clever' ways to address waste. Yet the scale of wasted animal lives is unknown, thus it is only a hopeful supposition that it will be less. Interestingly, there is ambiguity in Leonard's comment about whether regret about killing surplus animals is attached to economic loss, ethical concern for the animals, concern for the human emotional toll, or a combination of all three. What is clearer is that Leonard seems attached to the idealised relation between producer, seller, and consumer, leading to seamless accessibility to live animal commodities that meet a researcher's specifications at the time they want, perceived as potentially workable for commercial operators, but not universities. After all, and as we previously heard from Gretchen, researcher-customers 'don't want to waste time and therefore ... always want to have animals on the ground ready to use'.

And yet, the reality perhaps can be different. As we observed in our ethnography, there can be a process of negotiation between a potential purchaser and a commercial supplier whereby surplus animals may be offered at a discounted price. Alternatively, in cases where the potential purchasers' specifications couldn't *quite* be met, a to-and-fro might occur between the two parties (via specialist administration staff) where the supplier makes an alternative 'offer' with a view to meet the demand, even if not completely fulfilling the whole specification. This begs the question: Will the experiment be designed differently if the seller is persuasive enough about the price discount? Outsourcing to commercial suppliers was therefore not simply used as a strategy for shifting waste upstream in the supply chain. We thus found an idealised view of the centralised breeder's ability to simultaneously bypass the ethical costs of breeding and provide ease of access.

Who carries the emotional costs in outsourcing?

During our fieldwork at Leonard's research institution, we witnessed orders being placed for animals to arrive the next day. This is not an unusual situation; administrative staff responsible for overseeing orders reminded users that the maximum cut-off point was midday the day before delivery. In the 1998 Laboratory Animal Science Association (LASA) Taskforce on Surplus report, 16 different reasons for surplus animals were identified.[28] 'Breeding pressures', such as an inability to match supply and demand and 'trying to meet a variable customer demand and short notice orders', was the first item on this list. Indeed, from the perspective of the supplier, customers' demand for the availability of mice with fairly tight specifications was indeed a major contributing factor for surplus, and in turn for creating distress among staff. Remembering the situation in commercial breeders in the late 1970s, Jacqueline reflected:

> actually some of these were quite distressing because they were euthanised at weaning, so you literally took them away from their mum and actually said, 'I've just weaned 700 females, I know I only ever sell 300, I'm going to kill those 400.' So it was quite a big waste. But the industry outside the commercial breeder wanted that flexibility, they wanted to be able to phone you up and say I want 300 female mice, 18–25g ... So you had to be flexible, there was no computer system, ... especially as a young technician back in 1977, it definitely felt that way to me. (Jacqueline, ex-animal technician at a commercial breeder, interview, 2018)

This story is in the past, but our fieldwork shows that customer expectations about next-day availability of mice at short notice from commercial breeders is still a regular occurrence.

It is clear from how Jacqueline tells this story that there are negative affective costs involved in handling surplus: a team of animal technicians tasked with culling will still carry the costs of surplus even if it is outsourced to a supplier. Using outsourced mouse breeders who are down a phone line and off-site runs the risk of simply shifting the affective (if not economic) costs of breeding and killing surplus mice, as paradoxically they become less visible. It also may then avoid tackling aspects of surplus production through a pan-institutional lens of a culture of care for humans and animals. We suggest that knowing how much surplus there is across a supply chain can be the basis for productive concern for doing the right thing, as exemplified by this reflection from Lydia, a senior technician at a university:

> I understand that there's a lot of surplus that's produced with these external suppliers, but the researchers are able to get the [mice] cohorts that they need. And perhaps this is because it's out of sight out of mind, perhaps, that you don't really give too much thought about it. But also that, you know, in the same breath we're not producing them here, and having to kill them or put the onus on the technicians to kill that excess stock. So it is something that we do think about, but it is happening somewhere else. (Lydia, senior technician, university sector, interview, 2019)

This quote illustrates Lydia's grappling with the spatial ethic of efforts to reduce surplus, as she ponders how achieving specific goals around surplus reduction locally can have repercussions elsewhere. Yet we also sense an ambivalence between this tentative reaching out, and the recognition that it is a matter of where it happens, not if it happens. Nonetheless, reflections such as these represent useful and important starting points for caring about breeding and surplus throughout the supply chain, and an impetus to extend the reach of the 'culture of care' through the supply chain, going beyond the institution. Bringing attention to Lydia's role and her concern, and amplifying the experiences of Jacqueline and others in her line of work, might counter attempts to organisationally 'externalise' the affective costs of surplus by locating it elsewhere.

Throughout our analysis, we have amplified the human affective labour and associated anxieties of those close to culling practice. Ultimately, we found that outsourcing breeding per se should not provide assurance that surplus will stop, unless done in tandem with cultural shifts surrounding activities elsewhere in the mouse supply economy. Using outsourced mouse breeders does shift the problem away and makes less visible the affective costs of breeding and killing surplus mice. Indeed, it is unclear the scale of surplus animals produced by commercial suppliers. For instance, the Additional Statistics point to a third of animals 'bred, not used' being wild types, which one might speculate are procured from a commercial breeder,[29] but with no further detail. During our interviews, several participants expressed a desire for greater transparency around the quantity of surplus animals culled by commercial breeders. Equally, we are aware that some surplus rodents enter the pet and zoo animal trade as food, but again the scale is unknown.

How can care extend throughout the supply chain?

We propose that recognising the relational nature of surplus and, especially, becoming attuned to the implications for animals and people elsewhere in the supply chain can change the moral economy around making life. To do so would mean to engage forms of 'caring at a distance': a form of ethical consumption[30] that goes beyond regulatory requirements or narrow readings of ethical principles. Returning to Jacqueline, she shared with us that it was in the private sector that she learned to manage colonies and match supply and demand. Eventually, she moved to the academic sector, and there made use of her expertise to encourage improvements in breeding efficiency at her university. Her experience means she is very attuned to the implications of surplus, as demonstrated by this story that took place around Christmas when people take time off work:

> somebody made a flippant comment that 'it's alright for the commercial breeders, they just breed over Christmas and keep killing them all off'. I looked at this person and said, 'So if the commercial breeders stopped breeding over Christmas will you promise to not want animals until February, and beyond?' ... 'No, no I want them when I want them.' I said, 'Okay, you can't have both worlds. You cannot take the commercials to task about breeding over Christmas and having to kill them all, because that's what happens invariably, and then tell me you want animals on 2 January' ... And I was quite offended by that comment because I thought actually you don't understand what your demand sometimes does. (Jaqueline, now facilities leadership in the university sector, interview, 2018)

This phrase 'you don't understand what your demand sometimes does' poignantly illustrates the frustrations about a lack of care for breeding and the consequences for staff tasked with culling. Although there are various useful strategies that can be deployed to minimise the making of surplus life, a more dramatic change could perhaps emerge from nurturing a deeper awareness of the affective costs on human and animal lives associated with some researcher-customer expectations. That means, of course, revisiting the customer–breeder relationship, perhaps with a view to making more visible the full panoply of costs or experiences associated with dealing with surplus, across the supply chain.

Happily, we have found some evidence of a diversifying of practices (following Gibson-Graham) within the customer–breeder relationship. The practices we learnt about – even if piecemeal or restricted to particular institutions – do, by dint of their specialist status or other factors, suggest other ways of negotiating supplier–customer relations. For instance, one approach is to adjust the expectations of researcher-customers. Take Leon's statement below. He works at a large institutional breeding facility that carries out contract research work. Albeit not exactly a commercial supplier, they must contend with similar concerns in terms of business sustainability, and again in the quote below we read quite strong feelings of a desire to take a moral stand with a customer about mouse availability, rather than his institutional supply system being framed as a supermarket:

> So now you know from my perspective it'll be really sort of 'okay, what do you need? This will be the timescale that we can deliver them on because we're breeding to your demands' rather than having them, you know, I'd hate to think of them ever being considered as almost like, you know, we're a supermarket, where you come in and it's a case of 'oh I'll have one of those, one of those and one of those'. Because I'd rather say, 'Well actually no, those aren't available yet' because we breed to requirements rather than having a big colony waiting a while. Because that's when you get stock that are, which I hate the term, surplus to requirements. (Leon, facilities leadership, large breeding facility, interview, 2018)

The strength of feeling in what Leon says suggests to us that he may have first-hand experiences, not captured in our interview, that explain his forcefulness about rejecting the breeding facility as a supermarket representation. Perhaps the capacity to stand firm on slower supply chains is aided by developments in cryopreservation, as well as other innovations in colony management systems that hold, for some interviewees, the promise of greater efficiency and transparency in the management of breeding. And yet, more than this we have found feelings and changes in practices that, although still with problems, are starting to address animal surplus by more careful procurement practices. These practices in turn also work to lessen risks – to people and to animals – of suffering unnecessarily.

Conclusion

We conclude with three points. Firstly, we have discussed what practices and feelings surround the making of 'avoidable waste' animals. We have demonstrated that these differ between roles, given the 'emotional division of labour' in animal research that increasingly separates those who bear the emotional cost of caring for research animals from those who carry out experiments and assume the economic costs of research.[31] The separation between, on the one hand, practices of husbandry and care, and on the other those of experimentation and knowledge-making, can engender a lack of awareness on the part of the animal users of the affective, emotional resources 'spent' when animals become waste. We heard from the voices in the laboratory who are most familiar with the practical details that create waste, who convey concerns over the acceptability of waste-making practices, girded by the public release of the 'bred, but not used' statistics in 2018. We found confidence in the voices of people working within the industry to speak up about their concerns and experiences, and thereby to shape the social contract, more actively, around animal research. This is perhaps related to the shift to a 'culture of care',[32] which is reducing the tolerance of animal care staff for practices that appear care-less to how humans suffer from the unnecessary making and killing of research animals. It is indicative of how the 'culture of care' is enabling people working in the industry to speak openly of their ethical concerns, and drive change in the industry's resource economy.

Secondly, and connectedly, the chapter also demonstrates the recognition of expertise perhaps previously overlooked – not only that of the animal technician as carer, which has been discussed before,[33] but also breeding expertise located either in-house or out-sourced. We illustrate these two points with reference to the surplus reduction strategy of outsourcing, which involves social and affective aspects that connect the whole supply chain, from breeding animal technologists to end-users. Stories from our research participants about this strategy suggest that it is important to take a holistic view of the supply chain and think relationally about the distribution of priorities and practices, whether around efficiencies, science-making, or caring. We have shown the importance of thinking holistically about relations within the animal research economy, through adopting a cultural economies approach. In addition, we have pointed to the important role of highlighting affective practices into our writing about animal research.[34] To this point, the culture of care concept could be usefully extended beyond an institution to be a consideration across the distributed economy of research animal supply.

Thirdly, we have shown how concerns about surplus animals extend the application of the 3Rs beyond those animals directly experimented upon. Russell and Burch's original definition of the principle of reduction does not specifically call for the overall reduction in the number of animals used in research.[35] Instead, it specifies a decrease in inhumanity or distress, according to Tannenbaum and Bennett. Notably, Russell and Burch were pre-occupied with ensuring that sufficient quantity of animals were used to ensure that the experimental results are sound.[36] However, and as Tannenbaum and Bennett also observed, newer definitions *do* take reduction to mean minimisation.[37] The UK's National Centre for 3Rs (NC3Rs) define reduction as 'methods which allow the information gathered per animal in an experiment to be maximised in order to reduce the use of additional animals',[38] which can include experimental, statistical, and breeding practices. Yet, and again according to Tannenbaum and Bennett, it is interesting that *efficiency* was already a concern for Russell and Burch, in the sense of 'generating maximum scientific or medical results from expenditures of monetary and animal resources, facilities, and personnel'.[39]

Notably, in the latest NC3Rs definition of reduction, there is a return to Russell and Burch's emphasis on maximising scientific efficiency whilst using minimal additional animals that would include the 'bred, but not used' category. However, as we learn from our study, in practice there is a complex ethical spatiality surrounding how and where efficient breeding is located and visible. Breeding efficiency competes with other resources in efficient science-making practices, such as time. Proximity between customer and breeder appears to matter in both cultivating sensitivities about waste and doing something about it. As researchers studying this topic, we see practices leading to the unnecessary breeding of surplus animals as threats to the social contract of the humane use of animals.[40] A life spent in the laboratory is not considered a 'good life' for an animal, so it is important that human benefits are realised from laboratory animal use and breeding.[41] We would therefore recommend greater regulatory interest in the details of how surplus can be avoided, and closer scrutiny and transparency about the scale and location of avoidable waste animals.

Notes

1 https://assets.publishing.service.gov.uk/government/uploads/system/uploads/attachment_data/file/901224/annual-statistics-scientific-procedures-living-animals-2019.pdf [accessed 1 February 2023].
2 'How Many Animals Are Used in Research? | NC3Rs', www.nc3rs.org.uk/how-many-animals-are-used-research [accessed 5 December 2022].
3 Gail Davies et al., 'Animal Research Nexus: A New Approach to the Connections between Science, Health and Animal Welfare', *Medical Humanities*, 2020, DOI: 10.1136/medhum-2019-011778, 1–13, (p. 504). Overall the numbers of animals used are less than the number of procedures, as animals can be used in more than one procedure during their life, which is one further factor leading to the arguments around trends in animal use.
4 Robert G. W. Kirk and Dmitriy Myelnikov, 'Governance, Expertise, and the "Culture of Care": The Changing Constitutions of Laboratory Animal Research in Britain, 1876–2000', *Studies in History and Philosophy of Science*, 93 (2022), 107–122, DOI: 10.1016/j.shpsa.2022.03.004.
5 While in many EU countries, attention to surplus animals has been raised through the revised statistical reporting required by the EU Directive 2010/63/EU, it was an area of UK public interest much earlier. In 1995, questions in Parliament regarding the number of surplus animals produced at Porton Down led the Laboratory Animal Science Association (LASA) to establish a Task Force on Surplus Animals and to produce a report on the issue.
6 LASA, *The Production and Disposition of Laboratory Rodents Surplus to the Requirements for Scientific Procedures: A Report of a LASA Task Force Meeting Held on 12th June 1998*, 1998, www.lasa.co.uk/PDF/Surplus.pdf [accessed 3 December 2020].
7 Home Office, 'Additional Statistics on Breeding and Genotyping of Animals for Scientific Procedures for 2017', 2018, www.assets.publishing.service.gov.uk/government/uploads/system/uploads/attachment_data/file/678765/Additional_data_collection_2017_guidance_v2.pdf [accessed 10 July 2018].

8 Sara Wells, 'Increasing Transparency in Animal Research Numbers', *MRC Insights* (Medical Research Council, 2018), www.mrc.ukri.org/news/blog/increasing-transparency-in-animal-research-numbers/ [accessed 2 December 2020].
9 J. K. Gibson-Graham, 'Rethinking the Economy with Thick Description and Weak Theory', *Current Anthropology*, 55.S9 (2014), S147–153, DOI: 10.1086/676646.
10 Nicky Gregson and Mike Crang, 'Materiality and Waste: Inorganic Vitality in a Networked World', Environment and Planning A: Economy and Space, 42.5 (2010), 1026–1032, DOI: 10.1068/a43176.
11 Gregson and Crang, 'Materiality and Waste', p. 1028.
12 Katherine Perlo, 'Marxism and the Underdog', *Society & Animals*, 10.3 (2002), 303–318, DOI: 10.1163/156853002320770092.
13 Neil Smith, *Uneven Development: Nature, Capital, and the Production of Space* (Verso Books, 2010), p. 59.
14 Gibson-Graham, 'Rethinking the Economy with Thick Description and Weak Theory'.
15 Emma Roe and Beth Greenhough, 'A Good Life? A Good Death? Reconciling Care and Harm in Animal Research', *Social & Cultural Geography*, 24.1 (2021), 49–66, DOI: 10.1080/14649365.2021.190 1977.
16 Beth Greenhough and Emma Roe, 'From Ethical Principles to Response-Able Practice', *Environment and Planning D: Society and Space*, 28.1 (2010), 43–45 (p. 43), DOI: 10.1068/d2706wse.
17 The NVS is responsible for, monitors, and provides advice on the health, welfare and treatment of animals.
18 Sarah A. Moore, 'Garbage Matters: Concepts in New Geographies of Waste', *Progress in Human Geography*, 36.6 (2012), 780–799, DOI: 10.1177/0309132512437077; Gregson and Crang, 'Materiality and Waste'.
19 Gay Hawkins, *The Ethics of Waste: How We Relate to Rubbish* (Rowman & Littlefield, 2006); Gay Hawkins, 'The Politics of Bottled Water', *Journal of Cultural Economy*, 2.1–2 (2009), 183–195, DOI: 10.1080/17530350903064196.
20 Daniele Fanelli et al., 'Misconduct Policies, Academic Culture and Career Stage, Not Gender or Pressures to Publish, Affect Scientific Integrity', *PLOS ONE*, 10.6 (2015), e0127556, DOI: 10.1371/journal.pone.0127556.
21 Home Office, 'Efficient Breeding of Genetically Altered Animals: Assessment Framework' (London: Home Office, 2016).
22 Home Office, *Guidance on the Operation of the Animals (Scientific Procedures) Act 1986* (London: Home Office, 2014), sec. 2.4; European

Commission, *Commission Staff Working Document Accompanying the Document 'Report from the Commission to the European Parliament and the Council on the Implementation of Directive 2010/63/EU on the Protection of Animals Used for Scientific Purposes in the Member States of the European Union'* (Brussels, 5 February), 2020, p. 41, www.ec.europa.eu/environment/chemicals/lab_animals/pdf/SWD_Implementation_report_EN.pdf [accessed 11 October 2020].

23 Home Office, *Guidance on the Operation of the Animals (Scientific Procedures) Act 1986*, sec. 2.4.

24 Home Office, *Guidance on the Operation of the Animals (Scientific Procedures) Act 1986*, sec. 2.4.

25 *The Jackson Laboratory Handbook on Genetically Standardized Mice*, ed. by Kevin Flurkey et al., 6th edn, 1st printing (Bar Harbor, Me: The Jackson Laboratory, 2009).

26 Bronwyn Parry, 'Patents and the Challenge of "Open Source" in an Emergent Biological Commons or ... the Strange Case of Betty Crocker and the Mouse', *BioSocieties*, 2019, DOI: 10.1057/s41292-019-00158-4.

27 See Gail Davies, "Mobilizing Experimental Life: Spaces of Becoming with Mutant Mice', *Theory, Culture & Society* 30.7–8 (2013), 129–153, DOI: 10.1177/026327641349628.

28 LASA, *The Production and Disposition of Laboratory Rodents Surplus to the Requirements for Scientific Procedures*, p. 2.

29 Wild-type mice may more typically be sourced from commercial suppliers as they are a less specialised (often meaning a type of GA) mouse line. Specialised, bespoke research lines can be developed by a research team who then may choose to keep that line as a live colony of that mouse line within a research institution.

30 Clive Barnett et al., 'Consuming Ethics: Articulating the Subjects and Spaces of Ethical Consumption', *Antipode*, 37.1 (2005), 23–45, DOI: 10.1111/j.0066-4812.2005.00472.x.

31 Beth Greenhough and Emma Roe, 'Attuning to Laboratory Animals and Telling Stories: Learning Animal Geography Research Skills from Animal Technologists', *Environment and Planning D: Society and Space*, 37.2 (2019), 367–384, DOI: 10.1177/0263775818807720.

32 Home Office, *Guidance on the Operation of the Animals (Scientific Procedures) Act 1986*; M. Brown, 'Creating a Culture of Care', *NC3Rs News & Blog Online*, 2014, www.nc3rs.org.uk/news/creating-culture-care [accessed 30 July 2019].

33 Roe and Greenhough, 'A Good Life? A Good Death?'; Greenhough and Roe, 'Attuning to Laboratory Animals and Telling Stories'.

34 Davies et al., 'Animal Research Nexus', p. 8.
35 William M. S. Russell and Rex L. Burch, *The Principles of Humane Experimental Technique* (London: Methuen, 1959).
36 Jerrold Tannenbaum and B. Taylor Bennett, 'Russell and Burch's 3Rs Then and Now: The Need for Clarity in Definition and Purpose', *Journal of the American Association for Laboratory Animal Science: JAALAS*, 54.2 (2015), 120–132 (p. 128).
37 Tannenbaum and Bennett, 'Russell and Burch's 3Rs Then and Now', p. 128.
38 'The 3Rs | NC3Rs' www.nc3rs.org.uk/the-3rs [accessed 20 May 2022].
39 Tannenbaum and Bennett, 'Russell and Burch's 3Rs Then and Now', p. 123.
40 Davies et al., 'Animal Research Nexus'.
41 I. Joanna Makowska and Daniel M. Weary, 'A Good Life for Laboratory Rodents?' *ILAR Journal*, DOI: 10.1093/ilar/ilaa001.

13

Commentaries on distributing expertise and accountability

Edited by Pru Hobson-West

This chapter focuses on experts and looks at how various forms of expertise are performed in the animal research nexus. It features commentaries, including two from invited respondents to this chapter of the book, and one from the chapter editor. The first commentary is by Larry Carbone, a US-based veterinarian with a particular interest in the welfare of animals in laboratories. It focuses on the complexity involved in the doing of contemporary science, and the importance of ethnographic and collaborative methods in the social scientific study of animal research. The second commentary is from Ngaire Dennison, a UK-based veterinarian with previous experience as a government inspector of animal research facilities. Drawing on the book section on expertises, one of Dennison's key contributions is to stress the need for more complete and more careful conversations on animal research. The chapter editor's closing commentary then looks across all book chapters and commentaries, with the aim of foregrounding key themes. This final piece, by Hobson-West, identifies the importance of seeing expertise as a spatial activity, and a form of action that can be studied. However, this contribution also argues for the importance of reflexivity from those claiming their own forms of academic expertise.

13.1

Outsiders on the inside: citizens and scholars in animal research

Larry Carbone

In the novel that bears his name, Martin Arrowsmith, physician and scientist, escapes the pursuit of status and money in academic medicine for an isolated country home where he can conduct his bacteriology research on his own.[1] He joins a long line of scientists – Copernicus, DaVinci, Leeuwenhoek, Mendel – working in brilliant isolation and accountable to no one and nothing except pure scientific truth. He lives a life of science, free of administrators, ethics committees, safety committees, audits, inspections, mandatory training sessions, and the 1920s version of emails.

To a modern working scientist, this image is alluring – and nonsense. Where does Arrowsmith get his reagents, his cultures, his animals, his glassware? Who is funding his nights in the lab? Will he self-publish his treatises, or subject himself to peer review and possible rejection? Will he live and die in isolation with his beautiful data? Unlike Arrowsmith's reverie, doing real science in the twenty-first century is a complicated process, and even more so when it includes animal experiments.

The four chapters in this section bring to light two aspects of modern animal research that scholars have given relatively little attention. First, doing science is a many-step, cyclical process, so much more than the fun and exciting parts of running experiments and collecting data. Second, Arrowsmith and other lone scholars might do well to bring in as many eyes and hands as they can, for every step; it takes a village to do science.

I have spent my working life as a laboratory animal veterinarian, the US equivalent of a UK Named Veterinary Surgeon (NVS) and read these chapters through that lens, and through my obligation to try as much as possible to consider how this work affects the animals who might be in my charge. I know, too, the reflexive wariness of allowing outsiders into our animal laboratories. 'It will stress the animals', we say, but really, do we want yet one more set of eyes on how we handle our animals?

Chapters 10 and 11 in this section – Palmer's 'Field folk' and Davies, Gorman and King's 'Knowledge is power', respectively – raise the challenging question of what it is to do science, including animal-based science. Biologists invite bird ringers to participate in data collection, and medical researchers invite patients' input in basic preclinical research, following models of patient engagement in clinical research. In both cases, the invited outsiders push against the limits of their invited role.

In 'Field folk', Palmer asks what roles citizen scientists should have, with ornithology as her test case, but field botany could be an interesting contrast. Should experienced non-professionals be more than data collection volunteers? Can they not be project leaders themselves? Botanists and ornithologists alike would face questions of competing expertise: if the non-professional's decades of experience at the hands-on work exceed those of the professional scientist, is something wrong with the professional–amateur hierarchy? Or is it just right? I do not care, for instance, if my brain surgeon is as good at venipuncture as an experienced phlebotomist; I still want my brain surgeon directing my care.

The case illustrates the complexity of doing science, in which the scientist's work in hypothesising and analysing, knowing important questions to ask, and knowing if they have been successfully answered is every bit as important as data collection.

It also illustrates issues of quality control in researcher training: at what point does an experienced bird ringer qualify as sufficiently experienced to do the work on their own? Who should make that determination – the scientist legally responsible for the research project, or an experienced ringer? Is this an area for shared higher-level authority in the project?

Data-collection expertise is likely as important in field botany as in field ornithology, but the welfare of sentient animals adds urgency to the ornithologist's responsibility for competency in data collection – and in all other aspects of the work. I may trust that a team of experienced ringers can competently net and ring 30,000 (!) birds in two weeks with minimal injuries or deaths (the chapter does not give those details), but surely all those birds experienced some minutes of fright in the process. Somebody should be evaluating whether this huge number is actually necessary to answer a well-defined question, or should be reduced. We need accountability for potentially harming sentient animals, and perhaps a multidisciplinary group of research participants can best do that work.

Chapter 11, 'Knowledge is power', similarly explores citizen (patient) involvement in animal research, though in this case not for data collection but for input on important questions to ask, insight into what would constitute a successful outcome, and again, what kinds of accountability to bring to the harmful use of sentient animals. This chapter also shows the importance of giving others some control over their own lives. Control is a key element in developing humane animal housing. In this piece, patients also deserve control over the amount of involvement they may have in animal experiments – do they really want to watch or even see pictures of animals modelling the disease they themselves are suffering from? And in this instance, do they want some control of how animals are used and, indeed, of how they themselves are used, as poster cases of how important animal testing is?

Academic life perpetuates the Arrowsmith mythology of the scientist as a lone warrior. They develop a hypothesis from their knowledge of their field, conduct their experiments, analyse and publish their data. Modern science, animal research included, requires collaboration, often across many labs and many experts, but a scientist's status lies not in how well they collaborate but in how many lead-author publications they can list on their CV. In these chapters, I find myself wanting to change the reward and recognition systems, and to see all the bird ringers and patient participants promoted from the Acknowledgements to co-authors. And indeed, happy the day when the animals themselves receive more recognition than a sentence or two in a manuscript's 'Materials and methods' section.

In Peres and Roe's '"Bred, but not used"' (Chapter 12) and Anderson and Hobson-West's '(Dis)placing veterinary medicine' (Chapter 9), the outsiders visit the laboratories not to help, but to observe and ask (which itself can be a kind of help). I read these chapters on alert mode; part of the job of the NVS, at least in the US, is to mediate who gets access to information about the animals.

How would I advise my institution if these social scientist outsiders came asking about how we decide how many animals to breed (and cull), or how our vet work differs to that of 'real' vets in private companion animal practice? Forgive my vigilance. In the UK, the Concordat on Openness is driving a move toward increased transparency about animals in labs. In the US, laboratory animal use is much more of a guarded secret.

The fresh eyes of the ethnographer show insiders like me that things we take for granted are worthy of examination. The ethnographer's main audience may be interested outsiders, but as an insider, I found value in the insights of these two chapters. In fact, at one point Peres and Roe point out how making some animal statistics public has more of an effect on the insiders, who already have access to that information, than on outsiders.

Peres and Roe bring an anti-capitalist sense to their look at practices around (over)breeding and then culling rodents for experiments. One need not embrace all Marxist theory to see the wisdom in their analysis. What should be prioritised in deciding on the numbers of mice to breed? Scientists' desire to efficiently get a cohort of animals on demand (and perhaps, by implication, the public and patients waiting eagerly for the fruits of their experiments)? The emotional wellbeing of animal care staff tasked with culling the millions (at least, in the US) of unwanted animals that this approach generates? The unused animals who go to the euthanasia chamber?

As an NVS, I have frequently advised scientists to 'offshore' their animal breeding to the large commercial suppliers, to avoid overbreeding in-house that would result in excessive animal culling. I was wilfully blind to the likelihood that even this more efficient approach would also result in over-production, the culling of healthy animals, and the possible emotional toll on someone else's staff. Never did I think, or feel pressure to think, more globally about overproduction as a systemic issue, whose costs we should not just shunt to some other lab where we cannot see the results of our practices. I do not believe we would (over)produce lab dogs or monkeys this way. In the US, at least, we know that it will take weeks to fill an order for dogs or months to get some monkeys. Why should we take an 'animals on demand' approach to mice? Because they are smaller? Cheaper? Less of a public concern? My thanks to Peres and Roe for opening up these questions.

As a vet, I read Anderson and Hobson-West's interviews with NVSs the most eagerly. I found I wanted more. I wanted full-on ethnography: come into the labs and watch how what NVSs *say* maps onto what you see them *do*. Part of the enculturation of an NVS – and again, my perspective is US-based – includes knowing how to present the work to outsiders, when we must. Vets, like patients, are at risk of being propaganda tools for animal research, at risk of performing the role of the humane carer and healer. But this carries risks, I believe, for the actual care the animals receive.

In their interviews, Anderson and Hobson-West hear from an NVS who brings facial pain scoring ('grimace scales') from the lab to the private clinic, to improve pain management for companion animals. In my day, I similarly told friends in private practice that their then-current standards of pain management for abdominal surgeries (i.e., spays) would require special approval in the lab. Cross-fertilisation is good. Still, I worry.

I see facial/grimace-scoring as the NVSs' valiant effort to make up for the fact that vets in labs do not (and in current scales of efficiency in laboratory work, cannot) devote the same time and effort into animal patients' pain management as companion animal practitioners. In an interview, this sounds like a great example of how vets' practices in labs are actually better than what companion animal practitioners do. I longed to see Anderson and Hobson-West accompany 'Mia' through her days to see how often she and her scientists modify their practices in response to their (probably) quick facial scoring. I wanted Peres and Roe to keep pace during the vet's quick rounds through the animal rooms, to ask why we choose a particular animal-to-vet ratio in labs that makes this quick-assessment tool necessary. In essence, as a potential object of study for the social scientists' inquiries, I am pushing back and suggesting that we research insiders could be more involved collaborators, to each other's mutual benefit, as indeed the present project models.[2] I want a voice in how you study me.

Outsiders – government inspectors, accreditation site visitors, grant reviewers, and journal editors – have long had a role shaping vets' and scientists' animal lab practices. Since the 1980s, outsiders have also had a role in animal laboratories' animal ethics committees in many countries. Perhaps this role should be expanded; for example, Niemi says that unaffiliated technical experts on animal ethics committees could be a great complement to the non-expert outsiders most countries currently require.[3] This assortment of outsiders may make for better quality science and better treatment of animals, and a better ethical balance when good data seems to require animal harm. This selection of chapters shows the value of yet another cadre of outsiders, the social scientists watching and critiquing how we do what we do, despite insiders' reflexive resistance to their presence.

13.2

Moving forward: the need for more meaningful conversations around animal research

Ngaire Dennison

I think we stand at a crossroads in animal research. The question is how we move forward. The four chapters in this section reflect a move towards more openness and accountability and the recognition of the human and animal imperatives for a culture of care.

These chapters all demonstrate that there are significant challenges in having discussions around animal research, including: whether others will see you as a 'real vet' as a Named Veterinary Surgeon (NVS) working under the Animals (Scientific Procedure) Act (ASPA); that animals may be bred and not used and so are culled without having served a scientific purpose; that asking for patient opinions may create differences rather than consensus; and the difficulties in involving (skilled) members of the public in research. However, it is critical that we face up to these challenges and find ways to have open conversations, firstly to understand the issues and concerns of different people, and then to address them.

Chapter 11, by Davies, Gorman, and King, on patient involvement in research chimed particularly with me in its discussion of 'basic and pre-clinical' research because I believe that this is an area that needs particular societal discussion. The emphasis in this chapter, on not making assumptions but on being open to asking the difficult questions, is important if we are to find solutions that address different perspectives.

For example, the decisive vote in the European Parliament (15 September 2021) to phase out the use of animals in research, regulatory testing and education has been seen as an impetus to a move towards a ban on animal testing in Europe.[4] A move away from animal use for chemical testing is proposed in the US, where the Environment Protection Agency is prioritising the development of New Approach Methods (NAMS) and plans to reduce animal testing of chemicals by 30% by 2025 and to end testing in vertebrates by 2035.[5] Amazing strides are being made in non-animal alternatives such as organs on a chip,[6] but what about the basic research where we still need to understand biological processes, which organs are affected, and how different systems interact? At these early ('blue sky') stages, how do we work out the likely value of the outputs of the research and how these can be weighed up against the harm to the animals? How do we know what valuable information will be lost if those studies are not performed? These are some of the difficult questions that we need to find ways to discuss to allow meaningful engagement of all parties.

The issue of animal wastage described in Chapter 12 by Peres and Roe is an issue that has particular resonance in the context of the Covid-19 pandemic, where research programmes were shut down rapidly as people were told to stay at home to save (human) lives. The work of humanely killing the animals that could no longer be used because planned work was stopped fell mainly to the animal technicians, causing significant moral dissonance for many (when an individual's behaviour or their cognitions are in conflict with their moral values). Moral dissonance, which can be of immense detriment to an individual's health or welfare, is a significant issue in those working in the laboratory animal field.[7] It can be part of the reason that researchers, technicians, and NVSs can have difficulty in discussing their work with colleagues, families, or members of the public.

As an NVS myself, I found Chapter 9 by Anderson and Hobson-West of particular interest. This chapter discusses how the NVS role appears to exist 'at the intersection of a wide and conflicting set of expectations and assumptions', which is a statement that really resonates with me. As an NVS, I am proud of the work that I and my colleagues do. I always tell anyone who asks me the details of what I do, although it often takes quite a long time to explain – but I understand the reluctance of some others working in my field to do the same. It can be hard not to feel 'judged'. ASPA is an enabling Act – it is there to allow procedures to be done that may cause pain, suffering, distress or lasting harm to an animal for a scientific purpose. While my job is to advise on health, welfare, and refining procedures (and so to advocate for the animals), I deal daily with the moral dissonance of knowing that animals under my care may be harmed. I joined the profession 'to help animals' (however naïve that may sound) and the oath I took when I qualified as a vet was:

> I PROMISE AND SOLEMNLY DECLARE that I will pursue the work of my profession with integrity and accept my responsibilities to the public, my clients, the profession and the Royal College of Veterinary Surgeons, and that, ABOVE ALL, my constant endeavour will be to ensure the health and welfare of animals committed to my care.[8]

In my role as an NVS supporting research programmes, I often feel there is a conflict between my oath and my day-to-day work. Ensuring that people understand what I do and the passion I have to try and ensure the best possible welfare for the animals under my care – a concern shared by my colleagues, in particular the animal technicians – is important to me, even if those people disagree with research using animals.

The critical message from all four chapters is summarised for me in Chapter 11 where Davies, Gorman, and King write, 'We suggest it is important to continue to develop these conversations in more complete and careful ways'. They go on to say that it is essential to enable, 'everyone to have more open, honest, and meaningful conversations around animal research' (p. 283). This is true for the areas described in the chapters but is much more widely applicable for the use of animals in research.

The challenges of having care-ful conversations are both reduced and exacerbated in this web-based age of social media. Such platforms can reach huge numbers of people, but misinformation can be spread rapidly and can be difficult to counter. Individuals may be reluctant to voice their opinions and risk being put in the public pillory by a (likely) vocal minority of 'trolls' with strong views. We need to find ways out of the 'polarisation cycle'[9] to have these conversations and the social science approaches in these chapters may begin to give insight into ways that we might start these discussions.

13.3

Experts and expertise in researching animal research

Pru Hobson-West

Experts and expertise represent a key area for research in science and technology studies (STS). In a critical review, Grundmann goes back to the beginning, noting that the word expert 'has its root in the Latin verb *experiri*, to try. An *expertus* is someone who is experienced, has risked and endured something, is proven and tested'. However, in critiquing contemporary STS scholarship, Grundmann argues that 'experts are not only characterised by their embodiment of skills and experience. What matters is their performance'.[10] Or, to put it another way, expertise is a form of action. From a social scientific perspective, the animal research nexus (see Introduction) thus cannot be fully understood without being attuned to the question of who experts are, what they do, and, crucially, how they perform their various forms of accountability and expertise. In this brief commentary, I therefore look across the chapters included in this section from my perspective as an academic, and identify how these social scientific themes relate to key issues in animal research.

In terms of who animal research experts are, one option is to simply read this from the legislation. In the UK, the Animal (Scientific Procedures) Act (ASPA) articulates this in detail: indeed, the named roles themselves arguably make this particularly explicit – named individuals are assumed to have particular expertise or accountability. On one level, then, where institutional responsibility lies is 'relatively easy to identify'.[11] However, the chapters in this section of the book draw on qualitative empirical research to reveal the ways key actors are involved in navigating regulation's interpretive flexibility. By taking a nexus type approach, the studies are thus able to connect governance processes with the 'lived, embodied experience of those with regulatory responsibilities',[12] or, alternatively, to foreground those excluded from the legislation. As I will now briefly highlight, the analyses presented in the chapters also reveal the way in which expertise is spatially managed and narratively performed.

In characterising the relationship between regulation and expertise, Palmer (Chapter 10) points to the ways in which actors actively navigate regulation on a daily basis. In their case, citizen scientists 'are simultaneously excluded from positions of authority, yet also able to directly contribute to animal care and research practice'. As Palmer shows, 'expertise comes in multiple forms'. Indeed, the distinction between expert and non-expert is one long troubled by STS. Similarly, the chapter by Davies, Gorman, and King (Chapter 11) focuses on a form of expertise traditionally considered to be outside the boundaries of the regulation, namely patienthood. However, the authors track the increasing pressure for upstream patient engagement and the influence of other pieces of regulation (such as the 2001 Research Governance Framework for Health and Social Care), which, in turn, can create expectations for the way animal research is done. In Chapter 12, Peres and Roe also point to the importance of recognising expertise previously overlooked, in their case the work of animal technicians and animal breeders.

The chapter by Anderson and Hobson-West (Chapter 9) suggests that the constant navigation of legislation is also a key activity for the NVS. In the UK, NVS staff describe walking a careful line between ASPA and the Veterinary Surgeons Act 1966. This case helps highlight the difference between expertise and professional expertise, the latter being formally certified by a professional organisation. As the commentary by Dennison (in this chapter) reminds us, these distinctions are not technical or semantic matters. In their own biography as an NVS they report pride in their work, yet simultaneously identify a 'moral dissonance' toward the intentional harming of animals that ASPA allows, with the professional veterinary oath mandating a prioritisation of animal health and welfare.

The second theme that cuts across this book section concerns *expertise and space*, and the role of discursive labour in the creation of important insider/outside boundaries. For example, Anderson and Hobson-West show the way in which NVSs position their laboratory work as important but construct an intra-professional hierarchy whereby the NVS is 'not a *real* vet'. In terms of expertise, their analysis suggests that 'it is not just levels or depths of expertise that are made to matter, but rather where legitimate expertise comes from'. The insider/outsider distinction is also critical to the chapter by Palmer on Places Other than Licensed Establishments, where their physical space as 'outside' the laboratory is made to matter. The chapter is also valuable in showing that it is not just professionals that engage in this demarcation or 'boundary-work'.[13] Sticking with space and place, in their chapter on supply chains, Peres and Roe show the way in which moving ethics 'outside' has significant implications. This is picked up by Carbone (in this chapter), who welcomes the provocation to consider the emotional consequences of 'off-shoring' animal breeding. Overall then, this body of work confirms the spatial dynamics of expertise in the animal research domain.

The third theme I would like to highlight is the way in which expertise is performed in its narration. One of the benefits of qualitative work is that, rather than looking at job descriptions, we can explore how expertise is described or demonstrated and, crucially, attune ourselves to the struggles or inherent contradictions involved in performing expertise. Indeed, the Animal Research Nexus Programme has focused on writing and story-telling of various kinds.[14] The book continues this trend using various narrative-led research methods: Palmer (Chapter 10) draws on interviews, informal conversations, and participant observation to identify barriers to wider participation in science; Davies, Gorman, and King (Chapter 11) draw on a large and varied pool of interviews and analyse the diverse ways in which patients relate expertise to their lived and embodied experience; Peres and Roe (Chapter 12) draw on interview data and two periods of ethnographic observation to identify practices that lead to categories of animal waste and the affect that this creates for those handling 'surplus'; and in Anderson and Hobson-West's chapter (Chapter 9), veterinarians share their personal career narratives as 'refugees'[15] from or between the clinic and the laboratory.

This brief review leads me to the thorny question of how we narrate our own expertise as academics.

In a previous publication, Davies et al. argued that a nexus approach prompts us to 'apply the same principles of contingency, coproduction and reflexivity to our own role in the nexus, as we argue for in relation to our research data'.[16] Indeed, I regard such reflexivity as an ethical responsibility, and an essential part of what it means to work in areas of techno-scientific controversy, although I also accept that there is no single way to 'do reflexivity'.[17] However, in working out *how* to perform reflexivity in the arena of animal research, perhaps we can learn from the tone of the commentaries published here. Dennison acknowledges that there are 'significant challenges' in having 'care-ful conversations' around animal research, but they are still keen to articulate the 'difficult questions', and reflect on their own positionality; while Carbone reflects on their own experience as a veterinarian, recalling professional concern about letting 'outsiders' into the animal laboratory.

Speaking personally, participation in the Animal Research Nexus Programme has encouraged me to recognise, and perhaps make more explicit, the multiple identities I have in relation to animal research. At the very least, I am an academic researcher who studies animal research, a teacher of potential future animal researchers, a citizen with an interest in scientific governance, as well as a patient who sometimes consumes medicine developed using animals. In reflecting on these multiple hats, I am trying to acknowledge the 'performative essence of identity and the relationality of sense-making'.[18]

However, I am also aware of the irony that in the act of narrating this identity list here, others may judge that I am making a personal credibility claim – put bluntly, I am now performing the role of the so-called 'good, reflexive academic expert'. Carbone argues here that 'doing real science in the twenty-first century is a complicated process, and even more so when it includes animal experiments'. This claim also holds true for social science. It seems, then, that researching animal research does not only involve careful study of the way others exhibit expertise and accountability, nor even the creation of new opportunities for participation; as argued elsewhere in this volume, we also need to be reflexive about our own role (see Chapter 15 by McGlacken and Hobson-West). To help achieve this, perhaps it is worth going back to the definition of expert detailed at the start of this commentary: an expert has 'experienced, has risked and endured something'. What those of us who study the topic may risk is that our writing will never fully do justice to the complexity of the nexus, the messiness of our own identity positions, nor the experiences of the millions of research animals for whose lives and deaths we are somehow accountable.

Notes

1 Sinclair Lewis, *Arrowsmith*, revised edition, 1998 (New York: Signet Classics, 1925).
2 Gail Davies et al., 'Developing a Collaborative Agenda for Humanities and Social Scientific Research on Laboratory Animal Science and Welfare', *PLOS ONE*, 11.7 (18 July 2016), 1–12, DOI: 10.1371/journal.pone.0158791.
3 Steven M. Niemi, *Notes in the Category of C: Reflections on Laboratory Animal Care and Use* (London: Academic Press, 2017).
4 Goda Naujokaityte, 'Parliament Votes through Demand for Faster Phase Out of Animal Testing in Research', *Science Business*, 2021, https://sciencebusiness.net/news/parliament-votes-through-demand-faster-phase-out-animal-testing-research [accessed 28 November 2022].
5 US Environment Protection Agency, 'EPA New Approach Methods Work Plan: Reducing Use of Vertebrate Animals in Chemical Testing', 2020, www.epa.gov/chemical-research/epa-new-approach-methods-work-plan-reducing-use-vertebrate-animals-chemical [accessed 28 November 2022].
6 MPSCoRE Working Group, 'Organs-on-a-Chip: Report and Recommendations', 2021, www.nc3rs.org.uk/sites/default/files/2021-10/NC3Rs%20Organ%20on%20a%20Chip%20report%20and%20recommendations.pdf [accessed 28 November 2022].
7 Megan R. LaFollette et al., 'Laboratory Animal Welfare Meets Human Welfare: A Cross-Sectional Study of Professional Quality of Life, Including Compassion Fatigue in Laboratory Animal Personnel', *Frontiers in Veterinary Science*, 7.114 (2020), DOI: 10.3389/fvets.2020.00114.
8 Royal College of Veterinary Surgeons, 'Code of Professional Conduct for Veterinary Surgeons', 2012, www.rcvs.org.uk/setting-standards/

advice-and-guidance/code-of-professional-conduct-for-veterinary-surgeons/ [accessed 9 December 2021].

9 Alison Goldsworthy et al., *Poles Apart: Why People Turn Against Each Other, and How to Bring Them Together* (London: Random House Business, 2021), p. 55.

10 Reiner Grundmann, 'The Problem of Expertise in Knowledge Societies', *Minerva*, 55.1 (2017), 25–48 (p. 27), DOI: 10.1007/s11024-016-9308-7.

11 Gail Davies, 'Locating the "Culture Wars" in Laboratory Animal Research: National Constitutions and Global Competition', *Studies in History and Philosophy of Science Part A*, 89 (2021), 177–187 (p. 179), DOI: 10.1016/j.shpsa.2021.08.010.

12 Gail Davies et al., 'Animal Research Nexus: A New Approach to the Connections between Science, Health and Animal Welfare', *Medical Humanities*, 46.4 (2020), 499–511 (p. 505), DOI: 10.1136/medhum-2019-011778.

13 Thomas F. Gieryn, 'Boundary-Work and the Demarcation of Science from Non-Science: Strains and Interests in Professional Ideologies of Scientists', *American Sociological Review*, 48.6 (1983), 781–795, DOI: 10.2307/2095325.

14 Animal Research Nexus, 'Writing the Nexus. AnNex Newsletter Issue 7 – Autumn 2021', 2021, www.animalresearchnexus.org/sites/default/files/publications/other-files/Writing%20the%20Nexus_0.pdf [accessed 29 March 2022].

15 Alistair Anderson and Pru Hobson-West, '"Refugees from Practice"? Exploring Why Some Vets Move from the Clinic to the Laboratory', *Veterinary Record*, 190.1 (2021), e773, DOI: 10.1002/vetr.773.

16 Davies et al., 'Animal Research Nexus', p. 505.

17 Jonathan Ives and Michael Dunn, 'Who's Arguing? A Call for Reflexivity in Bioethics', *Bioethics*, 24 (2010), 256–265, DOI: 10.1111/j.1467-8519.2010.01809.x.

18 Renelle McGlacken and Pru Hobson-West, 'Critiquing Imaginaries of "the Public" in UK Dialogue around Animal Research: Insights from the Mass Observation Project', *Studies in History and Philosophy of Science*, 91 (2022), 280–287 (p. 286), DOI: 10.1016/j.shpsa.2021.12.009.

Part IV

Experimenting with openness and engagement

14

The Mouse Exchange: what can curiosity-driven public engagement activities contribute to dialogues about animal research?

Emma Roe, Sara Peres, and Bentley Crudgington

Introduction

The field of animal research has long been considered a controversial public engagement topic.[1] For many decades there has been a culture of fear from the activities of anti-vivisectionists,[2] though currently this threat is at a relatively low level.[3] In turn, this fear has created a culture of secrecy about practices of animal experimentation.[4] Steps to tackle the culture of secrecy, and to fulfil the ideals of transparent scientific experimentation, have encouraged the drive towards greater openness about animal experimentation. However, this goal has not yet been fully realised,[5] in part because efforts to engage with publics typically take the form of a knowledge-deficit approach in which experts convey information to publics under the assumption that greater knowledge will lead to greater support. Furthermore, negative feelings towards animal research – which are not only the legacy of animal rights campaigns and activism,[6] but also reflect a wider distrust in science and expertise[7] – have restricted publics' willingness to engage with animal research. A new approach to public engagements with animal research is therefore needed to achieve improved openness.

This chapter introduces The Mouse Exchange (MX), a public engagement activity that we propose helps address some of these issues. The MX was designed as an activity that contributed to, enriched, and explored findings from Roe and Peres's research into the supply, breeding, and biobanking of research animals (see Chapter 12).[8] This research lent itself to creating an engagement activity that broadened the focus beyond the animals used in experimental procedures to include all animals whose lives are involved with UK research, and to also consider their lives from breeding to culling or euthanasia. Initially, we were keen to understand what questions or concerns people involved in animal research had about their area of work that would help us forge research questions. Consequently, we took inspiration from participatory research methodologies in a more-than-human world,[9] and held a workshop at the Conference of the Institute of Animal Technologists in 2018. There, we gathered the thoughts of animal technologists – directly involved in animal breeding and care – about their understandings and experiences of the animal journey, to hear what they felt was important for them to know more about, and what they wished for others to know. This event helped us to frame, along with subsequent data collection, where the management of the production and use of animals continues to pose a challenge for animal research and those working with the industry.

The result of this development was a public engagement activity that approaches openness by shining a light on the making and supply of animals used in research, rather than on the experiment itself. Another key point of difference with traditional public engagement is that rather than provide information, we create a space where participants can experience becoming curious and creative. Through creative processes and informal conversations, the MX activity manages negative feelings like distrust, suspicion, and anxiety, which can be associated with animal research. Instead, the MX seeks to convey something of the emotional and ethical landscapes experienced by those working within animal research, which are complex and contingent.[10] For example, the MX aims to offer participants a mixture of: scientific curiosity; the rewards from caring for animals; the consequence of being moved by animal harm; and hopes from medical research that uses animals. Together, participants and facilitators feel a way into this animal research nexus, primarily through the activities of their hands and fingers, working with familiar objects, repurposed.

The chapter begins by describing efforts to achieve openness in animal research, including via public engagement with, and criticism of, this work. Drawing on these critiques, we conclude that openness is often narrowly framed, selective, and follows a problematic knowledge-deficit approach. We then set the scene by describing what the MX activity involves in practice, before discussing how we have been inspired by other performance art, and how MX facilitators generate talk during the activity. We then move to discuss particular aspects of the infrastructure around an MX Workshop – the biobank, the passport, the ear-punch, the Infinity Box, and the caging system – and what these can add to the activity. We conclude by reflecting on how the MX helps move beyond deficit-model approaches to public engagement around animal research, instead offering a valuable creative, curiosity-driven, participant-led approach.

Secrecy, caution, and public communication styles

The Concordat on Openness in Animal Research[11] has impressed openness as an important tool to develop public communication about animal research, but the dimensions of animal research that have been communicated have been selective.[12] The trajectory of animal research in the UK is one of institutional moves away from secrecy and towards 'openness': transparency is utilised to achieve social legitimacy.[13] Holmberg and Ideland's[14] study of public engagement strategies used by animal research institutions in Sweden identified two main problems, which we propose also apply to some extent in the UK. Firstly, there is a kind of 'selective openness',[15] where individuals feel they should manage the disclosure of their work. This finding echoes the argument, made by Wendy Jarrett of Understanding Animal Research, that some researchers involved in discussions in advance of the inauguration of the Concordat on Openness in 2014 were fearful that providing information to 'the public' would expose them to attacks from animal rights extremists.[16] Hence, the idea of doing public engagement can invoke fear and reticence from researchers. The history of controversy and the binary, adversarial nature of previous public communication could put members of the animal research community off from doing public engagement where they may be less in control of setting the terms and direction of conversations.

Secondly, Holmberg and Ideland find that those involved in animal research in Sweden align themselves with a deficit-model approach to public engagement, where the public feature as 'uninformed and misled'.[17] In practice, communication around animal research privileges a 'scientific witnessing'[18] over other possible ways of framing communication, which in a sense legitimates this controversial activity and prevents other ways of knowing and making sense of animal research. Recent evidence in the shape of a survey of the attitudes of Swiss animal researchers towards public engagement[19] backs this argument: Roten found that 72% thought that 'their main task was to educate the public', and 80% believed that 'if the public were more educated, it would be more positive toward science'. Conversely, 33% agreed or strongly agreed that 'the public may lack scientific knowledge, but it possesses a lot of relevant common sense and good judgement', and 19% similarly agreed that 'the public should have a say in the regulation of scientific activities and applications'.[20] Altogether, then, the concept of openness has begun to be performed with limitations to its scope and potential because of the wider context.

In this context, efforts to be more open about animal research have been limited in important ways. Communications aimed at achieving openness often take the form of institutional websites, newspaper articles, or media stories about the potential benefits to humans of a new scientific finding that involved animals.[21] While these communications counter the images and narratives about animal harms disseminated by animal rights organisations, they do not linger on what it was like for the animal taking part in the experiment, or how they live and are cared for in the laboratory. Consequently, these communications do little to eschew public anxiety about the experiment itself. A growing number of animal research institutions do, however, aim to give greater insight into life within the animal facility, via websites[22] and YouTube videos.[23] Yet the type of information that is conveyed is often carefully curated. Barney Reed from the RSPCA has been a vocal critic of the oblique and inaccurate language used in institutional websites, which implies that standards of animal welfare are of no concern.[24]

Furthermore, efforts to engage publics may focus too narrowly on ethical decision-making. Engaging laypeople in animal ethics committees[25] is a weak attempt to engage publics in animal research; these laypeople require expertise to be able to understand how to scrutinise paperwork, and there is no mechanism for the few people who hold these roles to disseminate their understanding more widely. Yet this approach is still advocated.[26] For example, this route is emphasised by a 2019 report on a two-day international expert workshop about how the current governance practices regarding openness and transparency could lead to better public engagement.[27]

Building on these criticisms, we argue that this view of openness as an element of ethical, democratic research culture has propagated a narrow vision of what one could be open about in relation to animal research. Openness efforts tend to focus on ethical decision-making, rather than the more mundane task of putting ethics into practice, including across the breeding, supply, and care for laboratory animals, which are the focus in the MX. Focusing on these other elements of work in the research laboratory also serves to counter the risk of controversy associated with focusing on animals' experience in the experiment. Rather, in the MX we make the research mouse the primary object of interest, putting the science and the experiment into the background. Through the tasks that participants are invited to perform, the MX puts people into the shoes of those who are practically involved in caring for animals used in research, such as administrators, breeders, and animal care technicians. The MX also provides the opportunity, should participants wish to take it, to learn more about the wider social world around animal research beyond the experiment, which may be difficult to find out about. Furthermore, the structure of the MX, with its privileging of participant-led, un-scripted dialogue, enables questions to arise that may otherwise be excluded if researchers (or facilitators!) hold all the power in determining the content.

In summary, our approach carefully tackles some of the ongoing challenges about engaging publics in animal research. It encourages a different culture of communication around animal research, a primary goal of the Animal Research Nexus Programme (AnNex). It proposes an alternative to the historical tendency for communication to be framed as a debate between supporting and opposing 'sides'.[28] We now describe what the MX activity involves.

Encountering The Mouse Exchange

> 'Have you ever wondered where lab mice come from?'[29]
> 'Do you want to make a mouse?'[30]

On the table are threads, scissors, and homely fabrics. Using these materials, we invite people to make a type of mouse that most of us have never seen: a research mouse. Through the collective work of the MX participants, research mice become day-time residents in unlikely places (see Figure 14.1).

Although details have changed over time as we iteratively developed the MX, the fundamentals have always been a set of tables with sewing equipment in the middle. At different events we have added our own enrichment for the mice to the activity. Beginning with cardboard houses and lab-grade treats sourced from colleagues, we progressed to try different things.

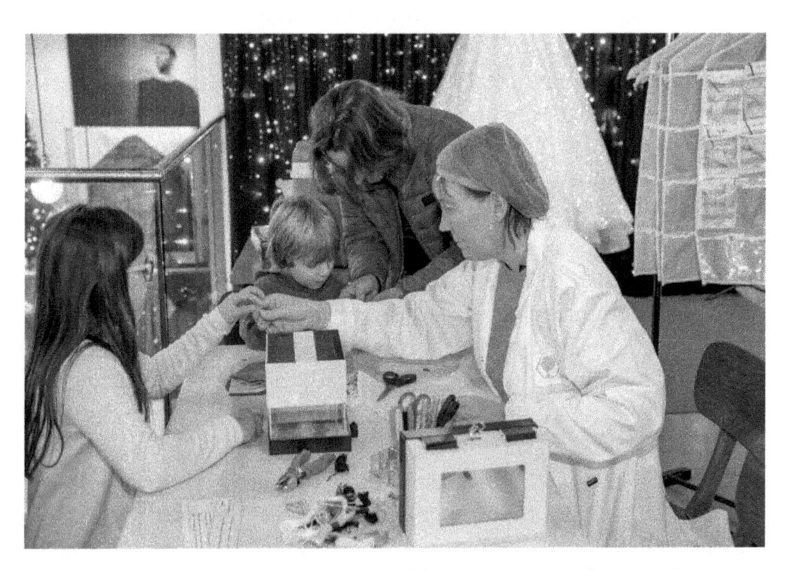

Figure 14.1 MX materials on tabletop (Source and copyright: University of Southampton).

We created a large mouse nest (Figure 14.2) on one occasion; on another, we wore lab coats. We have been in different contexts: university seminar rooms, academic conferences, science festivals, museums.

Allow us to set the scene. On a Saturday morning in November 2019, we are in a theatre. A table stands prepared with needlecraft materials – thread, felt of various colours, needles – and small white, black, or pink stuffed felt objects in the shape of pasties waiting to be picked up.

These felt objects represent the bodies of three of the most popular research mice strains: C57Black6 (black), BalbC (white), and nude (pink) mice (Figure 14.3). Passers-by and pre-registered

Figure 14.2 Mouse nest that mice and their makers can play with (Source and copyright: Bentley Crudgington).

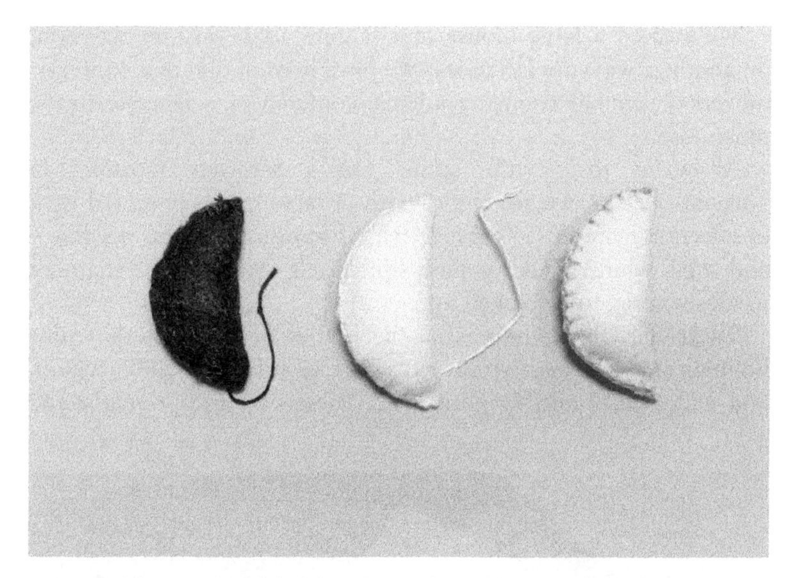

Figure 14.3 Stitched felt-fabric pasty-shaped bodies of the three strains of mice (Source and copyright: Bentley Crudgington).

participants are invited to take a seat to make a mouse. We brief the maker-participants on the basic plan to turn the felt pasty, now in their hands, into a laboratory mouse.

The invitation to make a mouse enables participants to play with developing a relationship with a research animal by making it. The process of making meanings and generating feelings begins with finding oneself caught up in a process of crafting and creativity as fingers and hands are put to work with fabric, needles, and thread. Participants get to experience the mouse taking shape by sewing on felt circles as ears, embroidering on a nose and eyes, sewing on strands of whiskers to the face and trimming them with scissors to a certain length and shape, and finally sewing on more thread, sometimes plaited, to make a tail. Within these moments, there is a shift at some point from thinking solely about how to do it, to feelings for and about the developing animal form as eyes, whiskers, tail, and ears are added. Through the act of creating, a sense of belonging and care develops for the thing forming in one's hands.

Crafting, like kneading bread,[31] slows down time, as it forces mindful attention to be brought to what one is doing with one's hands, emptying the mind of other things. Consequently, crafting invites unstructured conversations and remarks. A conversation when crafting is necessarily broken by reaching for a pair of scissors, asking how to do something, and yet all the while the making of a research mouse is intimate to the participant's and facilitator's fingers. This physical engagement is important; it creates another means of relating to the issue that is not grounded in thought alone. Participants sit with us without knowing what the mouse will be; the meanings and values are built into the process. It is designed to enable all to take part, not just those with a pre-existing view. In the later versions of the MX, what kind of mouse they will make, and the origin of the mouse, is decided through the selection of a chance card; will they make a brand-new mutant, or a mouse held in a biorepository as frozen embryos?

At the outset the MX was devised by thinking with Roe and Buser's 'becoming ecological citizen' methodology (BEC).[32] This approach was developed with artist Dr Paul Hurley,[33] who is part of the MX team and has contributed to its creative life. The original application of BEC was to food, which made it possible to engage very directly with its materiality, and to draw on participants' extensive embodied and other knowledges of food. In contrast, the MX had the added difficulty of overcoming the absence of actual laboratory mice, and existing knowledge of mice in research. The methodology involves taking two steps towards creating a space for engagement. Firstly, it involves 'facilitating sensory experiences that enable the agential qualities of [object of concern] to shape knowledge making'.[34] This is why the rich sensory experience of sewing a colourful, soft felt mouse is at the centre of the activity. Secondly, it aims 'to create a space where people can perform, or relate differently, in unusual manners to [the object of concern]'.[35] In this case, our object of concern is research mice. Hence, we turn to creating a comfortable space of curiosity coupled with a crafting activity to invite people into a relationship with research mice.

The crafting materials scattered on the table afford the transformation of curiosity into the deeply political act of creating a body and advocating for an animal's care through its documentation via the mouse passport. In the next section we discuss the nature and style of generating table talk through the activity and how MX facilitators can work against the knowledge-deficit model.

Facilitating The Mouse Exchange

The MX toolkit[36] provides guidance about the materials needed to set up an MX event and a guide for facilitators about how to set up the space and hold conversations that meet the aims of the MX. We encourage others to download the toolkit and to run their own MX. In this section we discuss in depth the thinking behind why the MX is facilitated in the way it is, and what type of participatory experience we are aiming for.

Participants assemble and take a seat at a table with felt-crafting materials laid out. At the table, hierarchies and power imbalances can be set aside. This conception of the table is informed by queer feminist performance artist Lois Weaver's work *The Long Table*, an 'experimental open public forum that is a hybrid performance-installation-roundtable-discussion-dinner-party designed to facilitate dialogue through the gathering together of people'.[37] It empowers by literally tabling or gathering excluded and included voices to speak on difficult and conflicting subjects in their own terms; certain responses or degrees of knowledge are not discarded as unacceptable. The table bridges the private domestic setting and the connected, yet distant, public domain. The MX table is a hospitable place for experimenting around what might legitimately be discussed. Participants come and go from it, and with that, experiences and viewpoints both overlap and differ.

As people assemble and take a seat, we are curious about what might have attracted these people to come to the table. Is it the appeal of making something of their own? The appeal of sewing, or an activity that can occupy their children and offer a rest for a little while? Or is it part of an educational experience, and if so, how will their expectations of learning be challenged by how the workshop is structured?

The materials laid out on the table are both familiar materials and unrecognisable objects – soft, felt pasties, a 'thing'[38] that can be assembled into becoming a research mouse. The pasties are taken into the hands of participants who are invited to add ears and a face to make a soft-toy research mouse. The conversation is initiated at this point; the facilitator does not have to direct but can let those gathered around the table make a mouse with their hands while engaging in curious chatter. Requests to please pass the scissors, thread, or the felt support conviviality and contact between participants. The practice of sewing together encourages an atmosphere with a mixture of talk about how to do something (sew a mouse nose) alongside reflections on the object taking shape in their hands, and issues related to the origins and lives of research mice. This approach works with Deleuze's statement that 'something in the world forces us to think', to talk, to feel.[39] Conversations happen simultaneously alongside the making. In this way the MX is not an output, but a process.

Instructions of what to do are conveyed through conversation and observation as there are no written instructions. Facilitators have an important role in shaping the conversation, ambiance, feelings, and thinking around the MX table by recognising that thought and talk are generated in relation to the context. To this end, participants should feel empowered to lead their own knowledge-creation, co-authoring content with those around them. Beyond a couple of opening questions from the facilitator – 'have you ever met a mouse?' or 'where do you think laboratory mice come from?' – participants should always take the lead when exploring the topic and directing what is, and what is not, spoken about. This approach is in opposition to a traditional public engagement audience member who is cast as needing to learn something to address their knowledge deficit.[40] Collectively, conversations do not crystallise but keep changing, since the outputs of knowledge and meaning making processes are not decided in advance. What takes place in the MX is the outcome of the work that participants and facilitators collectively perform and consent to.

Space is made for talk, but it does not have to be forthcoming. Facilitators are asked to let go of the need to control responses; rather, they should focus on supporting participants to not only become makers of mice but to make space for those who choose to take the opportunity to reflect and learn. Consequently, it is important to not talk at participants but respond to expressions of curiosity. The learning objectives are not a set of facts, figures, or ethical guidelines but just to sew a mouse and complete its passport. Participants are not in an audience on receive mode. Indeed, participants may opt to stay in their established habits of thinking, perhaps knowing little about research animal origins and lives, and it is then up to them whether they ask questions or share thoughts about the materials or anything else as they make a mouse. Participants may talk about the lives of mice they have encountered and move to the life of a research mouse if and when they are comfortable.

Finding ways to collectively enable participants to hold an interest in the lives of laboratory animals is important because it allows them to engage in the process of, rather than the products from, animals being used in science. In the MX, we achieve this by avoiding the head-on discussion of animal research as an 'object-issue'[41] to instead bring attention to the research mouse, its origins, and how to care for it. Animal research as an 'object-issue' has a rich patterning of emotions, disruptions, disagreements, and agreements that extend around it. Acknowledging this, the MX registers a need to support the inclusion of the multiplicity of affective, emotional, rational, historical, and ethical engagements that participants may have with animal research. Indeed, we found that making something tactile and tangible enables feelings towards the animal to develop; it equally allows issues and themes that arose in conversation to evolve into more meaningful concerns, rather than abstracted facts. Experience has shown that holding a felt research animal in one's hands, and completing a mouse passport, has taken participants through a process that can change their stake in animal research. Feelings surrounding the life experience of the mouse can be made and expressed that exceed objective facts about the research animal industry, its animal welfare standards, and binding ethical principles.

Along the way, we have, so far, held conversations that include: cats bringing in mice; imaginations of wild mice being captured for research (a common assumption when participants had never been asked to think about the subject before); identification of mouse models for a son's genetic condition; childhood memories of needlework; and feeling squeamish about mice. We have observed children playing with the felt mice in a display area decorated with enrichment used in lab mice cages, and we have learnt a lot about how to manage conversations to give people confidence to articulate what they are thinking and feeling, nurturing their attachment to the mouse they are making. Each time it is different, inflected by the occasion and the people who pass by. As a process, it is prone to evolution and mutation. We have identified new needs and developed other experiments that are adapted to new situations and new questions. The MX will also, we hope, find new tables, new participants, as a different set of facilitators learn how to set up and run their own MX, using the MX toolkit.[42]

The Mouse Exchange Infrastructure

We found curiosity was inspired by the infrastructure around making mice as we built more into the MX performance. For example, people were curious about the different colours of the mice and the different ways that mice could be sourced from a biobank or live colony, and what these different practices involved.

The biobank was enrolled into the performance when participants were invited to collect cold, 'frozen' embryos (in reality, the soft pasty-shaped mouse forms are kept next to a plastic ice pack) from our coolbox (the kind that is more commonly taken on a picnic), which performed as our biobank. Waiting for the embryos to warm up on the pretend heat pad creates a pause where we can begin to talk about the way mouse strains circulate within the animal research community, and to think about freezing down strains as a form of animal welfare.

Once the mouse is made, we ask mouse makers to complete a passport (Figure 14.4) for their mouse. This encourages participants to articulate and reflect on their participation and helps both to make meanings more concrete and to evaluate their experience. Drawing on the structure and purpose of the mouse passports recommended for genetically altered animals,[43] which commonly travel with mice, the passport enables makers to detail their mouse's specific care needs, which vary from strain to strain.

Figure 14.4 An early form of mouse passport and the mouse (Source and copyright: Bentley Crudgington).

The passport has evolved over time in the project from a simple, two-question prompt to a more detailed form that includes data like the mouse's name, place, and date of birth as well as information about their phenotype (what they look like), character, and instructions for their care, including who they want to care for them (Figure 14.5). We also ask makers about their hopes and expectations about the future of their mouse. What, then, does it look like to care for this mouse, now and in the future? Questions about character and phenotype, who cares for them, and what needs they have continue the work of thinking about individual animals' sentience and welfare. By collecting the passports that participants have created alongside their mice, we are putting together an archive that not only preserves the mice, but also – in a small, creative way – records the makers' engagement with their mice as beings to be cared for as well as scientific resources. Together, the mouse and the passport make up the primary units of the MX and embody something of the experience after the event is complete.

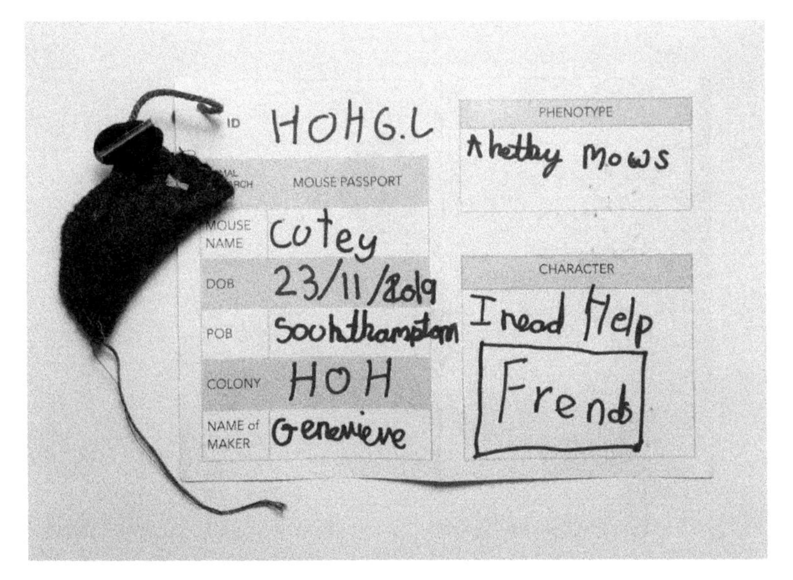

Figure 14.5 The current form of mouse passport (Source and copyright: Bentley Crudgington).

The ear-punch, a hole-punch stationery tool re-purposed, clips the ear for genotyping (to find out what genes are in this particular mouse) (Figure 14.6). Participants get to encounter how the work of caring for a research mouse does not allow for easy refusal of inflicting harm on them. Genotyping the mouse, and ensuring that the mouse and its passport remain together, depend on the maker's willingness to punch a hole in the mouse ear. The meaning of this act, however, is constructed by each maker, and is as unique as their mouse.

As the activity draws to a close – the mouse is completed and the passport has been written – participants are invited to put their mouse inside the Infinity Box and see their mouse multiplied into many animals. The Infinity Box uses mirrors and a light to show multiple ongoing reflections of their mouse (Figure 14.7). They watch as their single unique mouse becomes many indistinguishable mice; it provides a way to exemplify the sorts of practical ethical questions that may not necessarily be covered by regulation, but which arise when making life. As they watch their single unique mouse become a lineage, we unfold the idea of care from

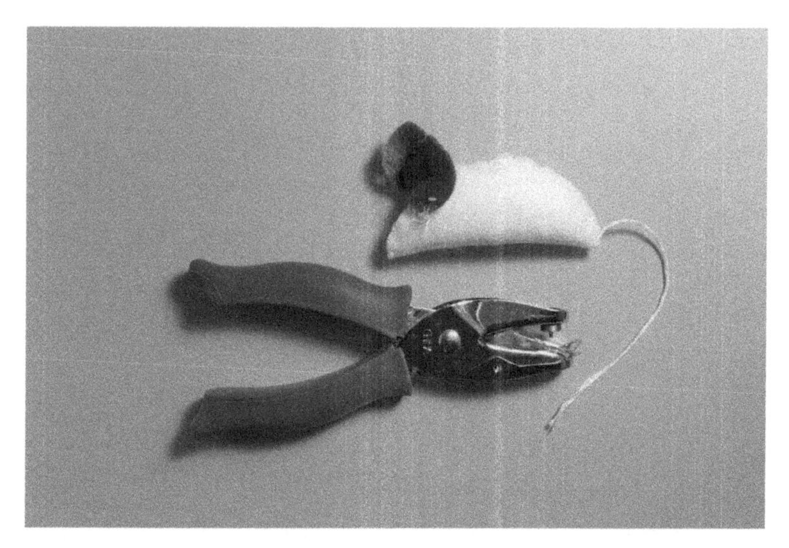

Figure 14.6 Hole-punch as ear-punch (Source and copyright: Bentley Crudgington).

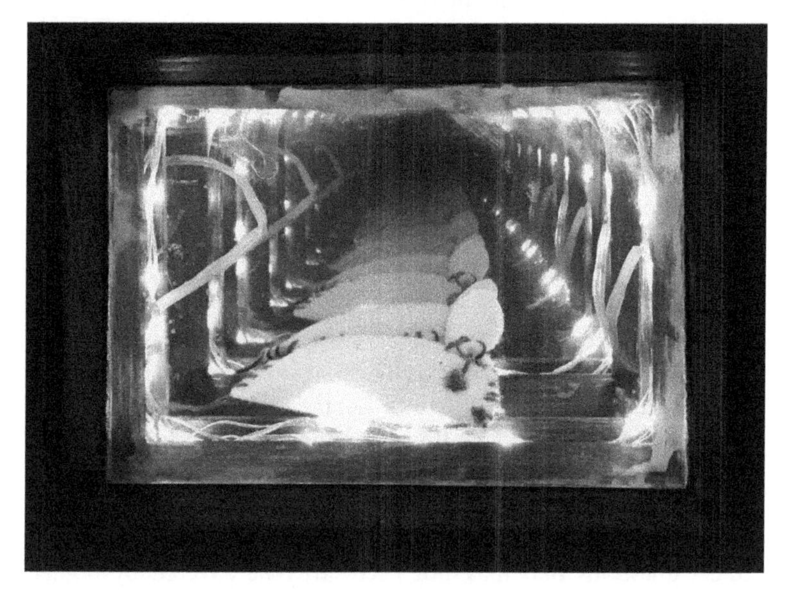

Figure 14.7 Image of a mouse in the Infinity Box (Source and copyright: Bentley Crudgington).

the individual to a colony. Mice in science are often lost in the multitude.[44] With the Infinity Box, we can start to ask about caring for an individual versus caring at scale. What kinds of responsibilities and care does a maker have for the future life of a creation? The Infinity Box multiplies our relationships over space and time.

Whereas the original MX offered people the opportunity to take their mouse home, swap it, or leave it in our care, later variants of the MX asked participants to leave them behind.

The completed mouse and passport are then left to live in our 'caging system', where rows of different felt mice start to stack up in a hanging shoe-storage rack (Figure 14.8). These are then photographed and added to a virtual online archive. This isn't easy: the invitation to leave the mouse behind brings a sense of loss and anxiety, of not knowing and trusting someone else with the responsibility to care. The anxiety is ours, as much as the makers': what does it mean for us to be custodians of these mice, and responsible for what happens in their future? These kinds of questions remain live as we seek to create a future for the MX as a toolkit for others to use.

Collectively, these items help to demonstrate how the assemblage that is animal research requires caring maintenance, and mandates specific forms of care for the animals intimately entangled with UK science. Moreover, they do so in a way that both recognises the animals' individual sentience, but also the collective expression of being one of a multitude,[45] one example amongst an expanding variety of mouse strains (Figure 14.9).

Conclusion

Public engagement suffers many of the same critiques as other form of participatory art: that it is trivial or trivialises[46] in that the science is dumbed down. However, as Helen Molesworth[47] argues, as one moves away from attempting to fill any knowledge deficit, public engagement can create situations where questions that ask, 'what if the world was different?' can be articulated and responded to. The MX crafting table can be read as a performance 'art form' for legitimising public discourse, locating itself as 'a conduit through

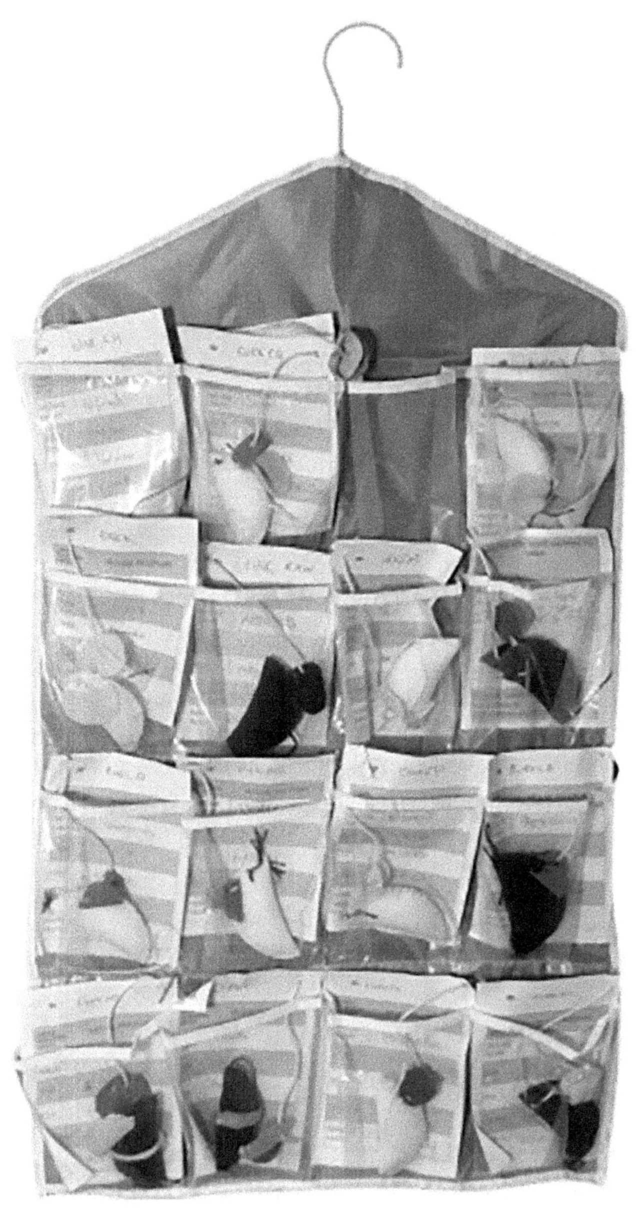

Figure 14.8 The shoe-holder as mouse caging system (Source and copyright: Bentley Crudgington).

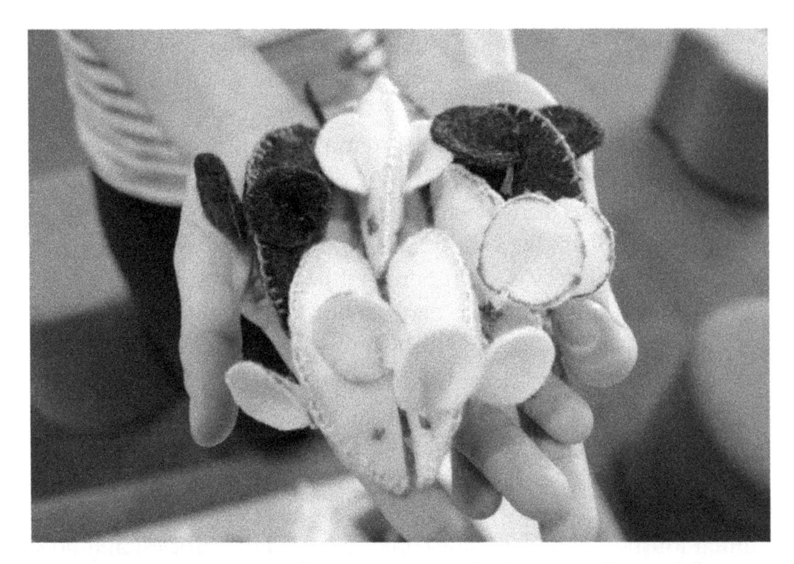

Figure 14.9 Variety of mice strains made (Source and copyright: Paul Hurley).

which to enter ideas into public discussions'.[48] This is a deliberate move away from many forms of animal research public engagement, where roles are predefined, and facts are neatly packaged up to be taken away. Instead, the MX is a feast for situated, spontaneous knowledge-making and transfer. In doing this, it continues the work of combining science and technology studies with performance and theatre studies.[49]

The MX adds something unique to forms of public engagement on animal research. It offers participants a space where they can engage with animal research on their own terms. Here, hesitancy about taking part because of associations of anxiety or controversy in relation to animal research can be allayed, and participants are invited to engage with curiosity and care. In this way, we have devised a method that seeks to engage participants' curiosity and create space for a range of perspectives including affective, emotional, and embodied engagements with animal research. The MX positions the participant not as audience or information deliberator and ethical decider, but as maker and carer through being provided with crafting materials and engaged in curiosity-led conversations that inspire care. We found that creating a space for curiosity, whilst crafting a felt mouse with stitches and thread in a curated performance space, could achieve a different way to support reflective conversations about the objects and animal supply chain infrastructure of animal research.

In part because it involved communicating research findings on the supply, biobanking, and rehoming of laboratory animals, the MX has worked to be open about what happens in spaces and practices around animal research that are not structured around ethical decision-making, but rather ethics in practice of making and caring for animal life. It never asks whether a particular experiment, or any experiments, should be carried out, but places participants as carers, which can be read as ethics in practice. While the introduction to the event does not presume an interest in any of these elements, the MX brings a focus on the practices that perform laboratory animal journeys, how animals come to be in laboratories, and how they are cared for while they are there. Doing public engagement that focuses on these practices is therefore a way to engage beyond either the science or the (deliberative) ethics of the use of animals in research. Rather, it shows how participants become immersed in a process of animal care through the practices of making, listening, and conversing around a table, which changes the way publics talk about ethics associated with using animals in research.

Finally, the activity's name gestures at the idea that making mice is a collective endeavour. It was also chosen to highlight other highly social aspects of this practice. Firstly, mice can be exchanged between collaborators. Secondly, the 'exchange' refers to the dialogues that we wish to encourage between participants. The MX is an experiment that we construct together. We have created this process together, and we offer it up for curiosity and future evolution.

Notes

1 Carmen M. McLeod, *Assuaging Fears of Monstrousness: UK and Swiss Initiatives to Open up Animal Laboratory Research*, *Science and the Politics of Openness* (Manchester: Manchester University Press,

2018), pp. 55–74, DOI: 10.7765/9781526106476.00010; Elisabeth H. Ormandy et al., 'Animal Research, Accountability, Openness and Public Engagement: Report from an International Expert Forum', *Animals*, 9.9 (2019), 622, DOI: 10.3390/ani9090622.

2 Mary Ann Elston, 'Attacking the Foundations of Modern Medicine? Anti-Vivisection Protest and Medical Science', in *Challenging Medicine*, ed. by Jonathan Gabe et al. (London: Routledge, 1994), pp. 196–219, DOI: 10.4324/9780203864319.

3 Unknown, *Animal Rights Extremism*, AnimalRightsExtremism.Info, www.animalrightsextremism.info/ [accessed 25 August 2021].

4 Tora Holmberg and Malin Ideland, 'Secrets and Lies: "Selective Openness" in the Apparatus of Animal Experimentation', *Public Understanding of Science*, 21.3 (2012), 354–368, DOI: 10.1177/0963662510372584.

5 Holmberg and Ideland, 'Secrets and Lies'.

6 Niklas Hansson and Kerstin Jacobsson, 'Learning to Be Affected: Subjectivity, Sense, and Sensibility in Animal Rights Activism', *Society & Animals*, 22.3 (2014), 262–288, DOI: 10.1163/15685306-12341327.

7 Harry M. Collins and Robert Evans, *Rethinking Expertise* (Chicago, IL: University of Chicago Press, 2007).

8 See also Sara Peres and Emma Roe, 'Laboratory Animal Strain Mobilities: Handling with Care for Animal Sentience and Biosecurity', *History and Philosophy of the Life Sciences*, 44.3 (2022), 30, DOI: 10.1007/s40656-022-00510-1.

9 *Participatory Research in More-than-Human Worlds*, ed. by Michelle Bastian et al., Routledge Studies in Human Geography, 67 (London; New York: Routledge, 2017).

10 Gail Davies et al., 'Animal Research Nexus: A New Approach to the Connections between Science, Health and Animal Welfare', *Medical Humanities*, 46.4 (2020), 499–511, DOI: 10.1136/medhum-2019-011778.

11 Understanding Animal Research, *Concordat on Openness on Animal Research in the UK*, 2014, https://concordatopenness.org.uk/ [accessed 26 August 2020]; Wendy Jarrett, 'The Concordat on Openness and its Benefits to Animal Research', *Laboratory Animals*, 45.6 (2016), 201–202.

12 Holmberg and Ideland, 'Secrets and Lies'.

13 Carmen McLeod and Pru Hobson-West, 'Opening up Animal Research and Science–Society Relations? A Thematic Analysis of Transparency Discourses in the United Kingdom', *Public Understanding of Science*, 25.7 (2016), 791–806, DOI: 10.1177/0963662515586320.

14 Holmberg and Ideland, 'Secrets and Lies'.
15 Holmberg and Ideland, 'Secrets and Lies'.
16 Jarrett, 'The Concordat on Openness and its Benefits to Animal Research', p. 202.
17 Holmberg and Ideland, 'Secrets and Lies', p. 366.
18 Joan Leach, 'Scientific Witness, Testimony, and Mediation', in *Media Witnessing: Testimony in the Age of Mass Communication*, ed. by Paul Frosh and Amit Pinchevski (London: Palgrave Macmillan, 2009), pp. 182–197, DOI: 10.1057/9780230235762_9.
19 Fabienne Crettaz von Roten, 'Animal Experimentation and Society: Scientists' Motivations, Incentives, and Barriers toward Public Outreach and Engagement Activities', *Society & Animals*, 30:5–6, 524–546, DOI: 10.1163/15685306-00001674.
20 Roten, 'Animal Experimentation and Society', p. 10.
21 *Covid: Immune Therapy from Llamas Shows Promise*, BBC News, 22 September 2021, section Science & Environment, www.bbc.com/news/science-environment-58628689 [accessed 22 September 2021].
22 *Animal Research*, University of Cambridge, 2013, www.cam.ac.uk/research/research-at-cambridge/animal-research [accessed 23 September 2021].
23 Agenda Life Sciences, *Discovery Park Vlog Series | Episode Two | Caring for Animals and Staff*, 2021, www.youtube.com/watch?v=bzlHBn3eXS4 [accessed 27 September 2021].
24 Barney Reed, 'Talking about Harms'. Presentation at Concordat on Openness Signatory meeting, May 2018. https://concordatopenness.org.uk/wp-content/uploads/2018/11/Talking-about-harms-Barney-Reed.pdf [accessed 14 June 2023].
25 Helena Röcklinsberg, 'Lay Persons Involvement and Public Interest. Ethical Assessment in Animal Ethics Committees in Sweden. The Swedish Transition Process of the EU Directive 2010/63/EU with Regard to Harm–Benefit Analysis in Animal Ethics Committees', 2015, 4.
26 Ormandy et al., 'Animal Research, Accountability, Openness and Public Engagement'.
27 Ormandy et al., 'Animal Research, Accountability, Openness and Public Engagement'.
28 Davies et al., 'Animal Research Nexus'.
29 Question from the facilitator opens the conversation as people sew a mouse.
30 Question on publicity material that invites participation in MX.
31 Emma Roe and Michael Buser, 'Becoming Ecological Citizens: Connecting People through Performance Art, Food Matter and

Practices', *Cultural Geographies*, 23.4 (2016), 581–598, DOI: 10.1177/1474474015624243.

32 Roe and Buser, 'Becoming Ecological Citizens'.

33 Paul Hurley, *Engaging Creativity and Research*, www.drpaulhurley.com [accessed 27 September 2021].

34 Roe and Buser, 'Becoming Ecological Citizens', p. 591.

35 Roe and Buser, 'Becoming Ecological Citizens', p. 593.

36 Bentley Crudgington et al., 'The Mouse Exchange Toolkit', *An Animal Research Nexus Report*, www.animalresearchnexus.org/publications/mouse-exchange-toolkit [accessed 15 December 2021].

37 *Long Table on Feminism Documentation*, LADA Live Art Development Agency, www.thisisliveart.co.uk/projects/long-table-on-feminism-documentation/ [accessed 23 September 2021].

38 Bruno Latour, 'Why Has Critique Run out of Steam? From Matters of Fact to Matters of Concern', *Critical Inquiry*, 30.2 (2004), 225–248, DOI: 10.1086/421123.

39 Gilles Deleuze, *Difference and Repetition* (New York: Columbia University Press, 1994), p. 139.

40 *Interdisciplinarity: Reconfigurations of the Social and Natural Sciences*, ed. by Andrew Barry and Georgina Born (London: Routledge, 2014).

41 B. Latour, *From Realpolitik to Dingpolitik – An Introduction to Making Things Public*, www.bruno-latour.fr/node/208 [accessed 23 September 2021], p. 5.

42 Crudgington et al., 'The Mouse Exchange Toolkit'.

43 D. J. Wells et al., 'Assessing the Welfare of Genetically Altered Mice', *Laboratory Animals*, 40.2 (2006), 111–114, DOI: 10.1258/002367706776318971.

44 G. Davies, 'Caring for the Multiple and the Multitude: Assembling Animal Welfare and Enabling Ethical Critique', *Environment and Planning D: Society and Space*, 30.4 (2012), 623–638, DOI: 10.1068/d3211.

45 Davies, 'Caring for the Multiple and the Multitude'.

46 Philip Moriarty, 'Confronting the Critics of Public Engagement', *Times Higher Education*, 2016, www.timeshighereducation.com/blog/confronting-critics-public-engagement [accessed 23 September 2021].

47 Helen Molesworth, 'House Work and Art Work', *October*, 92 (2000), 71–97, DOI: 10.2307/779234, pp. 95–96.

48 Molesworth, 'House Work and Art Work'.

49 Saul E. Halfon et al., 'Engaging Theatre, Activating Publics: Theory and Practice of a Performance on Darwin', *Engaging Science, Technology, and Society*, 6 (2020), 255–282, DOI: 10.17351/ests2020.403.

15

Labelling medicines as developed using animals? Opening up the topic of animal research

Renelle McGlacken and Pru Hobson-West

Introduction

The use of animals in science remains socio-ethically contentious and the UK has a particular history of debate and dissensus around the matter.[1] Public dialogue around animal research is often framed as polarised, with scientists pitted against animal rights activists and 'the public' constructed as a neutral entity[2] occupying the middle ground. Science–society relations around the topic have historically been strained, with communication on the topic regarded as a source of personal and institutional risk.[3] However, responding to characterisations of the bioscience sector as secretive, the UK advocacy organisation Understanding Animal Research (UAR) launched the *Concordat on Openness on Animal Research* in 2014. As a high-level agreement around openness on animal use in the bioscience sector, the Concordat was intended to mark a significant shift in science–society communication. Therefore, while not a public engagement initiative in itself, the Concordat aims to 'alert the research community to the risks of secrecy, and provide support for greater transparency, highlighting its benefits for science, animal welfare and communications'.[4] To date, 127 UK organisations, such as universities, commercial organisations and funders have signed up. Similar initiatives have occurred in other countries, such as the 2010 Basel Declaration (now Animal Research Tomorrow) in Switzerland, which amongst other principles around good scientific and ethical practice, called for signatories to 'Promote the dialogue concerning animal welfare in research by transparent and fact-based communications to the public'.[5] More recently, transparency agreements around scientific animal use have proliferated across the EU[6] and beyond, with New Zealand launching their own 'Openness Agreement on Animal Research and Teaching'[7] in 2021.

Although the UK Concordat has helped to make certain information on bioscientific animal use more accessible, its impact on fostering in-depth science–society dialogue arguably remains limited. In practice, current policy initiatives or enactments of openness around animal research have largely treated openness as an end in itself, with the release of annual statistics on national animal use, establishment of institutional webpages that outline how and why animals are being used, publication of non-technical summaries of licensed project applications, creation of virtual tours of certain research facilities,[8] and so on, often being presented as a fulfilment of the bioscience community's contribution to public discourse around animal research. However, McLeod argues that such initiatives 'are unlikely to be enough on their own to build greater trust between the AR [animal research] community and wider society'.[9]

The language used in the Concordat is itself illuminating. For example, the third commitment states that organisations will 'be proactive in providing opportunities for the public to find out about research using animals'.[10] This wording implies a *unidirectional* approach to communication (from 'laboratory animal science' to the wider 'public'). By contrast, in recent decades, social scientists and research funders have been keen to promote more reciprocal relations between science and society, based on an understanding of engagement as a two-way process. To give just one example, concepts like RRI[11] (Responsible Research and Innovation) demand continual investment in *reciprocal* dialogue to ensure that societal expectations of science and technology are valued and have upstream impact.

On the topic of animal research, Davies et al. previously ran an agenda-setting exercise with social scientists and animal research stakeholders to identify priorities for future research. The eventual list was substantial but included the following provocation: 'where are the opportunities for greater and meaningful public and stakeholder engagement in the policy and practices of animal research?' This question fed into our shared desire, as part of the Animal Research Nexus Programme (AnNex), to think carefully about how we might create or facilitate opportunities for shared discussion about animal research. What unites our work across AnNex is thus a commitment to 'open up' the topic, allowing conversations to develop without being tethered to specific areas such as regulation or traditional ethical positions. In other words, we are keen to 'resist the pressure to resolve dispute by removing difference, instead seeking to involve different voices and evolve the terms of the debate'.[12] As McLeod has contended, 'The biggest challenge to opening up AR remains how to provide these opportunities and spaces where there can truly be inclusive, co-productive and safe conversations'.[13]

In 2018, the team at the University of Nottingham began informal discussions about how we might translate this AnNex-wide aim into a more specific engagement activity that connected with our priority research theme of how publics interact with animal research. On this subject, the Nottingham strand of AnNex has studied writing on the topic of animal research collected by the Mass Observation Project (MOP) a national life-writing project working to record 'everyday life in Britain'.[14] We eventually decided to focus our engagement activity on the topic of medicine consumption, which we believed could attend to the complicated nuances of animal research, whilst also enabling participants to engage with it through their own 'worlds of relevance'.[15] As well as offering a way of centring conversations on animal research in lived experiences, medicine consumption also arguably represents a moment of direct complicity in scientific animal use and thus helps draw us closer to a topic that might be felt as abstract or as a polarised issue of right and wrong, us and them.[16]

Our interest in medicine consumption was also indirectly influenced by existing sociological studies of 'mundane' or 'everyday' technologies,[17] and by our work with the MOP and its emphasis on the 'ordinary'.[18] We were also encouraged by the success of the innovative The Mouse Exchange project (see Chapter 14, this volume), which foregrounds the everyday in material ways. Overall, we were attracted to the idea of seeing animal research not as a distant practice that only exists inside the closed walls of the laboratory, or less commonly 'in the field',[19] and as a matter for certified experts such as scientists, veterinarians, or ethicists, but rather as a practice entangled with medicine consumption in everyday spaces such as the home, shops, or pharmacies.

In reviewing the literature around animal research and medicine consumption, the published evidence suggests that this topic has been considered by some stakeholders, yet only in a very specific way. In brief, some have called for the labelling of medicines as 'tested on animals' to make the use of animals in medicine production clearer to publics and patients. In the next section, we introduce this policy proposal of 'labelling medicines as tested on animals', highlighting the relative lack of academic consideration of this topic in contrast to other domains such as food. Intrigued by this policy proposal, we ultimately decided to design a pilot engagement activity based on this idea. The following section describes our aims, and the 'methodology' we developed to structure our activities on this topic. We conclude by critically reflecting on our activities.

Labelling medicines

In 2013, Lord Professor Robert Winston, a prominent UK scientist with a significant media profile, proposed the Medicinal Labelling Bill.[20] The Bill, which was debated in House of Lords,[21] included provision to label medicines to make it clear that they had been developed and produced through animal use. If passed, the Bill would require medicine packaging to include a 'prominent' statement along the lines of 'This pharmaceutical product has only been made possible by the use of research in animals'.[22] Winston claimed that such 'legislation would demonstrate the widespread nature of the need for animal research and increase recognition of its importance for medical progress'.[23] Hence, they worked with the assumption that increasing the visibility of animal research via making the link with medicines clearer would lead to increased public support for animal research. Similar thinking has spurred initiatives such as UAR's Wellcome Trust-supported distribution of their leaflet 'Where do medicines come from?' in GP waiting rooms across the UK, aiming to 'outline the vital role animal research plays in the development of medicines and vaccines'.[24] Ultimately, the Medicinal Labelling Bill was not passed, with members of the

House expressing concerns around the impact on patient uptake of prescribed medications, misplaced expectations of what a label itself can achieve, and the regulation of medicine packaging being governed at the European level.[25]

Nevertheless, similar policy suggestions have continued to be made by other commentators in the field. For instance, researcher Khoo argues that if medicines were labelled as tested on animals, this may have advantages in 'opening healthcare to ethical consumerism through labelling and disclosure can [...] be seen as a means of respecting the ethical views of a significant minority of the population',[26] with such labelling assumed as enabling choice. More recently, in 2021, including labels on medicine packets was advocated by the US-based organisation Speaking of Research (SoR).[27] Writing for SoR, in a post discussing the UK's National Health Service (NHS) messaging around COVID-19 vaccines (i.e., advising that the vaccines contain no animal products), Bennett et al. advocate for 'a new way of labelling medications and vaccines'. Similar to other recommendations for such medicine labelling, the stated aim behind their proposal is to 'provide consumers with accurate and full information about the roles that animal research and testing have played in vaccines, medicines, treatments, and medical devices'.

Despite these policy proposals and the unease articulated in the House of Lords debate, we could not find published empirical work, or publicly available evidence of wider discussion about what the societal impact of labelling medicines in this way might be. The suggestion of such labelling occurred in Dignon's study of the views of healthcare professionals on animal research, with the idea associated with uncertainty about its potential consumer impact.[28] However, more broadly, what is lacking is published explorations of what publics, scientists, or other stakeholders would make of the suggestion to label medicines as 'tested on animals', or any critical exploration of the tacit assumptions underlying such a proposal. This led us to narrow our nascent interest in medicines, animal research, and the everyday, to the more specific question of labelling medicines as 'tested on animals'. Going back to the provocation put forward by Davies et al.,[12] our aim was to think through how this topic could be translated into a new opportunity for 'greater and meaningful public and stakeholder engagement in the policy and practices of animal research'.

Despite the lack of published research or engagement directly on medicine labelling, we were aware of existing research on food labelling which could potentially offer a useful point of comparison. Indeed, food labels can work to make animals and certain aspects of their lives (and deaths) visible or invisible. Going further, Evans and Miele point out that food labelling 'can function to "make animals matter", not only in a cognitive fashion but also in the very literal sense of intervening with the material and sensual qualities of animal food consumption practices'.[29] More than simply raising awareness, labels can thus work to 'gain political buy-in from diverse actors; provide a visual cue for consumers about ethical production practices and contribute to building new norms about production and consumption which may lead to further policy action'.[30] However, food labelling practices have also been criticised for 'greenwashing' and 'welfare washing'[31] production processes and reducing demands for systemic change to a single issue. Furthermore, where labels are used as part of industry self-regulation, Parker argues that labelling 'represents the further privatization of regulation, rather than a broader redistribution of power'.[32] Strategic use of labelling can thus be used to present processes and conditions of food production in sanitised and appealing ways whilst maintaining problematic practices. For example, discussing the 'Welfare Quality®' logo intended to represent an integration of animal welfare in the food quality chain, Cole argues that 'the instrumental relationship with non-human animals remains unchanged, in which their bodies remain as exploited commodities, but the relationship is instead presented as beneficent and caring'.[33] Scholars have also focused on the way in which labels can function as 'boundary-objects', 'serving as two-way translators or mediators' between social worlds, and have underlined the need to appreciate the complex way in which 'science and society are mixed up in mundane decisions and everyday encounters'.[34]

In summary, such insights from the food domain demonstrate that any kind of product labelling is neither straightforward nor neutral, and this is perhaps especially so when it concerns the lives and bodies of animals. In the case of medicine labelling, such labelling practices are likely to be further problematised by the lack of choice between medicines with and without animal use. Despite these potential concerns, we were struck by the recurrence of calls for medicines to be labelled as 'tested on animals', by various voices. We therefore decided to experiment with the idea by creating a format that allowed questions to be asked about what such a label might look like and what its societal consequences might be.

Experimenting with labelling

Planning the activity

Having decided to focus on medicine labelling, we discussed ideas about possible formats for a new engagement activity with an interdisciplinary mix of colleagues at the University of Nottingham and the wider AnNex team. After much discussion, we established the arc of an activity with six phases. The objective was to develop a small-scale activity that did not require specialist equipment or technology, or a particular kind of space.

Phase one involved us introducing ourselves as social scientists, explaining the format of the workshop, and contextualising the proposal made by some stakeholders to label medicines as developed and produced through animal use. However, from the start, we were keen to ensure that this opening phase was brief, avoided the style of an academic lecture, and also avoided casting ourselves in the role of expert or educator on the topic of animal research. In other words, we were mindful not to frame the activity as an opportunity for publics to 'find out' about animal research. However, we did seek to provide some broader context for the labelling exercise and decided to do this by introducing images of well-known existing logos that feature in food labelling, such as the Red Tractor[35] logo or the 'RSPCA Assured' logo from the Royal Society for the Prevention of Cruelty to Animals. In doing so, we were mindful of Michael and Brown's assertion that citizens (including ourselves as researchers) often understand unfamiliar technologies by reference to the familiar. As they claim, 'whatever the precise way people "grasp" a new phenomenon, it will involve drawing upon certain familiar cultural resources'.[36]

Phase two of the activity involved a facilitated session where participants were invited to use pen and paper to design their own label that could appear on a medicine packet. In actually drawing a label (rather than just discussing the idea of labelling), we hoped to foster an informal atmosphere in which there were no 'right' answers. Furthermore, we hoped that by discussing a hypothetical future or 'what might be' rather than the 'now' or 'what *is*', we might also help participants to feel comfortable in thinking and talking about this sensitive and challenging topic. As Dunne and Raby contend in discussing design as a mode of inquiry, such speculative engagements 'usually take the form of scenarios, often starting with a what-if question, and are intended to open up spaces of debate and discussion; therefore, they are by necessity provocative, intentionally simplified, and fictional'.[37] It is this ability of speculation and the hypothetical that we wanted to harness for discussions around animal research.

The third and most important phase was the group discussion. Here, we invited reflections on how it felt to design a label that informs about the use of animals in medicine development and production; what thoughts, problems, or questions this activity raised; and what sort of societal impacts were envisaged if their label were rolled out. To be clear, we were interested in the labels that were created, but our key aim was to foster discussion. In this, our intentions align with Ratto, who suggests that when 'using a shared process of making as a common space for experimentation' – the objects made (in our case, hypothetical labels) – 'are not intended to be displayed and to speak for themselves' but 'are considered a means to an end, and achieve value though the act of shared construction, joint conversation, and reflection'.[38]

Phase four entailed a brief summary of our wider research programme to give further context to our activity. Aiming to avoid the format of a traditional academic presentation, we did not include this in the opening of the activity, but nevertheless felt obliged to give participants an opportunity to challenge us on our wider activities or contact us at a future date.

Phase five invited participants to provide feedback on the session and (optionally) leave their label with us. Although we did not know whether participants would choose to do this, we applied for and received ethical approval from the School of Sociology and Social Policy at the University of Nottingham (number 2021–012). Finally, phase six of the activity consisted of discussion amongst our research team as part of a refinement process, giving feedback to colleagues on how the event ran, and identifying possibilities for improvement.

Running the activity

As noted, the desire to create new opportunities for creative engagement on animal research is shared across the AnNex team. However, the authors also work in an academic context that encourages staff to develop and embrace local opportunities for public and stakeholder engagement. This push for engagement can create pressures for staff, as they balance competing institutional incentives, scarce resources such as time, and personal and professional values.[39] In our case, we originally designed the six phases of the labelling engagement activity described above with the aim of trialling it via existing local initiatives.

In its first running, the labels activity was piloted at Pint of Science in May 2019 with 60 participants. This sold-out event was held in a music venue in Nottingham city centre and was open to all. Pint of Science is an international science festival and 'grassroots non-profit organisation' with a mission to develop 'a space where audiences are engaged with research; where walls are broken down and everyone has the opportunity to share their thoughts, questions, and ideas'.[40] Our Pint of Science event generated a good level of discussion despite the relatively large number of people. However, we are aware that the Pint of Science festival can be criticised for adopting a one-way engagement approach to public engagement. While the aims and scope are different, this criticism of being unidirectional is similar to criticisms of the Concordat outlined earlier, where research is assumed to be communicated by experts 'to' the public,[41] reinforcing notions of 'the public' as lacking relevant knowledge by concentrating on education rather than shared learning.[42] Such an approach conflicts with the National Co-ordinating Centre for Public Engagement's definition of public engagement as 'a two-way process, involving interaction and listening, with the goal of generating mutual benefit'.[43] This mode of engagement also does not fully address the question posed in the original Davies et al. piece, which called for 'stakeholder' as well as 'public' engagement.[44]

Following Pint of Science, we then ran the labels activity as an example of engagement around animal research at an international academic conference (the European Society for Agricultural and Food Ethics – Eursafe[45]) in Tampere, Finland, September 2019. Approximately 20 participants took part, consisting of academics with mixed disciplinary backgrounds, including philosophy, veterinary science, and bioethics.

When the COVID-19 pandemic hit, we considered whether it would be possible or desirable to run this activity online. Given our key interest in the quality of discussion, rather than any sort of outcome, we shared concerns voiced by professionals working in public engagement about the need to create 'engaging, productive or enjoyable' sessions in a virtual space.[46] Though mindful of the added challenges of facilitating such conversation and activities online, in November 2020 we ran the labelling exercise over Microsoft Teams as part of the ESRC Festival of Social Science. The festival was open to all and was specifically aimed at engaging non-academics with social science. While the event generated a good number of registrations, actual participation was much lower at approximately ten individuals. Although online, the format of the session remained more or less unchanged. However, given the lack of in-person contact, we were keen to develop a resource to allow participants to continue engaging with the project if they wished. We therefore used the opportunities provided by the ESRC Festival of Social Science to develop a small website outlining our activity with examples of labels produced by participants.[47] This website link was given to participants at the session, in case they wanted to find out more or keep in touch with the project team. Establishing a website also meant that other academics could read about our activity. The website includes some creative work by a UK artist, Kelly Stanford,[48] with an interest in animals and science.

In May 2021, we ran the labels activity with the University of Nottingham Animal Welfare Ethical Review Body (AWERB), the group responsible for the local ethical review of proposals to use animals in scientific experiments. The aim of the session was both to alert AWERB members to the activity we had designed and to run it as a potential tool to encourage broader ethical discussion in AWERBs.[49] Approximately 20 people took part online. In future we hope to explore other avenues and audiences for this activity and encourage others to run and adapt it. Indeed, one attendee at our Eursafe event was keen to adapt this activity for teaching as an alternative way into the topic of animal research and has so far used it with 150 biology undergraduates at Wageningen University, the Netherlands. This has encouraged us to reflect on the positives that can come from breaking down barriers between research, engagement, and teaching in this field.

Reflections on experimenting with labelling

This section includes some examples of the labels drawn by participants. Our aim is to critically reflect, not on the content of the labels produced and shared, but on the kinds of discussions that the activity promoted. Overall, this section demonstrates that the topic of labelling medicines can work well as a route into wider discussions around animal research. However, as a policy initiative it does not, as may have been assumed by some proponents, offer a solution to a perceived problem, but rather generates further difficult questions and conversations around how openness regarding animal research is navigated and enacted. In summary, this instance of opening up the topic of animal research succeeded in generating discussion of the wider themes of knowledge, power, and positionality.

Opening up knowledge

In the discussions that followed the drawing of labels (phase three), much of the conversation centred on what kinds of information could or should be included. In imagining what a label could look like, participants were keen to talk beyond the use of a simple declaration such as ‘animals were used in the production of this medicine'. Instead, detailed questions were opened up by participants around what sort of species are used in the production of a particular medicine, the number of animals used (and whether this could ever be calculated in practice), the status of animals used (i.e. as experimental 'models' rather than 'pets'), the severity of procedures, the role of regulation, the principles of replacement, reduction, and refinement (the 3Rs), and what care means in relation to how laboratory animals are treated and kept. This range of themes arguably confirms the value of a conceptual approach such as nexus, which demands attention to the intertwining of different aspects.[50] For instance, possibilities for openness and transparency are here enmeshed with other considerations around laboratory animal welfare, governance, and broader sociocultural values around human–animal relations.

For supporters of labelling as a policy proposal, the variety of themes discussed implies that for any labelling scheme to be meaningful, it would need to go beyond simple statements of declaration and offer information on the conditions and contexts of the animals involved. Indeed, several participants in different sessions suggested that a label could include a QR code to scan or website link (see Figure 15.1), to provide further information on both the background and specifics of animal use in medical research. At the very least, such conversations signal that enactments of openness around the use of animals in medicine development would require more careful attention than mandating vague declarations on medicine packaging, such as the statement, 'This pharmaceutical product has only been made possible by the use of research in animals', which was suggested by Winston.[51]

Furthermore, participant discussions about what to include and what to leave off a label remind us that, in highlighting some aspects of biomedical animal use and obscuring others, any policy decisions on what information might be included in medicine labels are far from arbitrary. Indeed, being necessarily selective,[52] such practices of openness do not only tell us something about the use of animals in medicine development and production but also cultivate (and foreclose) particular ways of relating to the issue. For instance, if the focus is on specific scales, such as the total number of animals used in procedures or the species of animals used, this frames the

Figure 15.1 Participant-designed label. The text reads 'I can use it so can you'. On the right of the label is an 'iScan' QR code (Source and copyright: University of Nottingham).

issue through quantification, and forms relations with research animals based on such numbers. By contrast, information about the severity of the procedures that animals have undergone or the practices of care in place at a particular research institution tell us about the lived experiences of the animals involved. Furthermore, as Palmer and Peres contend, focusing only on procedures prevents those animals 'bred, but not used' in an experiment from being accounted for.[53] Overall, then, the discussions that the label activity prompted encourage us to be mindful of the plurality of expectations toward openness in this area, and the impossibility of ever achieving 'complete' transparency.

Opening up power

Connected to the above discussion of what kinds of information medicine labelling around animal use could or should include, participants also discussed the impact that certain labelling choices might have on actions and agency in this area. For example, some participants suggested that labelling on the use of animals in medical research could enable individuals to make *choices* between different medicines, based on information such as numbers of animals used, or, as in Figure 15.2, which species had been involved in their production. Others considered the question of severity, or

Figure 15.2 Participant-designed label. The text reads 'Not tested on domestic animals' (Source and copyright: University of Nottingham).

even whether animals had been used at all or had been replaced by other models.

However, participants went further to discuss whether such choices were currently (or could ever be) possible, given the long and complex trajectory of research in developing medical treatments. In other words, questions were asked about the role of history and path dependency, which reveal the difficulty of providing specific information on animal use in medical development. During the sessions, participants also raised concerns about the impact that such labelling might have, for example around the potential risks of distressing patients and disrupting medicine compliance, particularly prescriptions. This echoes some of the concerns expressed during the Medicinal Labelling Bill debate highlighted earlier.

For us as social scientists, this discussion attunes us to the need to trouble the ethics of placing the responsibility of animal use on the shoulders of patients and medical consumers, an issue which echoes criticism of the 'responsibilisation' of consumers.[54]

The problematic nature of burdening individual consumers with the responsibility of governance is compounded in the case of medicines, where, unlike ethically motivated consumption in the food domain, there are no possible choices between conventional medicines with or without the involvement of animal use. Hence, as McGlacken has written elsewhere, 'consideration of the varying capacities that publics have to act on what they come to know is crucial. Without this, those who care about an issue yet feel unable to act on the moral and emotional trouble it evokes may feel it necessary to turn away altogether'.[55] Such insights into the potential negative consequences of current modes of communication around animal research emphasise that the pursuit of 'openness' should not be perceived as an end in itself.

For those interested in labelling medicines as a policy proposal, our reflection is that more information on animal use will not resolve societal concerns around animal research, nor remove the need for wider dialogue on the issue. Rather, we must also address the existing unequal distribution of power in decision-making around animal research, and whether patients and medical consumers could or should have the power to exercise choice in this regard. Indeed, participants in several sessions raised the question of whether and how patient interactions with medicine packaging are different when being prescribed medicines, rather than purchasing them in shops. Through the former route, patients are unlikely to view a medicine's packaging or information leaflet until the medication has been prescribed and collected from the pharmacy. Thus, the visibility of information around animal involvement in medicine production may be even more complicated in practice.

Opening up positionality

As highlighted in the first section of this chapter, rather than framing the topic as about regulation, ethics, or welfare, our aim instead was to provide an opportunity for participants to discuss any thoughts or questions that might emerge around the provocation of labelling medicines as developed and produced involving animals. Indeed, in introducing the activity, we were careful to stress that we as researchers were not advocating for or against the policy proposal of such medicine labelling. Given the controversial nature of the topic, we were also mindful from the start of our own positionality and the need to present this carefully. However, the discussion confirmed that participants themselves were equally aware of the importance of positionality in the animal research debate and interested in considering different positions and perspectives.

In summary, participants described how different kinds of labels could be used to convince different audiences of different things. For instance, one label could be designed to emphasise that animal use helps certify the safety of medicines, as in Figure 15.3, which includes the text: 'you are safe because of them'. Other labels could

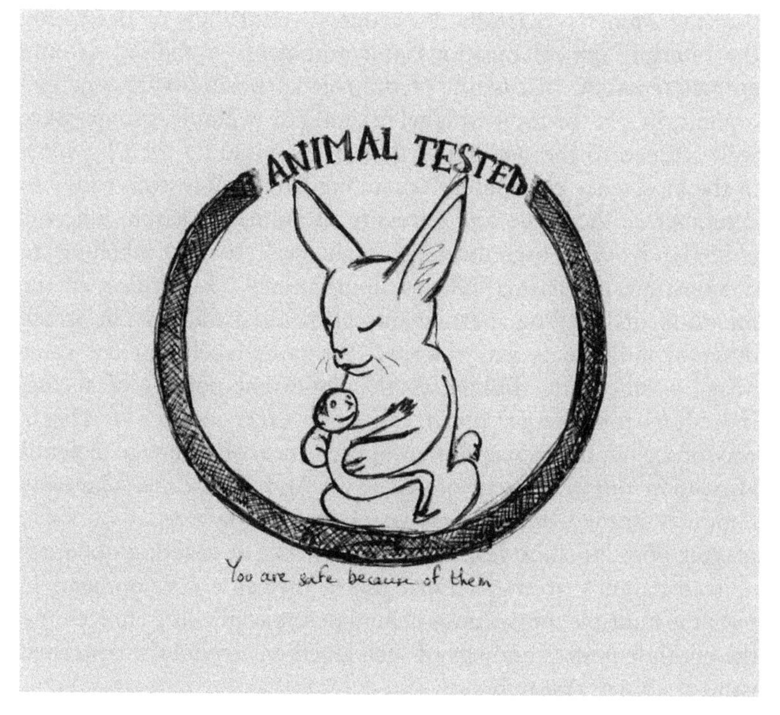

Figure 15.3 Participant-designed label. The text at the top of the image reads 'ANIMAL TESTED' and below reads 'You are safe because of them' (Source and copyright: University of Nottingham).

be designed to emphasise the 'humane' treatment of animals in laboratories or, as the example below might also suggest through its imagery, utilise the rhetoric of 'help' or 'sacrifice' to present the role of animals in a particular way. As social scientists have previously pointed out, such rhetoric is more than euphemistic, with the metaphor of the 'sacrifice' sanctioning the killing of research animals by enabling their transformation into scientific data[56] and by imbuing their lives and deaths with both personal and external meaning.[57]

Overall, the activity generated lots of discussion around the expectations that different individuals and groups might have for such labelling. However, rather than being prompted by us, for example by listing the stakeholders involved in animal research, summarising the regulatory framework governing scientific animal use (the Animals (Scientific Procedures) Act 1986), or describing the multiple ethical positions and arguments mobilised around animal research, discussion of the role of positionality emerged organically via the topic of label design. For example, groups were well attuned to the possibility that labels could be used by some in the bioscience community (including Robert Winston) to try to demonstrate the value and necessity of animal research, whereas animal advocacy organisations might seek to use labelling to demonstrate the harms inflicted upon animals. Regardless of the intention of labelling, participants also noted the way in which different individuals may interpret the same label – or the same image – differently, linking to the important politics of seeing. Indeed, as John Berger put it, 'although every image embodies a way of seeing, our perception or appreciation of an image depends also upon our own way of seeing'.[58] And, as Donna Haraway famously argued, there is no vision 'from nowhere'.[59] As such, playing with labelling reveals the complexity of enacting openness or transparency around animal use in medicine development, in both the multiple motivations behind possible labelling choices and the multiple understandings of such labelling around the contested issue of animal research.

Conclusion

This chapter began by acknowledging the recent push towards more openness and transparency in the animal research domain and the way in which this openness is sometimes restrictively enacted. We argued that enactments of openness have sometimes treated openness as an end in itself and that wider societal dialogue around animal research remains limited. While focusing on everyday medicine use was not envisaged at the start of AnNex, we gradually became interested in this topic as a possible route through which to make the topic of animal research more tangible, local, or mundane. We were also intrigued by the policy suggestion from some individuals and campaigners that the role of animals in medical research and testing could be acknowledged through the act of labelling.

Ultimately, our engagement with the topic of labelling remains small-scale, playful, and experimental. Asking participants to imagine what such a label could look like, and using drawing as a way of stimulating discussion, the labelling activity provided a space to raise and grapple with some of the complexities of both the scientific and socio-ethical aspects of animal research. The drawing exercise was designed with the hope of bringing nuance to the polarised 'for'/ 'against' positioning that often dominates dialogues around animal research. Using drawing to generate collective discussion, our rationale aligned with methodological approaches such as Guillemin's combination of both visual and word-based research methods, which is said to 'offer a way of exploring both the multiplicity and complexity that is the base of much social research interested in human experience'.[60] Overall, the labelling provocation functioned as a focal point around which wider discussion about animal research occurred.

Furthermore, the chapter has also started to illustrate the likely messiness of policy proposals to label medicines as 'tested on animals' in practice. This messiness is in part due to the practical complexity and feasibility of labelling medicines as 'tested on animals', given the non-linear histories of medical research. The proposal also creates ethical complexity, raising considerations of the multiple ways in which such labels might be employed and interpreted, as well as the dangers of providing 'knowledge' without attending to questions of power and capacity to act. In conclusion, we consider that the proposal to label medicines provokes further questions about knowledge, power, and positionality.

Finally, our involvement in this experimental activity raised several reflexive questions for us as social science researchers. We are not public engagement specialists, nor did we restrict our activities to a narrow definition of 'the public'. Indeed, previous work has focused on how staff working in animal research imagine the public,[61] and we have also written elsewhere about the problematic ways in which this category is deployed in general in the animal research field.[62] Nevertheless, we did feel conscious of being seen to be somehow 'promoting' the idea of medicine labelling, even unintentionally. This anxiety prompted some useful discussions between us, and with AnNex and wider colleagues, about the precise role for social scientists in organising such engagement activities in areas of technoscientific controversy. In aiming to generate discussion for its own sake, producing questions and uncertainties rather than resolutions, we would hope that this kind of role can help to build dialogue around controversial issues rather than transmitting knowledge in one direction.

However, we also recognise that this aim of building dialogue is itself not neutral and is instead perhaps part of our own political response to a perceived pressure to 'pick a side', or to put forward a clear and consistent argument 'for' or 'against' animal research. Indeed, our involvement in this activity has prompted us to consider and reflect on the opportunities that we as academics arguably have to hold space for multiple perspectives and voices. As such, our own role in trying to provide a space for discussion is itself a form of action or intervention, regardless of whether this has implications for policy. As Davies et al. have argued, 'dialogue events that do not seek to influence policy could (1) provide opportunities for empowering individuals for further involvement, (2) be viewed as personally beneficial, or (3) be part of a gradual step by step change in science and society'.[63] To be clear, aiming to stimulate or facilitate dialogue does not necessitate that we stand 'outside' of discussions and debates and pursue a false ideal of academic objectivity or neutrality, but instead means embracing the capacity to explore and experiment, without being strictly tied to specific positions, as is often the case for stakeholder groups or policy-makers. Finally, it also requires us to be reflexive about our role and positionality, and thus be willing to critically reflect on our own contribution to the animal research nexus.

Notes

1 Nuno Henrique Franco, 'Animal Experiments in Biomedical Research: A Historical Perspective', *Animals*, 3.1 (2013), 238–273, DOI: 10.3390/ani3010238.

2 Renelle McGlacken, 'Exploring Everyday Relations with Animal Research: A Sociological Analysis of Writing from the Mass Observation Project' (unpublished PhD thesis, University of Nottingham, 2021), www.eprints.nottingham.ac.uk/id/eprint/66576 [accessed 1 February 2023].

3 John Illman, 'Animal Rights Violence Spreads Fear Through U.K. Research Community', *JNCI: Journal of the National Cancer Institute*, 97.21 (2005), DOI: 10.1093/jnci/dji394, 1565–1566; Emma Marris, 'Animal Research: Grey Matters', *Nature*, 444.7121 (2006), 808–810, DOI: 10.1038/444808a.

4 Annabella J. Williams and Hannah Hobson, 'Concordat on Openness on Animal Research in the UK Annual Report', in *Concordat on Openness on Animal Research in the UK: Resources, Annual Reports, and Concordat Documents*, 1–53 (p. 9), https://concordatopenness.org.uk/resources [accessed 10 January 2022].

5 Basel Declaration Society, 'Basel Declaration: A Call for More Trust, Transparency and Communication on Animal Research', www.basel-declaration.org/basel-declaration/download-the-declaration/ [accessed 10 January 2022], p. 2.

6 European Animal Research Association, *Transparency Agreements in Europe*, www.eara.eu/transparency-agreements [accessed 12 February 2021].

7 Australian & New Zealand Council for the Care of Animals in Research and Teaching, *Openness Agreement*, www.anzccart.org.nz/openness-agreement [accessed 12 February 2021].

8 Lab Animal Tour, *360° Laboratory Animal Tours*, www.labani maltour.org/ [accessed 12 February 2021].

9 Carmen M. McLeod, 'Assuaging Fears of Monstrousness: UK and Swiss Initiatives to Open up Animal Laboratory Research', in *Science and the Politics of Openness: Here Be Monsters*, ed. by Brigitte Nerlich et al. (Manchester: Manchester University Press, 2018), DOI: 10.7765/9781526106476.00010, p. 70.

10 Understanding Animal Research, *Concordat on Openness on Animal Research in the UK*, https://concordatopenness.org.uk/wp-content/uploads/2017/04/Concordat-Final-Digital.pdf, 1–12 [accessed 8 January 2021]. p. 8.

11 Richard Owen et al., 'A Framework for Responsible Innovation', in *Responsible Innovation: Managing the Responsible Emergence of Science and Innovation in Society*, ed. by Richard Owen et al. (Wiley, 2013), pp. 27–50, DOI: 10.1002/9781118551424.

12 Gail Davies et al., 'Animal Research Nexus: A New Approach to the Connections between Science, Health and Animal Welfare', *Medical Humanities*, 46.4 (2020), 499–511 (p. 503), DOI: 10.1136/medhum-2019-011778.

13 Carmen M. McLeod, 'Assuaging Fears of Monstrousness: UK and Swiss Initiatives to Open up Animal Laboratory Research', in *Science and the Politics of Openness: Here Be Monsters*, ed. by Nerlich et al., p. 69.

14 Mass Observation, *Mass Observation Project*, www.massobs.org.uk/about/mass-observation-project [accessed 8 January 2021].

15 Camille Limoges 'Expert Knowledge and Decision-Making in Controversy Contexts', *Public Understanding of Science*, 2.4 (1993), 417–426, DOI: 10.1088/0963-6625/2/4/009.

16 Mike Michael and Lynda Birke, 'Accounting for Animal Experiments: Identity and Disreputable "Others"', *Science, Technology, & Human Values*, 19.2 (1994), 189–204, DOI: 10.1177/016224399401900204; Elizabeth S. Paul, 'Us and Them: Scientists' and Animal Rights Campaigners' Views of the Animal Experimentation Debate', *Society & Animals*, 3.1 (1995), 1–21, DOI: 10.1163/156853095X00017.

17 Sarah Nettleton et al., 'The Mundane Realities of the Everyday Lay Use of the Internet for Health, and Their Consequences for Media Convergence', *Sociology of Health & Illness*, 27.7 (2005), 972–992, DOI: 10.1111/j.1467-9566.2005.00466.x; Kate Weiner and Catherine Will, 'Thinking with Care Infrastructures: People, Devices and the Home in Home Blood Pressure Monitoring', *Sociology of Health & Illness*, 40.2 (2018), 270–282, DOI: 10.1111/1467-9566.12590.

18 Pru Hobson-West et al., *Mass Observation: Emotions, Relations and Temporality. Workshop Report*, in Animal Research Nexus:

Publications, www.animalresearchnexus.org/publications/mass-observation-emotions-relations-and-temporality 1–9 [accessed 12 January 2021]; McGlacken, 'Exploring Everyday Relations with Animal Research'.

19 Alexandra Palmer et al., 'Animal Research beyond the Laboratory: Report from a Workshop on Places Other than Licensed Establishments (POLEs) in the UK', *Animals* 10.10 (2020), 1868, DOI: 10.3390/ani10101868.

20 UK Parliament, *Medicinal Labelling Bill*, Publications & Records, https://publications.parliament.uk/pa/bills/lbill/2013-2014/0011/lbill_2013-20140011_en_2.htm#l1g1 [accessed 4 February 2021].

21 UK Parliament, *Medicinal Labelling Bill [HL]*, Hansard, www.hansard.parliament.uk/Lords/2013-10-25/debates/13102540000357/details [accessed 4 February 2021].

22 UK Parliament, '*Medicinal Labelling Bill*', Publications & records, https://publications.parliament.uk/pa/bills/lbill/2013-2014/0011/lbill_2013-20140011_en_2.htm#l1g1, [accessed 4 February 2021].

23 Robert Winston, 'Animal Experiments Deserve a Place on Drug Labels', *Nature Medicine*, 19.10 (2013), 1204, DOI: 10.1038/nm1013-1204.

24 Understanding Animal Research, 'Watch and Read – Where Do Medicines Come From?' www.understandinganimalresearch.org.uk/news/research-medical-benefits/watch-and-read-where-do-medicines-come-from/ [accessed 8 February 2022].

25 UK Parliament, '*Medicinal Labelling Bill [HL]*', Hansard, wwwhansard.parliament.uk/Lords/2013-10-25/debates/13102540000357/details, [accessed 4 February 2021].

26 Shaun Yon-Seng Khoo, 'Justifiability and Animal Research in Health: Can Democratisation Help Resolve Difficulties?' *Animals*, 8.28 (2018), 1–12 (p. 5), DOI: 10.3390/ani8020028.

27 Allyson J. Bennett et al., *Is It Vegan or Not? A Proposal to Clearly Label Medications*, www.speakingofresearch.com/2021/04/22/is-it-vegan-or-not-a-proposal-to-clearly-label-medications/ [accessed 15 January 2021].

28 Andrée Dignon, '"If You Are Empathetic You Care about Both Animals and People: I Am a Nurse and I Don't like to See Suffering Anywhere". Findings from 103 Healthcare Professionals on Attitudes to Animal Experimentation', *Journal of Health Psychology*, 24.5 (2019), 671–684 (p. 678), DOI: 10.1177/1359105316678307.

29 Adrian B. Evans and Mara Miele, 'Between Food and Flesh: How Animals Are Made to Matter (and Not Matter) within Food

Consumption Practices', *Environment and Planning D: Society and Space*, 30.2 (2012), 298–314 (p. 303), DOI: 10.1068/d12810.

30 Christine Parker et al., 'Can Labelling Create Transformative Food System Change for Human and Planetary Health? A Case Study of Meat', *International Journal of Health Policy and Management*, 10 (2020), 923–933, DOI: 10.34172/ijhpm.2020.239.

31 Kristian Bjørkdahl and Karen Victoria Lykke Syse, 'Welfare Washing: Disseminating Disinformation in Meat Marketing', *Society & Animals* (2021), 1–19, DOI: 10.1163/15685306-BJA10032.

32 Parker et al., 'Can Labelling Create Transformative Food System Change for Human and Planetary Health?'

33 Matthew Cole, 'From "Animal Machines" to "Happy Meat"? Foucault's Ideas of Disciplinary and Pastoral Power Applied to "Animal-Centred" Welfare Discourse', *Animals*, 1.1 (2011), 83–101 (p. 93), DOI: 10.3390/ani1010083.

34 Sally Eden, 'Food Labels as Boundary Objects: How Consumers Make Sense of Organic and Functional Foods', *Public Understanding of Science*, 20.2 (2011), 179–194 (p. 192), DOI: 10.1177/0963662509336714.

35 Red Tractor, *Red Tractor Website*, www.redtractor.org.uk/ [accessed 12 January 2021].

36 Mike Michael and Nik Brown, 'The Meat of the Matter: Grasping and Judging Xenotransplantation', *Public Understanding of Science*, 13.4 (2004), 379–397 (p. 380–381), DOI: 10.1177/0963662504044558.

37 Anthony Dunne and Fiona Raby, *Speculative Everything: Design, Fiction, and Social Dreaming* (Cambridge, Massachusetts: MIT Press, 2013), 1–240 (p. 3).

38 Matt Ratto, 'Critical Making: Conceptual and Material Studies in Technology and Social Life', *The Information Society*, 27.4 (2011), 252–260 (p. 253), DOI: 10.1080/01972243.2011.583819.

39 Veronica Heney and Branwyn Poleykett, 'The Impossibility of Engaged Research: Complicity and Accountability between Researchers, "Publics" and Institutions', *Sociology of Health & Illness*, 44: S1 (2021), DOI: 10.1111/1467-9566.13418, 179–194.

40 Pint of Science Festival. About us https://pintofscience.co.uk/about/ [accessed 9 August 2023].

41 Praveen Paul and Michael Motskin, 'Engaging the Public with Your Research', *Trends in Immunology*, 37.4 (2016), 268–271, DOI: 10.1016/j.it.2016.02.007.

42 Renelle McGlacken and Pru Hobson-West, 'Critiquing Imaginaries of "the Public" in UK Dialogue around Animal Research: Insights from the Mass Observation Project', *Studies in History and Philosophy of Science*, 91 (2022), 280–287, DOI: 10.1016/j.shpsa.2021.12.009.

43 National Co-ordinating Centre for Public Engagement, '*What Is Public Engagement?*' www.publicengagement.ac.uk/about-engagement/what-public-engagement [accessed 15 January 2021].
44 Davies et al., 'Developing a Collaborative Agenda for Humanities and Social Scientific Research on Laboratory Animal Science and Welfare'.
45 European Society for Agricultural and Food Ethics, www.eursafe.org/ [accessed 15 January 2021].
46 National Co-ordinating Centre for Public Engagement, '*Online Engagement: A Guide to Creating and Running Virtual Meetings and Events*', www.publicengagement.ac.uk/meaningful-engagement-online-events, 1–12 [accessed 15 January 2021].
47 Labanimallabels, '*Labelling Animal Research?*' www.labanimallabels.co.uk/ [accessed 15 January 2021].
48 Kelly Stanford, '*Kelly Stanford Science Communicator and Artist*', www.kellystanford.co.uk/ [accessed 15 January 2021].
49 Penny Hawkins and Pru Hobson-West, 'Delivering Effective Ethical Review: The AWERB as a "Forum for Discussion"' in *RSPCA: Animals in Science: Reports and Resources: Ethical Review*, 2017, www.science.rspca.org.uk/-/ethical-revi-1 [accessed 15 January 2021].
50 Davies et al., 'Animal Research Nexus'.
51 Winston, 'Animal Experiments Deserve a Place on Drug Labels'.
52 Tora Holmberg and Malin Ideland, 'Secrets and Lies: "Selective Openness" in the Apparatus of Animal Experimentation', *Public Understanding of Science*, 21.3 (2012), 354–368, DOI: 10.1177/0963662510372584.
53 Alexandra Palmer and Sara Peres, 'Present, Not Used (Part 1): A Spectrum of Visibility', in *Animal Research Nexus:* Blogs (2020), www.animalresearchnexus.org/blogs/present-not-used-part-1-spectrum-visibility [accessed 20 January 2021].
54 J. Littler, 'What's Wrong with Ethical Consumption?' in *Ethical Consumption: A Critical Introduction*, ed. by Tania Lewis and Emily Potter (London; New York; Routledge, 2011), 27–39; Markus Giesler and Ela Veresiu, 'Creating the Responsible Consumer: Moralistic Governance Regimes and Consumer Subjectivity', *Journal of Consumer Research*, 41.3 (2014), 840–857, DOI: 10.1086/677842.
55 Renelle McGlacken, '(Not) Knowing and (Not) Caring About Animal Research: An Analysis of Writing From the Mass Observation Project', *Science & Technology Studies*, 35.3 (2021), 2–20 (p. 15), DOI: 10.23987/sts.102496.
56 Michael E. Lynch, 'Sacrifice and the Transformation of the Animal Body into a Scientific Object: Laboratory Culture and Ritual Practice

in the Neurosciences', *Social Studies of Science*, 18.2 (1988), 265–289, DOI: 10.1177/030631288018002004.

57 Arnold B. Arluke, 'Sacrificial Symbolism in Animal Experimentation: Object or Pet?' *Anthrozoös*, 2.2 (1988), 98–117, DOI: 10.2752/0892 79389787058091.

58 John Berger, *Ways of Seeing* (London: Penguin, 1972), p. 8.

59 Donna Haraway, 'Situated Knowledges: The Science Question in Feminism and the Privilege of Partial Perspective', *Feminist Studies*, 14.3 (1988), 575–599, DOI: 10.2307/3178066.

60 Marilys Guillemin, 'Understanding Illness: Using Drawings as a Research Method', *Qualitative Health Research*, 14.2 (2004), 272–289 (p. 273), DOI: 10.1177/1049732303260445.

61 Pru Hobson-West and Ashley Davies, 'Societal Sentience: Constructions of the Public in Animal Research Policy and Practice', *Science, Technology, & Human Values*, 43.4 (2018), 671–693 (p. 589), DOI: 10.1177/0162243917736138.

62 McGlacken and Hobson-West, 'Critiquing Imaginaries of "the Public" in UK Dialogue around Animal Research'.

63 Sarah Davies et al., 'Discussing Dialogue: Perspectives on the Value of Science Dialogue Events That Do Not Inform Policy', *Public Understanding of Science*, 18.3 (2009), 338–353 (p. 341), DOI: 10.1177/0963662507079760.

16

Building participation through fictional worlds

Bentley Crudgington, Natalie Scott, Joe Thorpe, and Amy Fleming

Figure 16.1 Biocore logo (Designed by Joe Thorpe (2018), copyright: Bentley Crudgington/The Lab Collective).

Act one: invitation

'Hello, if you are here for Biocore please follow me.'

You have arrived at the venue and been told someone will come to collect you shortly. Others gather too. In all there are about thirty, some in small groups and some alone. Someone new appears. You are led away from the open space and invited somewhere different. You follow others along corridors that display Biocore marketing materials. There are offers:

> Putting your voice into biomedical innovation.

And promises:

> Safer tomorrow – today!

One by one you enter a new space. You are warmly greeted and offered hand sanitiser by someone in a Biocore-branded lab coat.

> 'Welcome to Biocore – please take a seat.'

You take a seat at one of three tables. Each has a tablet connected to a monitor. Both display the powder blue Biocore logo, which

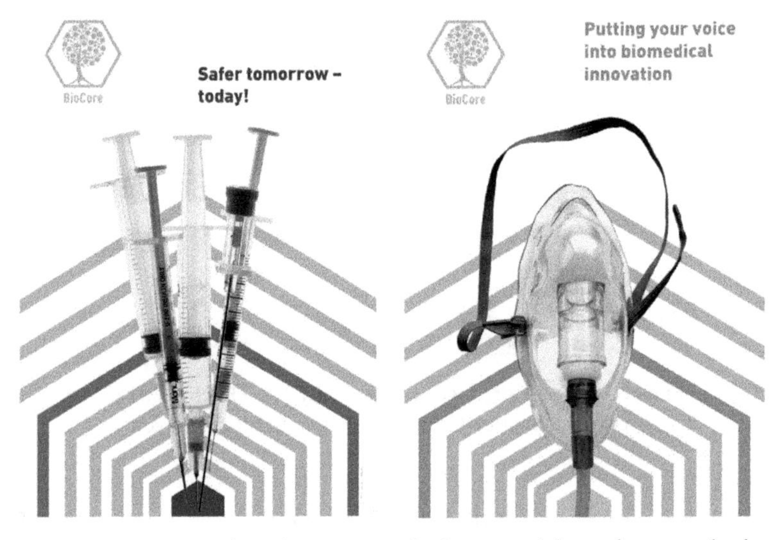

Figure 16.2 Examples of Biocore marketing materials used as a method for immersion. Visual signposts guide participants from venue foyers/waiting areas to the performance space. Designed by Dmitriy Myelnikov and Bentley Crudgington (2018). (Copyright: Dmitriy Myelnikov and Bentley Crudgington).

also appears on a large screen at the front of the space, where the person who welcomed you now stands.

> 'Hello everybody. Welcome to Biocore; thank you so much for joining us today for this session. So, I suppose I'd better start off by telling you a little bit about who we are. Biocore was formed in 1973 as a small lab based in Nottingham; it started as a response to the growing public concern of the use of animals as part of medical research, and we responded to this by placing welfare and ethical practice within our core company values. Since then, we have grown to become a global company with 37 research facilities in over 15 countries, thousands of researchers, vets, and technicians as well as working with ambassadors like me. And who am I?
>
> I am Dr Agnes Arber, Ethical Ambassador for Biocore. My job is to make sure that the policies and processes we work with today comply with the ethical standards that we've set across the organisation and I undertake most of the public-facing work we do in the UK. We could engage with you over Twitter, but I prefer a more face-to-face approach. Our public consultations are becoming very popular and important to the work we do; working with the public can have a positive and lasting input into Biocore's decision-making processes.
>
> And this is where you come in; though we are living through a pandemic, this does not mean that standards have slipped.
>
> This will not be the last pandemic and Biocore is looking to further future-proof our processes, and make sure that you, the public, feel more empowered to help us make decisions when developing future vaccine candidates.
>
> Good is never good enough.
>
> Today you'll be undertaking a simulation to find an animal model that will help us find a safe and ethical pathway towards developing future antivirals.
>
> Remember: 'Better Welfare equals Better Science.'
>
> We are asking you to consider three things during the simulation:
>
> Cost: you have eighteen million pounds to spend – use it wisely.
>
> Success: we want to create an experiment that has a high probability of discovering beneficial interventions.
>
> And last, but certainly not least – Harm: look carefully at the welfare of animals, researchers, Biocore, and of course, you, the public.'

Dr Arber explains that you will work as groups and use the Biocore Artificial Intelligence System, AL, to input your decisions. You are guided to the tablet on the table. The newly elected group spokesperson taps the button.

We are ready to begin.

What follows is a negotiation of what success could look like, what harms are made in pursuit of success, and how you might measure these together. Unaware, you are already deeply involved in a harm–benefit analysis. AL asks for your decisions via text messages, sometimes directing you to information they have placed in the document store, to emails that have arrived from other members of Biocore staff, or even to recent social media activity. Each time you input a decision new information arrives. Sometimes, this calls for celebration, sometimes regret, always complexity.

AL asks if you are ready for your progress report. You have been aware of the hum of conversation from the other groups, but it is not until Dr Arber asks your spokesperson to share your results with everyone that you all look beyond your own table.

At the end of each round, the groups share their total economic spend, harm rating, and likelihood of success, as calculated by AL. They also share what decisions they have made and why. Sometimes, other groups made a different decision to you and sometimes the same decision but for different reasons. Dr Arber reminds you:

> 'You will be in competition, however, don't let this affect your ethical compass!'

You return to your group discussions and to AL and make uneasy but urgently ethical progress. You complete another round of decision-making and report your scores to the other teams. There is one more round; you have a feel for them now. Only one research proposal will be taken forward by Biocore.

There are scores to settle.

Collectively each proposal is considered in turn. Is a higher harm level mitigated by a higher chance of success? What is the ethical cost of increasing success by 5%? Is the proposal trustworthy and to whom? Does your success feel different to someone else's? Are their harms unacceptable to you? What feels different out in the open? At what distance can you care about this?

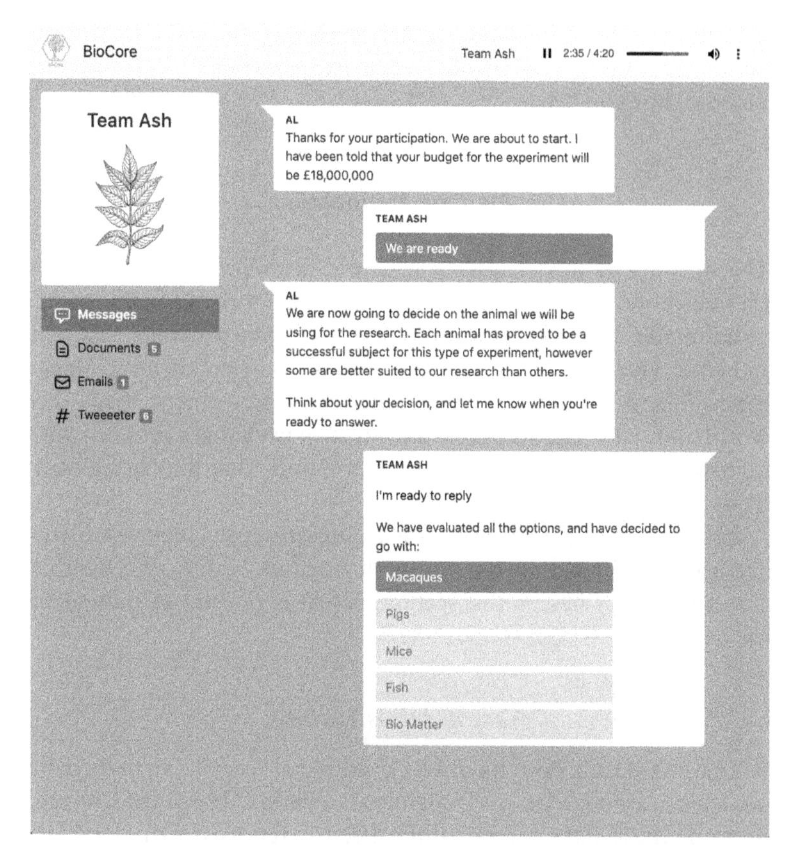

Figure 16.3 Screenshot for the Biocore ethical review interface. Participants use the app during the performance to input their decisions. New information responding to that decision is released in various formats, e.g., documents, emails, or Tweeeeter (a social media app). New information is highlighted by red icons in the information menu (see left panel below team leaf icon). AL, Biocore's AI assistant, communicates with participants via SMS (see central panel). App content and function were originally designed by Thorpe, Scott, Fleming and Crudgington. The functionality was refined and the standalone app was built by software designers Katja Mordaunt, Kris Sum and Nick Wade in collaboration with Thorpe, Scott, Fleming and Crudgington (2020). (Copyright: Bentley Crudgington/The Lab Collective).

A final decision is made. Dr Arber thanks you for your participation. It is up to you whether what happened at Biocore stays at Biocore, but you leave.

Act two: beginnings

The opening of this chapter is a hybrid account of the immersive theatre show, *Vector*. It combines quotes from the current script,[1] visual materials used in the show, and narrative accounts that aim to evoke *Vector* and the world of Biocore via two affective trajectories. The first is how we, as makers and performers, made this world and why. The second is the affective exchanges between participants who activate the Biocore world and what they bring back from travel into this world.

Vector is an experiment in care as infrastructure, immersion and agency, and ethical encounters with anxious bodies. We explore care and agency here in this section, before exploring encounters in the next.

Care as infrastructure

Animal research has its own 'culture of care', harm–benefit analysis frameworks, and openness agendas. However, as yet, none of these have meaningfully been extended to think about how to care for participating citizens.[2] The ambiguity of who is responsible for citizens taking part in conversations about animals in science stitches care anxiety across the animal research landscape.[3]

We wanted to experiment with the ways in which conversations about animal research take place and to think about who cares for these conversations rather than who controls them. We felt this was an approach vitally lacking in current openness agendas and frameworks. We worked to radically reimagine the procedural, cultural, and physical architectures of animal research as care-full infrastructures.

Vector is modelled on the operation of an Animal Welfare and Ethical Review Body (AWERB), a regulatory requirement under the Animals (Scientific Procedures) Act 1986. Their duties include: improving science and animal welfare, promoting a good culture of care, and improving public accountability via robust governance. However, *Vector* is also nothing like one of these. Its reviewing and reviewed bodies perform differently. The question is: what are they performing in, and why does this matter?

Previous experiments with the form of citizen juries and ethical review panels suggest that the barrier to public engagement with complex and controversial issues is not the amount or quality of information available, but the lack of an accessible framework in which to explore choices and responses to information.[4] Responding to this insight, we developed *Vector*: an immersive experience that uses elements of performance, game, and integrated technology to engage the public in issues around animal research (see Figure 16.4). *Vector* took the concept of the AWERB as a way of playing with the context in which information is offered and how it is contextualised and by whom. Drawing on theories of affect in theatre and examining the performance roles and power held by performers and audiences, we attempt in this chapter to move

Figure 16.4 Team Elm participants during a *Vector* performance at HotBox Live, Chelmsford, as part of the British Science Association Festival (September 2021) (Copyright: Matthew Kaltenborn).

beyond thinking in terms of the *public* towards the transformative concept of the *participant*.

To begin, we visited sites where animal research ethics are enacted, and reviewed them through the lens of performance. First was an AWERB of a high-profile UK facility, and second were the new laboratories and animal housing of a research group that had changed locations and introduced a new species into their new research institute.

Site one: AWERB

We were permitted to attend and fully participate in one institutional AWERB. The main body of the meeting was a pre-submission project licence review from a new Principal Investigator (PI). We were sent the documentation prior to the meeting and were able to ask questions throughout the research presentation and review process. As guests, we performed a gentle frame analysis,[5] with particular interest in the effects of non-human actors such as research animals and other more-than-human assemblages such as legislation, regulation, cultures, the concept of 'The Institute' and where and how their boundaries were imagined, and, critically, how the public were invited to or prevented from crossing these boundaries.

Site two: the new laboratory

Here, we were given a full tour of the new space by a different PI, with an overview of the research group's current work, and an introduction to what the new space would allow them to do. We toured the offices and laboratory spaces and met students and staff members who talked us through the move, the new space, and new equipment. We also toured the new animal housing and research areas and spoke to animal technologists and support staff. We were particularly interested in how the new space helped them think about the future: how proceduralised care processes had travelled with them; how these were shared, re-contextualised, and integrated into the existing care infrastructure of the new host institution; and in what ways a 'new' species helped or hindered these processes. After each site visit, we had a collective, gently facilitated, debrief session to review what we had discovered, how we felt about it, what other questions it prompted, and what we wanted to do about it.

We separately conducted further desk research and spoke with colleagues and peers before reconvening at a two-day planning session. We talked through scenarios and approaches and reached some critical moments of intent. We knew we wanted participants to shape the outcome of the experience through their own agency and decision-making and were drawn to a re-imagining of an AWERB. We realised that using a digital interface in the performance would allow us to bring in information from a variety of sources and formats. As a nod to the principles of replacement, reduction, and refinement (the 3Rs), we decided to base the performance around three decision-making 'sprints', which all aimed to improve animal research. However, we were aware that what was refined, replaced, or reduced would be very different in our speculative fiction than those proposed by the 3Rs framework.

We wanted to actively resist offering resolutions or merely 'problematising' scenarios, but instead to make full use of narrative and performance structures to deepen affective participation. Beginning in 2018 we used a 'here but not now' scenario of an emerging viral zoonotic pandemic that required the rapid development of new vaccines. There were three main reasons for this decision. The first is the scale. The size and complexity of a pandemic holds space for multiple nuanced storylines. Secondly, it allowed us to use a familiar trope to lure people in and then make it unfamiliar because, as Martin Crimp proposes, letting the unfamiliar into familiar situations can deliver the kind of truth that we have to work to discern, not a message that is offered up for us to digest.[6] Thirdly, it instils a sense of urgency that can be utilised to drive decision making in a way that advances the experience without shutting things down.[7] Gradually we mapped out a decision matrix that gave us the structure to make an interactive digital information cascade, where each decision was weighted and triggered the release of new information. For example, participants might receive a summary of the latest scientific findings from the PI, get a personal email from an Animal Technologist, or receive a fretful request from Jenny from PR. The information received is dependent on the previous decision.

The structure and content were reviewed by the Animal Research Nexus (AnNex) Public Engagement Subgroup before extensive 'play testing' with peers, stakeholders, and publics. Each play test included a post-show discussion with participants and a makers' debrief. Any questions raised but unaddressed in these sessions were sent out to members of the animal research community, and their responses and reflections passed back to the participants. *Vector* premiered at Niamos Radical Arts Centre, Hulme, Manchester, in March 2019 (Figure 16.5).

Figure 16.5 *Vector* premiere, Niamos Radical Arts Centre, Hulme, Manchester, March 2019 (Copyright: Bentley Crudgington).

Immersion and agency

Casting a citizen in any role is political. The role shows how you imagine them, and the role defines what they are allowed to imagine. Many roles allocated to citizens restrict their agency by setting limits on how much they can change things and how much those things can change them. The act of casting is an aesthetic choice. It is ideological. It is never passive. If we are to re-imagine how publics engage with animal research, we need to move beyond strategic opacity that escorts visitors at surface level through curated discovery trails and move towards affective immersive acts of creation. While one can perform a play without an audience, *Vector* requires bodies to immerse themselves in a world of their own creation.

We looked at what was lacking in citizenly care in current public-facing animal research agendas by paying attention to and honouring the ways in which citizens could care for their own participation in animal research. We then asked: could immersive theatre elevate publics from passive audiences to active participants who decline deferring[8] and immerse themselves into fully lived moments, despite the uncomfortable knowledge[9] around which they cultivated strategic ignorance?[10]

Immersive and theatre are two words with many definitions, like public and engagement, or patient and involvement, animal and research. Worlds are built on how words like these are understood. Affect, like immersive and openness, is a tricky word that has found its way into the vocabularies of many disciplines but has resisted a stable definition. Contemporary performance scholars, such as Patricia Clough, trace its emergence from the thinking of Deleuze and Guattari,[11] Spinonza,[12] and Bergson,[13] and propose 'affectivity as a substrate of potential bodily responses, often automatic responses, in excess of consciousness'.[14] Others, such as Brian Massumi[15] and Rhoda Blair[16] suggest it precedes emotions and language, it is contingent and ineffable. In these worlds affect becomes lively and unpredictable, 'We cannot consciously opt to be or not to be affected.'[17] Current openness agendas seek to educate the emotions out of concerns, to reduce the ineffable to the knowable, to settle necessity in rational and predictable terms.

We can explore other 'anti-affect ontologies' with Deleuze. In the so-called 'deficit-model',[18] for example, an expert has discovered a truth and can pass that knowledge down to the non-expert. This system relies on correct ways of thinking, feeling, and responding to knowledge; it is about how a citizen should act once properly informed. Deleuze argues that ways of knowing are not discovered but created. We argue that current openness agendas act as anti-affect agendas that prevent immersion.

To understand audience agency, affect, and immersion within our approach to immersive theatre we used the following plot of agency-affect, which asks how much I am affected by affecting the thing I am engaged in (see Figure 16.6).

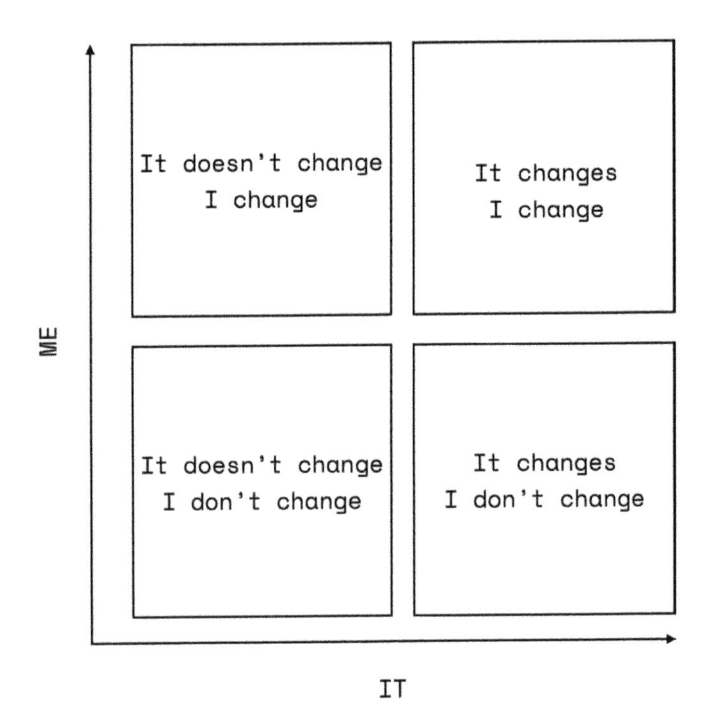

Figure 16.6 Axis of agency/affect. Created by Bentley Crudgington (2020). (Copyright: Bentley Crudgington).

Immersive theatre should occupy the top right quadrant by giving agency to participants to change both their own and others' objective, subjective, and affective experience – something Adrian Howells would call a 'cathartic revolution' and commitment to mutual transformation.[19] The art historian and proponent of socially engaged art criticism Grant Kester's dialogical aesthetics[20] identifies that affective dialogical exchange requires reciprocal openness that is sensitive to the singular subjectivities of those involved. This framing calls for more participant-led, context-dependent work that sees the role of artist/performer as a facilitator rather than content provider. Curator and art critic Nicolas Bourriaud offers a system of relational aesthetics, which involves 'art taking its theoretical horizon as the realm of human interaction and its social context, rather than the assertion of a private symbolic space'.[21] Immersive theatre can be considered as an inter-personal performance, which brings together dialogical and relational aesthetics in the praxis of reconstituting social relations by and for the participants that take seriously their complex and contingent identities which they may or may not wish to act out for us.

Immersive theatre offers methodologies for citizenly co-creation, where participation is not contingent on how one should act, but radically open to how one *might* act. *Vector* offers a mechanism for articulating good care as an intervention rather than a factual evaluation or judgement of practices[22] by resisting re-presenting normative moral obligations in favour of co-creating thick, impure involvement in a world where the question of how to care needs to be posed.[23]

Act three: experiments

Vector is designed so participants can appreciate that the world is produced through knowledge practices, and that participating in them comes to matter. How, then, do you design within worlds that people think are not for them? How do you ask people to participate in knowledge practice that might produce things they do not want to know?

As an icebreaker, in the introduction, participants are asked to introduce themselves and state how they feel about animal research. Word choice is important as 'feel' gives permission for emotions to be included in the discussion. By introducing the emotional ecology of the group, difference is given space and taken into the next phase of analysis and decision-making.

The first decision AL asks a team to make is which model organism they wish to conduct their research on. It is intentionally a difficult task – it needs to be. The models available are macaques, pigs, mice, fish, and biomatter (subdivided into human-derived stem cells and biobank material). AL reminds participants that each model organism has previously been successfully used for this type of research, and that factsheets are available in the document store. Each factsheet contains a brief description of the organism with its advantages and limitations, how many model organisms you can purchase, their cost, probability of success, and a harm rating.

The original version of these factsheets included a photograph of each model organism in a laboratory setting. We invited theatre makers, gamers, creatives, and activists to a 'play to break' session to assess the robustness of the world we had built, to see if it held up to scrutiny, experimentation, and a cohesive experience. It did not. The photographs on the factsheets seemed to form an impenetrable wall that the participants could not move past. It appeared that having even representations of research animals in the room was simply too disruptive.

The images were not of procedures, nor did they show any evidence of harm or suffering. Yet it was not what these images showed but what they represented. Their singularity had moved the focus from an animal 'model' to this particular animal subject. Our testers had specific questions about the individual animals that could not be answered within our *Vector* world. It was interesting to discover that the assumed richness and honesty we thought a picture could bring had the opposite effect. During their questioning, the pictures became fixed and flat and began to feel dishonest and opaque. As we had noted before, becoming transparent is a process that can obscure more than it reveals.

What could this tell us regarding what people want to know about animal research, how they want to experience it, and how they have learnt to analyse data in the public realm? More care is required in the thinking of those who dismiss citizens' calls for more information by saying all the information is already in the public domain. Such thinking fails to consider the structural barriers that prevent or put off 'publics' from entering certain 'domains', how it feels if they do, and how that information is activated and contextualised by the participant. Public or not, nothing comes without its world.

It is important to remember that providing access to information is not the same as revealing the process responsible for its production. This feels particularly vital when striving to create a new culture of communication around animal research. Images of research animals do not end up in the public domain without bias or design and therefore are not impartial data points. If our aim is to create a space for more nuanced discussions around animal research, perhaps it was self-defeating to use imagery that can be viewed as 'sanitised public relations' material.

We did not procure and photograph these organisms for this explicit purpose but re-homed them from various online image repositories under creative commons licences. What was not initially clear was whether the organisms or our experiments were unsuitable. After considerable discussion we decided to replace the photographs with stylised vectored images, which resolved the issue for future participants but not for us.[24]

We tried to justify the decision to remove the images as a matter of accuracy. Even though the primary aim of an AWERB is to discuss the details of animal research there are rarely any pictures of animals in the proceedings. We could, and did for a time, legitimately claim that omitting these images was a more accurate and transparent representation of the process we are modelling. But that never felt like vindication; it felt like an excuse. But why?

There is a long history of ocularcentrism in Western cultures, a perceptual and epistemological bias ranking vision over other senses and its intimate association with truth. It is possible to find traces of ocularcentrism in animal research engagement and communication strategies, and its etymology and epistemologies: openness, transparency, witnessing, oversight, exposé, and so on. This is understandable given the historical and ongoing contested visual culture of animal research, where the 'truth' images contain are used to justify actions in both pro- and anti-animal research arenas.

We argue that using images as settled representation of truth in order to control the narrative by offering up cinematic solutions to ethical problems is another hallmark of an anti-affect agenda. Anti-affect agendas act to say, 'the truth that you object to in this picture is false, you do not *understand* this image', with the subtext being 'therefore, your concern and emotional responses are illegitimate too'. We began to work on a visual culture for Biocore that could care for the full affective richness of images of animal research.

During the Covid-19 pandemic we experimented with delivering online versions of the experience. Many things were different online, but we noticed that teams took much longer to make their first decisions and select a model organism, which unbalanced the pace of the show. We restructured the introduction and Dr Arber now introduced each of the model organisms before participants were divided into teams. To make this more dynamic and aesthetically engaging we introduced five-second video clips of each model organism, which faded into the vectored image as we talked about the species. These videos are not shot in labs and there is nothing to suggest imminent experimentation, but they do re-instate the liveliness of the animal body. Perhaps this was the initial problem. Our original images were not integrated into the wider narrative and as such their liveliness was reduced to a commodity, which failed to avoid the ramifications, outlined by Chris Shilling, of alienating the body from the results of labour.[25] The videos and their integrated and personalised introductions gave the organisms agency and restored their bodies' ability to be both active and acted upon as the basis for intersubjective or empathic exchanges with audience members.[26] We did not want to concentrate on the technologies of power and subjectivation of the spaces of animal research themselves, but to focus on the narratives offered to the participant with which to activate them. Our experimental body is the social body, for which there are currently no legislative guidelines regarding husbandry or harm.

Did we fall victim to ocularcentrism? With critical reflection, we felt it was important to include images because of a desire to have the animals in the room with us, to make sure our framework was not an empty cage, rather than a higher desire to create a truth, or honestly depict animal research. It was finding a way of allowing the animal body to be activated through discussions, and to affectively remind participants, and ourselves, what was at stake.

This was not the only experimental procedure relating to model selection. The second is an issue we referred to as 'compro*mice*', where, despite dedicating a lot of time to discussing other model options, particularly macaques and biomatter, teams defaulted to selecting mice. Is 'compromice' a problem? Yes and no. How to approach this question depends on the interpretation of *Vector*'s aims.

There are numerous routes to take through *Vector*. However, if each team selects the mouse as their model organism, then the diversity of options and decisions, and therefore insights, that are shared with the wider group are reduced.

This became another ethical experiment we wrestled with; who is the author of the experience? On one hand it was imperative that participants retained agency over their decisions but on the other we knew the worlds being created were small and *Vector* could offer so much more. This initial decision is critical, but the crucial element is not *which* model is selected but more that a model is *selected together*. If that decision is taken from the group or compromised in some way, it erases ownership and makes participants less complicit in the consequences that follow. Since complicity is an affective manifestation of agency and therefore critical to immersion, solving this issue was central to our methodology and crucial to the success of the experience.

The structure of *Vector*, with its three teams and three rounds of decision-making, was designed to bring as many stories into the experience as possible. Each team would benefit from how others explored the world and made decisions in it. Therefore, as developers, we decided the performance, participants, and overall experience would be enriched by each team selecting a different model organism. This would open new storylines that posed different ethical decisions across the worlds of animal research. We thought critically about how to achieve this without participants feeling like choices had been taken away or that they were being influenced to act in a certain way.

We revisited all our post-performance debriefs and looked for new insights. In a post-show debrief, one participant had mentioned that their team really wanted to use biomatter, but they also wanted to win, and it seemed just too experimental: 'That question mark was just too big!'[27]

It turns out this was quite literal as the original biomatter factsheet did not give a likelihood of success as a percentage, like all the others, but instead had a question mark (see Figure 16.7). We replaced the question mark with a range of 0–100% and added a bullet point to say as biomatter was novel and experimental Biocore could not yet give a precise rating (see Figure 16.8). Changing this one detail seemed to transform a concern for the unknown into an intriguing opportunity in participants' imaginations. Since this edit, biomatter has been selected by at least one team in each performance.

Next, we decided to try to offer more context to the controversial macaques. We selected one of the host institutions of the AnNex and used their most recently published animal research figures to add the 'number of animals used by Biocore, 2020' to each model organism factsheet. We knew this would be zero for macaques – we

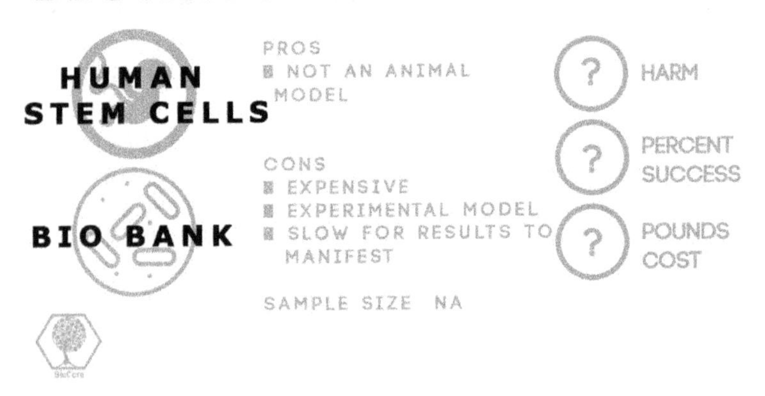

Figure 16.7 Original Biocore biomatter profile card, including '?' to represent the unknown potential of biomatter (2019) (Copyright: Bentley Crudgington/The Lab Collective).

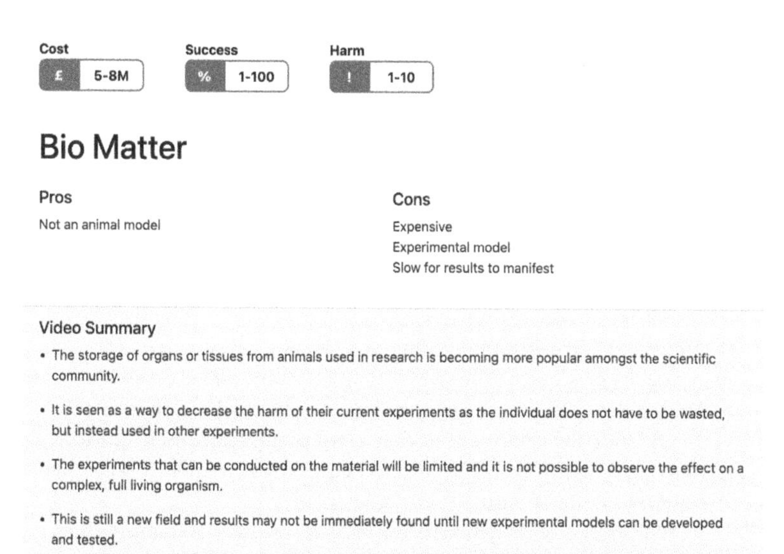

Figure 16.8 Updated Biocore biomatter profile card based on participant feedback. The question marks have been replaced with min–max scales. While these still represent an unknown and not-yet-quantified performance rating, the inclusion of a scale subtly introduces the potential of success (Copyright: Bentley Crudgington/The Lab Collective).

added into the introduction for this species that previous use of macaques and similar non-human primates had been phased out across all Biocore labs, but considering the complexity and urgency of the pandemic Biocore thought it important to now offer this option to the public.

To tackle the 'compromice' issue we introduced new storylines and some additional hard programming that links specific outcomes to specific teams. If, for example, you play as team Oak and select mice for your model organism you will receive an email to say that the mice strain used in the lab is not susceptible to the virus. However, if you play the same as team Elm, the email says the mice are highly susceptible to the virus and therefore mice cannot be used without breaching severity ratings. There is a prohibitively expensive option of breeding a brand-new mouse strain or participants can change their animal model. Each team is given two options of alternative animal models, one with a lower harm rating than mice and one higher, e.g., Team Elm can move to macaques or fish, and Oak to either pigs or biomatter. This now prevents all but one team using mice without compromising participant agency.

Intermission: Covid-19

> 'Which brings us to why we're here today – as you've probably seen in the news a global crisis is unfolding. A new zoonotic disease, Xenna, is marching its way across the world killing humans and animals.
>
> Declared a pandemic, what makes Xenna so dangerous is that you may not even know you have it – symptoms take up to fourteen days to manifest from being infected either by our spiky friend the mosquito or contact with an infected person.
>
> You'll present with symptoms such as fever, muscle pain, swollen glands, and lethargy – symptoms that in fact seem flu-like, but which end in haemorrhagic fever, and ultimately death.
>
> There have been widespread protests across the UK to apply pressure to the government in finding an immunisation for this dangerous disease, particularly as the UK's first case was identified in Slough last week.'[28]

Vector was researched and developed in 2018 and premiered and toured in 2019, all pre-Covid-19. The above is an excerpt from Dr Arber's welcome statement. What had been a brilliant narrative device in early 2019 felt almost unethical by 2020. Dr Arber's words that open this chapter are from our current post-Covid script, which we have used in physical and online versions of *Vector*. We moved away from the narrative of developing a vaccine for a current outbreak to future-proofing Biocore's research ethical approval processes. In 2018, we knew there would be a pandemic of some form, and we know there will be others in the future.

It feels reductive to say Covid-19 changed everything, but it certainly deeply affected how we felt about this work. It felt too close for a long time, and we questioned if we could return to it. The pandemic itself became another actor-participant we needed to work out how to care for. We took a break from the work until late 2020, when we began again the iterative process of replacing, reducing, and refining content, through play testing and reflective practice.

It is important to remember that *Vector* used a pandemic narrative as a tool to get people talking about animal research; it was never our intention to use animal research as a tool to discuss pandemics or vaccine design. In centring agency we knew that *Vector* would be shaped by what people brought to the table and that now, having had this experience, they would bring new things, or think and feel differently about the existing things. Previously, in post-show de-briefs we had asked, 'If this experience had not been based around developing a vaccine would you have played differently (e.g., cancer, dementia)?'

> Yes – would have spent more time researching as not a pandemic.

> Maybe – I think the idea of impending DISASTER is compelling and conducive to extreme decisions.[29]

However, we never imagined asking: Would you have made different decisions if you had not just lived through this?

It is tempting to reflect on this prospect. The two quotes above are indicative of the pre-pandemic responses to this question, which all relate to the size and scale of the task at hand, suggesting urgency gave permission to transcend certain hesitancies. When asked (in post-show debriefs and surveys) 'What, say, should the patients who potentially benefit from research have?' there was a suggestion that what matters to the individual or smaller collective may not matter at scale, and respondents questioned where expertise on ethical decision-making around animal research is derived from.

> They should have less say than the general population. Their judgement is compromised by their position so they are likely to choose options that are less acceptable to society as a whole.

> I think it is more important for experts to decide on this stuff.[30]

We know that cancer and dementia also operate at tremendous, urgent scales, and we also know from spending the last eighteen months apart, it only takes a kiss or someone else's distant inhalation to result in your diagnosis or to install a new caring responsibility in your life. It is unclear as to why urgency was imagined so asymmetrically. Perhaps it is, as Paul B. Preciado notes while reflecting on the pandemic via the works of Foucault, 'Tell me how your community constructs its political sovereignty and I will tell you what forms your plagues will take.'[31] Anne Boyer writes 'the history of illness is not the history of medicine – it is the history of the world – and the history of having a body could well be the history of what is done to most of us in the name of a few'.[32] Your biopolitical compass, of what you find acceptable to have done to other bodies in order to construct your own, and those you care for, is set by the material and temporal intimacies within which you experience these scales.

As we reframed the narrative and edited some of the content, we cannot make extensive claims about participants' pre- and post-Covid-19 decision-making. There were still broad differences in the decisions made and the reasons behind them, with one new addition. It was only in post-Covid-19 shows that teams explained they ruled out certain species for reasons of trust. They felt the public would be less trusting of a vaccine developed in fish, which risked it failing as a product in the field. It could be that in intensive media coverage of 'anti-vax/vaccine critical/hesitant' voices – such as those questioning, or rejecting, vaccines developed using 'novel' technologies at pace – alerted our actor-participants to the risk of operating outside of societal imaginaries, which can put acceptability and validity in jeopardy, when, for some, necessity is already in question.

The pandemic saw us move *Vector* online; while we had previously had multinational attendees at in-person shows, it felt different to have actor-participants situated in and speaking from different places. In one show, we had an actor-participant from a small town in northern Italy and one from Berlin, Germany, together discussing one of the biomatter storylines. They had received an email from Biocore confirming that their choice to use embryonic stem cells had been unpopular with the public and asking if they would consider switching to a different research model. The team rejected this request on the basis that 'pro-life protesters' were 'extremists' and that these views were not taken seriously.

> This isn't America.[33]

The Berlin actor-participant added further contextualisation by saying that currently in Berlin veganism has more influence than religion. They explained that it is deemed to be rational and a default position to be against the use of animals, adding that they felt, for the average citizen of Berlin, it would be more controversial to move from embryonic stem cells to a research animal, rather than the other way around. It is interesting to see how new 'extremists' are created and by whom and how their imagined irrational approaches push them off the affect axis to become unstable outliers. In another example following a request to swap to a model organism of 'lower sentience', one actor-participant's response perfectly encapsulated our aims in creating *Vector*.

> It does not solve the controversy – it just moves it elsewhere.[34]

As long as animals are used for research they will pose an ethical issue to citizens and society, including those who conduct the research and those who benefit from it and all those inbetween. What we attempted, and feel we succeeded in doing, was demonstrating that such ethical decisions are never resolved, but transformed, transferred, and translated into and onto new obligations and caring responsibilities.

The nexus is not infinite, neither does it end neatly.

Act four: ending well

We have attempted to make a case for immersive theatre as a methodology for citizenly co-creation, where actor-participants are not restricted to how they think they should act but offered radical hospitality that allows them to ask how *might* I act? The question of how I want to be in this world is quite different from how am I allowed to be in this world. We have proposed that *Vector* was an experiment with care as infrastructure, with agency and immersion, and with ethical encounters with anxious bodies. We facilitated this through inter-personal performances that brought together dialogical and relational aesthetics, through a praxis of reconstituting social relations that welcome participants to become immersed and complicit in uncomfortable knowledge on their own terms. We have explored how care is required when inviting citizens to participate in knowledge practices that might produce things they do not want to know, and that it matters less how we theorise the care we seek to culture and more how that care is experienced.

Some worlds are bigger than others. Immersive theatre does not have to be done at scale but can offer both the researcher and the engagement professional mechanisms for caring for citizenly encounters. It asks you to consider who you are casting in what role, what are you asking them to perform, and if your staging and scripting allows for experimentation or only delivery. Immersive theatre should prevent you from pre-defining what your outcomes will be but instead ask what might be created together, and avoids the unethical practice of demanding a citizen publicly destroys their view of the world in order to participate in yours. We feel these reflections and methodologies could be transformative across the engagement practices in science and technology studies and are not just limited to animal research.

To quote from our own script, *Vector*

> 'critiques utilitarian frameworks used to authorise animal research, unmasks the complex ways in which ethical responsibilities are distributed and enacted around structured decision-making processes, and reveals how increasing demands for openness and translational research extend responsibilities for care across professional roles. It enacts the inter-relations between scientific research, human health, and animal welfare that are held together through ethical practices and social norms embodied in governance, regulation, and care.'[35]

It is our nexus. However, this does not mean it is complete.

In recent iterations we end the show by displaying a QR code and explaining that there are many AWERBs who do not have lay (or, as we say, citizen) members, and if people are interested in applying what they learnt at Biocore to other ethical bodies they can scan the code to be introduced to their local AWERB.

The page they are directed to reads:

> We are thrilled participating in Vector inspired you to get involved with your local Animal Welfare and Ethical Review Body (AWERB).
>
> 'The most ethical' will always depend on who was in the room.
>
> Unfortunately, there are currently no direct ways for you to do this. There are no publicly available contacts, no sign-up sheets, and no referral process.

We could tell you what percentage of people scan that code but until the problem of who is invited to become immersed is solved, it would be redundant.

Notes

1 *Vector* script was written and edited by Amy Fleming, Natalie Scott, Joe Thorpe, and Bentley Crudgington. Last edited June 2021. More about *Vector* is available at https://vector.incidentallyb.com/ [accessed 1 February 2023].
2 Carol Nicholson et al., 'Postmodernism and the Present State of Integrative Studies: A Reply to Benson and His Critics' (Association for Interdisciplinary Studies), 1987, www.our.oakland.edu/handle/103 23/4023 [accessed 9 May 2022]. Renelle McGlacken and Pru Hobson-West, 'Critiquing Imaginaries of "the Public" in UK Dialogue around Animal Research: Insights from the Mass Observation Project', *Studies in History and Philosophy of Science*, 91 (2022), 280–287, DOI: 10.1016/j.shpsa.2021.12.009.
3 Kjetil Rommetveit and Brian Wynne, 'Technoscience, Imagined Publics and Public Imaginations', *Public Understanding of Science*, 26.2 (2017), 133–147, DOI: 10.1177/0963662516663057.
4 COPUS (Committee on the Public Understanding of Science), 'To Know Science is to Love It? Observations from Public Understanding of Science Research', 1998, *PUS*.
5 Erving Goffman, *Frame Analysis: An Essay on the Organization of Experience* (Cambridge, MA: Harvard University Press, 1974), pp. ix, 586.
6 V. Angelaki, *The Plays of Martin Crimp: Making Theatre Strange*, p. 15 (Basingstoke: Palgrave Macmillan, 2012).
7 Crudgington, Bentley, and Gail Davies. 'Publics, Pandemics, and the Performance of Ethical Review', *The Polyphony Blog* (15 May 2020), https://thepolyphony.org/2020/05/15/publics-pandemics-and-the-per formance-of-ethical-review/ [accessed 9 May 2022].
8 Matt Hills, *Fan Cultures* (London; New York: Psychology Press, 2002).
9 Steve Rayner, 'Uncomfortable Knowledge: The Social Construction of Ignorance in Science and Environmental Policy Discourses', *Economy and Society*, 41.1 (2012), 107–125, DOI: 10.1080/03085147.2011.637335.
10 Renelle McGlacken, '(Not) Knowing and (Not) Caring About Animal Research: An Analysis of Writing From the Mass Observation Project',

Science & Technology Studies, 35.3 (2021), 2–20, DOI: 10.23987/sts.102496.

11 Gilles Deleuze and Félix Guattari, *A Thousand Plateaus: Capitalism and Schizophrenia* (London: Bloomsbury Publishing, 1988).

12 Benedictus de Spinoza, *A Spinoza Reader: The Ethics and Other Works* (Princeton, NJ: Princeton University Press, 1994).

13 Henri Bergson and Mabelle L. Andison, *The Creative Mind: An Introduction to Metaphysics* (New York: Dover Publications, Inc, 2010).

14 *The Affective Turn: Theorizing the Social*, ed. by Patricia Ticineto Clough and Jean Halley (Durham, NC: Duke University Press, 2007).

15 Brian Massumi, *Parables for the Virtual: Movement, Affect, Sensation* (Durham, NC: Duke University Press, 2002).

16 *Affective Performance and Cognitive Science: Body, Brain and Being*, ed. by Melissa Trimingham (London: Bloomsbury, 2013).

17 Matthew Reason and Anja Mølle Lindelof, *Experiencing Liveness in Contemporary Performance: Interdisciplinary Perspectives* (London: Taylor & Francis, 2016), p. 85.

18 Robin Millar and Brian Wynne, 'Public Understanding of Science: From Contents to Processes', *International Journal of Science Education*, 10.4 (1988), 388–398, DOI: 10.1080/0950069880100406.

19 *It's All Allowed: The Performances of Adrian Howells*, ed. by Deirdre Heddon and Dominic Johnson (London, Bristol: Live Art Development Agency and Intellect Books, 2016).

20 Grant Kester, 'Conversation Pieces: The Role of Dialogue in Socially-Engaged Art', in *Theory in Contemporary Art Since 1985*, ed. by Zoya Kucor and Simon Leung (London: Blackwell, 2005), pp. 76–100.

21 Nicolas Bourriaud et al., *Relational Aesthetics* (Dijon: Les presses du réel, 2002).

22 Annemarie Mol, *The Logic of Care: Health and the Problem of Patient Choice* (London: Routledge, 2008), DOI: 10.4324/9780203927076.

23 'Matters of Care' [n.d.], University of Minnesota Press, www.upress.umn.edu/book-division/books/matters-of-care [accessed 9 May 2022].

24 Vectors are a form of digital visual image comprised of geometric shapes, such as points, lines, and curves. They differ from cartoons or sketches as they are simplified abstracted images that aim to convey the recognisable essence of the subject/object rather than a realistic representation.

25 Chris Shilling, 'Sociology and the Body: Classical Traditions and New Agendas', *The Sociological Review*, 55.1_suppl (2007), 1–18, DOI: 10.1111/j.1467-954X.2007.00689.x.

26 Nick Crossley, 'Body-Subject/Body-Power: Agency, Inscription and Control in Foucault and Merleau-Ponty', *Body & Society*, 2.2 (1996), 99–116, DOI: 10.1177/1357034X96002002006.
27 Anonymous quote about selecting biomatter, from *Vector* participant, gathered during post-show debrief using a 'pot luck' survey. Oxford Festival of Ideas, Old Fire Station, October 2019.
28 Excerpt taken from the original *Vector* welcome introduction, delivered by Dr Arber, pre-Covid-19.
29 Anonymous quotes from *Vector* participant gathered during post-show debrief using a 'pot luck' survey. Manchester, Niamos Radical Arts Centre, March 2019.
30 Anonymous quotes from *Vector* participants gathered during post-show debrief using a 'pot luck' survey. Oxford Festival of Ideas, Old Fire Station, October 2019.
31 Paul B. Preciado. 'Learning from the Virus', *Artforum*, May–June (2022). www.artforum.com/print/202005/paul-b-preciado-82823 [accessed 4 July 2022].
32 Anne Boyer, The Undying (London: Penguin, 2020).
33 Anonymous quote from *Vector* participant gathered during online show. Berlin Science Week, November 2021.
34 Anonymous quote from *Vector* participant gathered during post-show debrief using a 'pot luck' survey. Oxford Festival of Ideas, Old Fire Station, October 2019.
35 Adapted from Gail Davies et al., 'Animal Research Nexus: A New Approach to the Connections between Science, Health and Animal Welfare', *Medical Humanities*, 46.4 (2020), 499–511, DOI: 10.1136/medhum-2019-011778.

17

Commentaries on experimenting with openness and engagement

Edited by Emma Roe

This chapter focuses on experimentations with openness and engagement with animal research. It presents three separate commentaries – two from invited respondents to this chapter of the book, and one from the chapter editor. The first commentary is from Bella Lear, a social researcher and science communicator who works to drive and support change in the animal research sector. Her commentary charts changes to the openness agendas in animal research from the perspective of someone closely involved with those changes. She reflects on how the three chapters in this section create new points of entry to discussions about animal research, which can add dynamism to debates. The second commentary is from Louise Mackenzie, an artist who experiments with the imaginative possibilities of extending animal welfare and care to all manner of organisms. This commentary brings artistic practice into conversation with the three chapters, arguing that honesty and truth are at stake in how openness is performed, for whom, and for what purpose. The section editor's closing commentary looks across all of the book chapters and commentaries, with the aim of identifying key themes. In this final piece, Roe identifies how the contributors have created activities where participants lead in how, where, and when they engage with animal research, rather than being presented with a preformatted vision or version of animal research. These build to reveal the contours of the animal research industry's contemporary culture of both openness and closedness.

17.1

Changing openness agendas in animal research

Bella Lear

Scientists whose research involves the use of animals – and the ethical, legislative, and animal welfare teams who work alongside them – have changed the ways that they communicate in recent years. As a former researcher and science communicator who worked with the sector to drive and help implement that change, I have witnessed the enormous shifts in perspectives and approaches to the soft skills that underpin this animal use. Care, communication, and resilience are now seen quite differently, following a paradigm shift in how those involved in animal use talk about their work. I have been privileged to be part of that change, and here I discuss how different the working practices around animals in science once were, and the rapid changes that allowed organisations to support the innovative deep-engagement strategies developed by the Animal Research Nexus Programme (AnNex). My reflections are based on hundreds of visits to research facilities, in the UK and beyond, and on my conversations, both formal and informal, with the researchers, technicians, managers, communicators, and senior leaders of research organisations, large and small, commercial and public. Over the years, I tried to persuade them, sometimes more successfully than others, to try a new approach to talking about their research.

Prior to the 'new openness', animal research communities subscribed to a widely held belief that discretion was the better part of valour, whereby avoiding a potentially unpleasant or dangerous situation was the sensible thing to do. It followed that the complex values associated with research animals' interactions with human societies meant public expectations were best met through legislation, which was developed to represent 'society's voice'. In practice, this meant that many institutions kept details of their animal research on a need-to-know basis.

Animal facilities were windowless buildings, hidden away in basements, on top floors or in service-yards. Multiple layers of security were needed for access, and the use of cameras inside them was strictly forbidden, other than as part of the research protocol. Staff, including researchers, were trained not to tell anyone where they worked or what their jobs entailed. Engagement with the public was avoided at all costs, and researchers were forbidden to even showcase their work through science festivals or on their own webpages. In one research university a team of PhD students was refused permission to take a display stand to the local science festival over last minute concerns about security and reputation; others only admitted vetted participants to carefully managed public events, which were planned to avoid awkward questions or disruption. Support and administrative staff, students and – in some cases – senior managers at UK universities had no idea that the animal facility existed, and I attended events where senior figures proudly, and incorrectly, declared that their institution did not use animals in its biomedical science. At the time it all seemed to make sense, as talking about these sensitive topics would draw unwanted attention to a clear 'PR own-goal'.

This changed with the adoption of openness and transparency initiatives by the UK life sciences sector,[1] which followed years of work by the Science Media Centre, RDS, Coalition for Medical Progress, the Wellcome Trust, and others to encourage those who used animals in science to discuss it publicly.[2] This 'new openness' was pioneered by the well-documented media campaign of the University of Leicester, as it responded to protests at the building of their new animal facility, culminating in a public opening of the building in 2012.[3] The Concordat on Openness on Animal Research in the UK, initiated in 2012 and finally launched in 2014, used underpinning public dialogue work to consider public expectations around openness and animal research.[4] It asked what people wanted and felt that they should know about this challenging topic; it also aimed to show the leaders of research organisations how they could communicate more effectively without catastrophe. It supported the aims of governance bodies such as Research Councils UK, National Co-ordinating Centre for Public Engagement, and Sciencewise[5] in creating a policy change that facilitated public engagement with research.

Communication with wider publics, beyond those who worked with animals, was the focus from the outset, but within the first few years it became clear that conversations within organisations were equally critical. Internal discussions encouraged those involved with animal research to think again about how and why it was conducted. They increased the visibility of animal research within an organisation, so that it needed to be more accountable. The positive impacts of this were wide ranging, including higher quality applications for animal technician positions (now openly advertised) and reported improvements to animal welfare related to greater investment and oversight.

Participation in public discussion, with its potential to mitigate the risks around a controversial topic, was of course an important motivator for driving the change, but contrary to the assumptions of some, the Concordat was not developed to shape or dictate public acceptance, nor to win approval for the use of animals in science. Rather, the Concordat's aim was to offer organisations ideas and practices, enabled through high-level commitments, to help them demonstrate their motivations and considerations when dealing with ethically complex research. Many signatory organisations committed to a considerable overhaul of the information they provided publicly through their websites, taking information that had in the past been confined to intranets and making it fully accessible. Links to these collected 'Concordat websites' were made accessible through a single, easily located webpage. In addition, these websites were required to be easily located by an individual browsing or searching for them.[6]

An important aspect of these commitments was that they should support the media with access to reasonable and balanced information about animal research, so that they could, in turn, present a fair perspective to the public. Public engagement activities, which focus on two-way, often deliberative activities involving 'outsiders' and non-specialist audiences, are another important aspect of these public-facing initiatives. However, it requires strong groundwork in institutional transparency along with fresh thinking to initiate programmes that are truly innovative and engaging, so these are often difficult to fully realise.

The 'openness' shaped through the UK Concordat has given rise to similar *transparency agreements* across the EU and worldwide, with New Zealand recently launching their version, and Australia and the US both working to develop similar codes of practice. Like the UK's Concordat on Openness, these agreements are essentially sector focused, and the key actors are the research institutions themselves and the staff that work within them. Unlike the Concordat on Openness, these later transparency agreements are not founded in deliberative process, and most take the Concordat on Openness as their starting point.

There is now substantial information about the use of animals in research in the public domain but, while the landscape has shifted, barriers still remain. People need to be motivated to find out about research in the first place; they may need to ask the 'right' questions, and feel empowered to enter an unfamiliar, scientific space. Members of the public reviewing Concordat signatories' websites about animal research reported to me in my role at Understanding Animal Research (UAR) that the language is challenging, and the expected level of readers' education is high. Even when publics are motivated to question animal research, the moral complexities of the issues, and the narratives of both science-focused and active disinformation campaigns make this area challenging to engage with.

Animal research is often said to be one of the most intensely regulated areas of research, and that regulation brings established, deeply considered values and ethics, supported by rehearsed narratives. The existing utilitarian framework that underpins them focuses on the benefits to (human) knowledge traded against the sacrifice made by animals in support of our society. Animal welfare has a rich history and has developed in support of these ethics, demonstrating not only the importance of knowing about anatomy, physiology, and the disciplines they underpin, but also the value of the animals to researchers and the need to care and provide for them. These values and narratives have become accepted and embodied by the research community, so that they have become internalised and represent deep feelings for many. Developed over many years to address societal concerns, the constructed narratives to attend to animal welfare serve real purpose: they ensure that those carrying out research on animals give due attention to the ethical landscape in which they work.

Yet, despite the benefits of greater openness, it can also present risks that the public debate around this complex moral issue becomes static. Social benefit from public discussion about animal research will be limited if the script is already written, and the research community fails to move with an evolving society, listening to their thoughts and feelings on a complex and shifting ethical issue. However, breaking free of the usual conversations to allow the possibility of something new is as challenging as it is important.

The chapters in this section present three ways of creating new points of entry to discussions about animal research. Each one brings an intention to step beyond the existing narrative around animals in science and introduces new perspectives, inviting novel conversations on an old theme. Taking openness and transparency around animal research as a starting point, they show how shifting to different frameworks for engagement can allow access to complex moral and ethical perspectives, which challenge the familiar framing of animals used in research. Each project expands the notion of 'openness' beyond a policy focus, and into a more conceptual discussion that reimagines aspects of society's relationship with research animals. They explore deep questions about the ethical frameworks that shape existing research practices and provide the space to reflect on moral complexities and inconsistencies without criticism.

In The Mouse Exchange (Chapter 14), conversations about how animals are used in research are provoked as participants engage in crafting felt mice. On first hearing about this programme, the researchers, technicians, vets, and others who make up the core of the community working on animals in science seemed incredulous. They found it hard to imagine how a crafting activity would communicate the practices, realities, and even ethical decisions of their workplaces. The concept felt epistemologically and ideologically distant from their work and ways of communicating. This work was certainly different, inviting public participation with a low threshold, yet enabling discussions of complex emotional spaces that even the most seasoned of practitioners find it challenging to articulate.

Taking a different approach to previous engagement exercises around the use of animals in research, The Mouse Exchange did not start from openness-as-resilience. The aim was not to show why animals are important to science, or to counter circulating misinformation. Instead it allowed any interested parties to participate in or influence polarised discussions about rights or wrongs, good or bad, caring or callous approaches to animals. This starting point required a new approach, and participants were invited to shape and create something that they cared for.

The physical task of crafting a 'mouse' and creating a story that imagined it as a being gives participants a small taste of what it is to care for an animal. As with laboratory mice, participants are also introduced to the idea of intentionally creating an animal with a specific purpose, and what that means for the human creators and carers. The activity invites complex questions about human–animal relationships: how much of the compassion and care we feel for animals when we connect with them is co-created by the human and animal, and how much is the person investing themselves into an object they care for? To what extent is our connection with other animals anthropomorphised? What does it mean to create a living being with a specific purpose?

These are complex questions that are important to all of us, and which are considered, but not answered, by our current ethical frameworks for working with animals. Through these and other ideas of performing animal care, The Mouse Exchange enables public audiences to participate in some of the more challenging discussions that take place among the animal welfare community, yet with general audiences.

In 'Labelling medicines as developed using animals?' (Chapter 15), the authors open up a long-standing deliberation about what is the best way to illustrate personal connection to the subject of animal research. While Lord Professor Winston proposed his Medicinal Labelling Bill in 2013, it was a subject discussed at length among advocates of animal research, and indeed has been used to teach undergraduate research ethics at several universities. Obtaining or taking a medicine, whether it is prescribed or purchased from a pharmacy, is a moment of direct involvement with medical research, yet polls show that few people are aware of the requirement for animals to be used in testing medicines. Patients were provided with direct information about the use of animals in medicine development in a programme developed by the Coalition for Medical Progress, which was later repeated by UAR and the Wellcome Trust to place information leaflets about drug development in pharmacies and GP surgeries, though in practice providing public information through a leaflet-drop proved difficult to evaluate.

The activity described here uses the familiar notion of consumer labelling, and the ethical discussions it inspires about choice, coercion, intention, and positionality, through a creative group activity that provided a focus for participants from different publics to consider what labelling might look like. This gave participants an entry point to consider animal research and its role in society, particularly its relationship with the production of medicines, and where consumers and their ethical perspectives fit into how this is done. The activity saw both researchers and participants realising that providing a neutral statement or comment on such an ethically complex issue is impossible. Attempts at simplicity and neutrality led to further questions and a need for more extensive information, drawing public participants into discussion about the purpose and value of consumer labelling, who benefits from it, and the role such labels play in decision-making.

In 'Building participation through fictional worlds' (Chapter 16), performance art was used to allow groups of public audiences to experience the deliberations and decisions made by an Animal Welfare Ethical Review Body (AWERB) in a way that completely changed their access to, and experience of, the discussions. Researchers were able to create a new type of ethical review, embedded in a fictional scenario, and ask how and why the performative contexts of an AWERB matter to its function and the outcomes it provides.

Since the development of the Concordat on Openness, signatory institutions have been encouraged to not only provide information about animal research in an accessible way, such as on a public-facing website, but also to arrange opportunities for those outside the immediate concern of working with animals in science to join ethics committees, participate in discussion, and provide a public voice to the oversight process. However, while there is plenty of public information available, as well as institutions willing to provide tours, events, and discussions with their AWERBs, there is little appetite among those outside the research community to engage with this topic. The key barrier to public engagement is arguably, as the authors state, not the information provided, or its availability, but its accessibility. Additionally, researchers who use animals, and who seek to engage audiences beyond academia, either through public engagement initiatives or through more formal processes such as seeking their input to AWERB discussions, are limited by the familiarity of didactic knowledge transfer. They themselves work within a profession immersed in formal teaching and learning, making these methods a natural starting point for communications that follow established narratives.

This initiative created a performance experience in which audiences were invited to address research ethics within a creative space as 'invested agents' rather than 'aloof observers'. Altering the way that participants access and engage with AWERB-type discussions gives an opportunity to imagine things differently in a fictional space. Radically changing the approach to consider how animals can, or should, be used in biomedical research helps to question existing assumptions, providing valuable insights for understanding, training, and policy deliberations.

In conclusion, relatively recent changes to the communication of animal research have moved an uncertain and concerned sector from silence to a recognition that communication with those beyond their immediate professions is not only possible, but desirable. Many researchers and institutional representatives want to tell their side of the animal research story. They hope to show the care they take to minimise harms and support the benefits of their research, hoping that when publics understand their motivations better they will be sympathetic to the research. Indeed, public dialogues[7] have shown that audiences care deeply about the motivations of scientists when considering the potential harms and benefits of research, particularly when the subject matter seems ethically contentious or inaccessible. In welcoming new audiences into these discussions, the three engagement programmes outlined here begin the process of developing new critiques and deliberations of how and why animals are used in research.

17.2

Can I be honest? Querying kinship and communication in animal research

Louise Mackenzie

Our deep societal entanglement with animal research is already a foregone conclusion.[8] This is a difficult reality for many of us, but one that we don't have to think about regularly, due to the relative secrecy within which animal research takes place. As a child in the 1980s, I grew up during a period of extreme animal rights activism in the UK (the chapters in this section adopt a UK-centric perspective on the whole, and therefore I will too). It was difficult not to be aware of how animals suffered at the hands of humans, with public demonstrations and television advertising campaigns showing graphic, bloody imagery,[9] and activist violence towards humans escalating to the placement of explosive devices at researchers' homes.[10] This undoubtedly shaped my own, and many of my generation's, attitudes towards the use of animals by humans. I remember feeling angry and conflicted. How could humans do these things to animals? And to each other?

Changes in legislation followed, and while reporting on animal research in the UK is publicly available, national levels of awareness are certainly lower than they were towards the end of the twentieth century. This is where the authors of the chapters in this section step in. In an academic culture of interdisciplinarity and collaborative working, researchers are finding new ways in which to engage the public in the still-vital questions around animal research. The chapters in this section turn to creative strategies, drawing from design, craft, gaming, performance art, and theatre to explore different approaches towards opening up animal research to the public in the UK.

For the authors who have contributed to these chapters, practices of animal research are tacitly accepted and understood, to the extent that the question is not, as Patricia MacCormack asks, whether they should exist at all[11] but rather, as Jacques Derrida enquires, how they can exist well.[12] Through years of specialism and increased mass production, most of us have become so distant from processes that work with the visceral materiality of the animal that we can no longer relate to the animal in the food that we eat, nor less see any connection between our medical care and the humble mouse. While we may acknowledge that animal research exists, it is still so far removed from our day-to-day experience that we have perhaps never contemplated the many facets that comprise animal research, from production or breeding, to purchasing, maintenance, and ultimately disposal.

As an artist, I don't have to sit on the fence when reflecting upon animal research. In fact, my research playfully jumps off the fence and dives deep into the imaginative possibility of extending animal welfare all the way down to single-celled organisms. If we could start here, with the utmost respect for the smallest motes of life, it might be possible to reimagine working practices and definitions of care. I like to think that we have a deep, ancestral kinship with all forms of life – from the smallest living organisms, free-floating on the air or in the ocean, to the great variety of plant and animal species with whom we share the planet.

I first identified this sense of kinship whilst making *Oltramarino*,[13] where my research into human uses of micro-algae led me to understand that chloroplast-bearing single-celled organisms were responsible for the Great Oxidation Event around 2.5 billion years ago and thus ultimately responsible for all current forms of life on Earth. Kinship is an interesting word: conceptually it offers a breadth of scale and at the same time encompasses a sense of duty. It implies a form of dependency akin to a familial bond, something that feminist science scholar Donna Haraway describes as a 'mutual, obligatory, non-optional, you-can't-just-cast-that-away-when-it-gets-inconvenient, enduring relatedness that carries consequences'.[14] It is this enduring nature of kinship, and how it extends beyond the face-to-face relationship, that brings me to my question in the context of these chapters, *can I be honest?*

Being honest in the face of animal research is not straightforward. Who needs to be honest and with whom? One might assume that the parties involved are the animal research community and the public. As McGlacken and Hobson-West identify in Chapter 15, 'enactments of openness around animal research have largely treated openness as an end in itself', with an increasing number of institutions sharing details of their animal research online and in published reports. But as all the authors in this volume attest, there is a need to move beyond one-way information sharing as fulfilment of an obligation and instead approach openness through the generation of dialogue. Scientists, recipients of the benefits of animal research, their extended networks of family and friends, and research animals all share a kinship here. Being honest about animal research demands a conversation amongst all those whose enduring relationships have consequences for one another. The requirement for honesty, therefore, is multi-directional and multi-layered.

Can I be completely honest with you, reader? My own artistic relationship to animal research is complex. In the research project, Evolution of the Subject, I chose to learn about genetic modification as an artist, to understand this technology from two perspectives: my own as an artist learning how to manipulate life genetically and from the perspective of the organism subjected to genetic modification. I became interested in the subject when I realised that genetic modification was no longer the preserve of scientific research, but extended to the realm of artists[15] and I felt compelled to understand it better. While I wanted to explore what it meant to manipulate life at the level of the gene, I knew that I did not want to involve animals in my research. My preferred choice was to modify my own cells (*in vitro*), but this was not within the scope of my collaboration and therefore we decided upon a micro-organism, the laboratory workhorse, *E. coli*. I did not anticipate quite how attached I would become to these tiny organisms, nor – paradoxically – how casually I would disregard their lives after spending years working with them in the laboratory.

I share this with you as I believe that without this experience, I could not have fully understood the implications of using another form of life as a resource. Ultimately, this is what all forms of life used in research are: resources for human use. This is still an uncomfortable truth, as the 2021 annual report from the UK Concordat on Openness on Animal Research acknowledges: 'accurate communication of harms done to animals in research remains a difficult topic for the research community'.[16] Which leads me to another question – what exactly are we being honest about? There is undoubtedly greater openness in terms of how animal research is reported, but what information is being shared and to what end?

The chapters in this section all deal with the question of transparency, which, in the context of animal research, has various meanings dependent upon whether it is used by animal researchers, the government, or animal protection advocates.[17] Each chapter offers a different approach for engaging with the public, which moves towards a greater transparency around animal research, but equally, each acknowledges that doing so serves to increase the complexity of 'how openness is navigated and enacted' (McGlacken and Hobson-West, Chapter 15). Two of the chapters discuss approaches that involve creative activity as a stimulus for conversation: the design of a label that declares medicine as tested on animals (McGlacken and Hobson-West, Chapter 15) and the crafting of a felt laboratory mouse to open up discussion around the making and supply of animal research models (Roe, Peres, and Crudgington, Chapter 14). The third chapter draws on gaming theory, live performance, and immersive theatre to develop an experiential world in which participants take on the role of the member of an animal welfare review body (Crudgington, Scott, Thorpe, and Fleming, Chapter 16).

While the approaches taken by each project seem distinct, there are interesting parallels. Each brings aspects of performance into our ethical relations with animal research. It is the nature of this performance that brings contrasting results. In the medicine labels project (Chapter 15), the performance of designing a medicine label maintains a certain remoteness from the question of the animal and thus introduces wider debates around who has the power to create, supply, and consume medicine. In The Mouse Exchange (Chapter 14), the performance of stitching together a fictional mouse focuses the work (and therefore the ensuing ethical discussion) squarely on the subject of the animal. The immersive theatre experience, *Vector* (Chapter 16), widens debate again through the performance of actors in a fictional world that encompasses not only a host of animals, but an array of narrative choices to make with regards to the care of each. In every case, there is performance of an action that mimics the real, rather than contends directly with it, thus distancing from the live-ness of animal research.[18]

The concept of the 'mundane' in the medicine labels project brought to mind the work of artist Mierle Ladermen Ukeles, whose focus on the role of mother and maintenance worker as art activity in the 1970s highlighted the unpaid aspects of this routine labour.[19] Ukeles practised domestic activities as art, drawing attention to the ways in which her roles as woman and mother were not assigned equivalent value to other forms of labour. Key to this was the process of enacting the work. Thus, it is the live or lived-ness of Ukeles' actions that confronts us with the truth revealed through this work. I translated my own experience of using life as resource into an interdisciplinary workshop, which has shaped much of my work since. Key elements to this approach were a 'lived experience' of genetic modification, a speculative reflection on the process, and an intention to engage scientists with the public and not the other way around. I invited a mixed group of scientists, artists, and members of the public to join me in performing genetic modification upon live *E. coli*.[20] This liveness was important in generating honest engagement with the subject matter of my research.

What I had not expected was the honesty that would be revealed through the second part of the workshop. After I had guided participants through my approach to working with, caring for, and ultimately modifying these organisms, I invited them to enter an imaginary scenario where they were interviewed by the future kin of the organisms that they had modified. In a dark space and filmed under a spotlight, participants were interviewed as if they were themselves perhaps under the microscope. From behind a screen, multiple disembodied voices asked them questions. The responses, which I developed into the short film *Zone of Inhibition*, were revealing.[21] Participants, perhaps most notably the scientists, were able to dissociate from their day-to-day role and engage in imaginative conversation with these unseen future kin. Something about being freed from convention brought a freshness to their responses. When being asked to reflect on what they had done, the answers were surprisingly from the heart, encompassing a spectrum of views on our relationship with the use of life as resource.

This returns me to the question I began with, *can I be honest?* This question lies threefold for me in the context of openness around animal research. Firstly, given our cultural disconnect from animals that we use to better our lives in so many ways, how can we rephrase Derrida's question of 'how to eat well' in the context of animal research? Are researchers ready to be wholly honest with the public about how animal research is performed? What does it mean to open discussion around the number of animal welfare facilities, the number of animals, and the extent to which they suffer? Who wants this level of honesty and for what reasons? These questions are necessary to challenge the prevailing approach to openness. As the authors of these chapters have acknowledged, it is not enough to want to engage the public with information that already exists. The authors in this volume take the next step, by finding ways in which to creatively engage the public in animal research, which has led to valuable dialogue.

The second aspect of this question, then, focuses on honesty in the context of representation. In choosing to perform openness, what do we mask by not offering the real but instead a representation of it? How does this inform the outcomes? By contrast, what truths are revealed through allowing members of the public, or indeed the scientific community, to be freed from their assumed roles through imaginary scenarios?

Finally, and perhaps most importantly, who are we being honest with? Which publics need to be engaged and why? To draw from the field of participatory arts, 'In short, whose interests are being served by [the] project?'[22] This seems the logical next step. By broadening who we have honest conversations with to include scientific researchers, government advisors, animal laboratory technicians, and others who are physically involved in animal research, the honest truth of animal research can become as multi-layered and multi-dimensional as it needs to be. The question which then remains is, how can we handle this truth?

17.3

Are we asking the right questions about openness?

Emma Roe

There has been a recent shift in focus around communication about animal research following the establishment of the Concordat on

Openness in Animal Research in 2014.[23] Institutions involved in animal research were invited to commit to being more open about their use of animals and many signed up, changing conversations within the industry. The Concordat has also altered the ways in which social scientists can engage with animal research. I am one of the lead social scientists in the Animal Research Nexus Programme and led the team behind The Mouse Exchange (see Chapter 14), which received an Openness in Animal Research award in 2020.[24] The Mouse Exchange is one of the ways that we, and the other authors in this section, have been experimenting with alternative methods of engaging new audiences and evolving conversations around animal research. The Concordat is often cited as a vital context for the increasing experimentation in engagements around animal research in these chapters. Lear explains in her commentary (this chapter) how the Concordat sought to support signatories in communicating their motivations and considerations when dealing with ethically complex research. This has been achieved by providing resources to scientists and increasing the information available to the public. In this commentary, I want to put this legacy into dialogue with the thinking about animals, openness, encounters, and experimental forms of enquiry in the social sciences that have developed in parallel.

There has been a rapid growth of social science and humanities research on the complex relationships between human and animal lives. This gives particular consideration to the human response-ability for the quality of the lives we give animals when this is bound up, often from their very creation, with human interests. These studies have driven methodological innovation around how to co-design research with sentient animals,[25] how to speak of animals in a way that does justice to their species-specific experiences of curiosity, ambivalence, or disinterest in us. These studies also remind us that we might learn something useful if we pay attention to the location of disinterest and ambivalence towards various animal roles amongst humans. In 2016, we worked to develop a collaborative agenda for social science and humanities research into animal research, which identified the following question related to public attitudes and engagement in animal research: 'Where are the opportunities for greater and meaningful public and stakeholder engagement in the policy and practices of animal research?'[26] The chapters in this section illustrate playful, speculative, and provocative approaches to addressing this question, broadening thinking about engagement and openness, for whom and how, informed by emerging perspectives from the social sciences. The chapters explain what it means to create spaces where the public can be heard as they are invited to play roles in animal research: to participate as maker and carer of a research mouse; to act as members of an animal welfare and ethics review board; or to perform as designer of speculative drug packaging that is open about the use of animals in drug testing and development.

Yet, when experimenting with openness and engagement in relation to animal research, it also seems important to acknowledge the diversity of species, strains, sub-strains, and individual animals at the centre of animal research, and to reflect on *their* ongoing experiments that involve engaging with humans, and where their vulnerability – *their* openness as one might put it – finds them. This is something Despret[27] and Haraway[28] each recognise in relation to animals – what if we asked animals the right questions? If we are asking ourselves whether we are asking the right questions when we open up public engagement with animal research, perhaps we should also consider more closely animals' openness to engagement with us.

Research animals in their highly variable form are often genetically manipulated in their making, and are set tasks or given treatments that often end with them being killed for dissection, tissue extraction, or in mass cullings. While these animals live in a highly controlled environment, abstracted from their ecological niche in the wild, this gains them a life free from predation and disease in the wild, but they are vulnerable, alongside the humans that care for them. Faced with the peculiarity of research animals, where can we learn how to ask the right questions to animals embedded within animal research?

In Western cultures, we learn to ask questions about animals in contexts that find it acceptable to treat animals as a resource to meet a variety of human needs: to farm and eat; to love and care for as a pet; to enthusiastically advocate for their conservation as wildlife; or to experiment on as a research animal. These are associated with variable expressions of concern. People can care and not care, know and not know.[29] Explaining how this happens in practice and what changes the outcome is related to the context of the encounter between humans and animals. Questions raised in the chapters in this section of the book include how the form and context in which the public currently gets to know animal research affects how they respond to laboratory animals, and what questions they want to ask. These questions about context are important given the dominant motivations for openness and transparency. The terms differ conceptually but are often blurred in practical discussions about opening up animal research. Three discourses or frames have been identified around the concept of 'transparency' in animal research.[30] These can be summarised as an ethical responsibility demanded by animal protection groups; a secretive industry's counter-move to misinformation; or a branding strategy of the animal research industry by science funders and government as part of their accountability processes, to build trust in science.

The stories that are often put into the public realm tend to reflect these interests, for example, in telling stories about eye-catching scientific outputs that involve the use of animals. What are absent are stories that address how we might be open to responding to the inherent vulnerability and diversity that comes from being a research animal in the first place. The current status quo in the UK is that both the animals bred for use in research and animals in research are hidden away from mainstream society; could or should this be otherwise? While there is a biosecurity rationale for separation and a lingering security concern, do both animals bred for research, and animals in research, need to remain hidden within innocuous buildings and basements? Is the context the same for all species used in research? Could a more varied selection of representations become mainstreamed in the case of research animals, and if so with what consequences for their future? We could compare here farm animals, who are increasingly housed indoors, yet lived historically alongside humans, leaving a legacy of friendly and concerning representations in the public sphere – from children's books, films of farm animal adventure, and petting farms, to super-dairy farms and intensive chicken farms, with rivers being polluted by their waste.

Techniques and technologies are also used to frame openness, often building in a particular and sometimes limiting vision of both how to engage and why someone would be curious about and interested in animal research. Hobson-West identifies 'public' participation in animal research governance as being constructed through public opinion surveys, which denote this 'public' as politically neutral in contrast to those who are members of 'social movements'. Indeed, Hobson-West argues that the influential MORI survey actively homogenises and depoliticises the audience.[31] There is little attempt to differentiate by gender, social status, culture, or experience, and no sensitivity to creating opportunities for people to show how, where, and when animal research matters to them as evaluative beings negotiating ethics and morals.[32] Indeed, one could instead suggest that strategic ambivalence to animal research[33] is normalised through imagining a public who appear to not care enough to investigate, or to find out more, or to seek out the gentle flow of information into the public domain through the internet, laboratory open days, science engagement festivals, or news items. Or a public that does not choose to expose their vulnerable selves to presumed painful truths about animal research. A recent study of the Mass Observation Project archive argues there is a relationship between certain assumptions about what the public thinks and wants, and how in turn this influences changes to the practice of animal research.[34] There are consequences to the style and form of engagement.

The chapters in this section demonstrate an interest in both innovating and challenging what the everyday experience of the 'openness' agenda could be, in part by opening up the interpretive possibilities of 'openness' itself. They bring being open about animal research into everyday life, experimenting practically with invitations to engage with animal research as a route to continuing the evolution of the conversation. The chapters discuss outputs that aim to deliver public engagement that can explore the 'changing ways in which scientific practices, research governance and public imaginations connect the, often divergent, domains of science, health and animal welfare'.[35] The authors of these chapters have experience of working with a variety of different industry stakeholders, and often have closely consulted with the industry, yet stand apart from it. The commentaries, drawn from Understanding Animal Research and a practising artist, add different perspectives. Lear (this chapter) describes, from her insider position as public engagement lead within Understanding Animal Research, what led to the birth of the Concordat, with a prevailing sense that there is still an uneasiness about further widening engagement. In contrast, Mackenzie (this chapter) holds the question, 'Can I be honest?' throughout her personal reflections on being an arts practitioner curious about kinship with experimental life, from animal through to the cellular form. She uses the refrain 'Can I be honest?' to speak of what unsettles her about how the industry operates. Readers are left asking themselves about the honesty of their own position about animal research – are they colluding with it – and, more pertinently, could engagement as openness go further with practising honesty?

Overall, the experiments in this section involve taking materials to participants and seeing what they build and what questions they ask, rather than offering them an existing vision of animal research about which to ask questions. The chapters challenge the contours of contemporary cultures of openness by both promoting activities that engage those less familiar with the workings of the animal research industry, and rehearsing the deliberations about why some technologies of the everyday seem impossible within the research industry. The activities shape possibilities for conventional and innovative modes of engagement, which could create new avenues for participants to demonstrate when, rather than how or if, animal research matters to them.

Notes

1 David Payne, 'Defending Science by Opening Up: Lessons from Understanding Animal Research', *Nature Jobs Blog* (2017), http://blogs.nature.com/naturejobs/2017/01/20/defending-science-by-opening-up-lessons-from-understanding-animal-research/ [accessed 14 December 2022].

2 F. Fox, *Beyond the Hype: The Inside Story of Science's Biggest Media Controversies* (London: Elliott & Thompson Limited, 2022).

3 £16 million medical research facility opens at University of Leicester, www.youtube.com/watch?v=abvgpZiCiX8 [accessed 15 June 2023].
4 Sciencewise and Ipsos MORI, Openness in Animal Research Dialogue Report, 201, https://sciencewise.org.uk/wp-content/uploads/2019/05/OinAR-Dialogue-Report.pdf [accessed 15 June 2023].
5 Sciencewise Guiding Principles, https://sciencewise.org.uk/about-sciencewise/our-guiding-principles/ [accessed 15 June 2023].
6 Concordat on Openness website portal, https://concordatopenness.org.uk/list-of-signatories [accessed 15 June 2023].
7 Darren Bhattachary et al., *Synthetic Biology Public Dialogue* (BBSRC, EPSRC), 2010, www.ukri.org/publications/synthetic-biology-public-dialogue/ [accessed].
8 Gail Davies et al., 'Developing a Collaborative Agenda for Humanities and Social Scientific Research on Laboratory Animal Science and Welfare', *PLOS ONE*, 11.7 (2016), 1–12, DOI: 10.1371/journal.pone.0158791.
9 Janice Li, 'Dumb Animals: Lynx's Campaign against the Fur Industry', *V&A Blog* (2019), www.vam.ac.uk/blog/projects/dumb-animals-lynxs-campaign-against-the-fur-industry [accessed 15 June 2023].
10 Understanding Animal Research, 'AnimalRightsExtremism. Info', n.d., www.animalrightsextremism.info/animal-rights-extremism/history-of-animal-rights-extremism/violent-extremism/ [accessed 13 December 2021].
11 Patricia MacCormack, *The Ahuman Manifesto: Activism for the End of the Anthropocene* (London: Bloomsbury Publishing, 2020), www.bloomsbury.com/uk/ahuman-manifesto-9781350081093/ [accessed 15 June 2023].
12 I refer here to Patricia MacCormack's ahuman philosophy, which advocates for the abolition of all animal use by humans and contrast this with Jacques Derrida's question of the animal other, 'the question is no longer one of knowing if it is "good" to eat the other ... nor of knowing which other ... the living or the nonliving, man or animal, but since *one must* eat ... how for goodness' sake should one *eat well*?'. See Derrida, Jacques, 'Eating Well or the Calculation of the Subject', in *Points ... Interviews, 1974–1994*, ed. by E. Weber, pp. 255–287 (p. 282) (Stanford, CA: Stanford University Press, 1988).
13 Louise Mackenzie, 'Oltramarino', Newcastle 2013, www.loumackenzie.com/oltramarino [accessed 15 June 2023].
14 Steve Paulson, 'Making Kin: An Interview with Donna Haraway', *LA Review of Books*, 6 December 2019, www.lareviewofbooks.org/article/making-kin-an-interview-with-donna-haraway/, no page [accessed 15 June 2023].

15 Joe Davis, 'Microvenus', *Art Journal* 55.1 (1996), 70–74, www.jstor.org/stable/777811; Eduardo Kac, 'Genesis', 1999, www.ekac.org/geninfo.html; Eduardo Kac, 'GFP Bunny', *Leonardo*, 36.2 (2003), 97–102, DOI: 10.1162/002409403321554125 [accessed 15 June 2023].
16 A. J. Williams and H. Hobson, 'Concordat on Openness on Animal Research in the UK', 2021, https://concordatopenness.org.uk/wp-content/uploads/2021/12/Concordat-Report-2021.pdf [accessed 15 June 2023].
17 Carmen McLeod and Pru Hobson-West, 'Opening up Animal Research and Science–Society Relations? A Thematic Analysis of Transparency Discourses in the United Kingdom', *Public Understanding of Science*, 25.7 (2016), 791–806, DOI: 10.1177/0963662515586320.
18 Arguably, the medicine labels project contends with the reality of medical labels. In this case, it is the material of the label itself that is deliberately distanced from the animal.
19 *Mierle Laderman Ukeles: Maintenance Art*, ed. by Patricia C. Phillips et al. (New York: Prestel Publishing, 2016).
20 Louise Mackenzie, 'Transformation', ACSUS Art & Science, 2017, www.ascus.org.uk/transformational-thinking-through-making-with-life/ [accessed 15 June 2023].
21 Louise Mackenzie, Zone of Inhibition: Relating to the Single Cell through Speculative Performance Practice', *PUBLIC* 30.59 (2019), 56–59, www.publicjournal.ca/59-interspecies-communication/ [accessed 15 June 2023].
22 François Matarasso, *A Restless Art* (London: Gulbenkian Foundation, 2019), p. 104.
23 Understanding Animal Research, *Concordat on Openness on Animal Research in the UK*, 2014, https://concordatopenness.org.uk/ [accessed 26 August 2020]. The Concordat was initially funded by central government but now requires an annual subscription of a few thousand pounds.
24 'Openness Awards 2020', *Concordat on Openness on Animal Research in the UK*, 2020, https://concordatopenness.org.uk/openness-awards-2020 [accessed 22 November 2022].
25 *Participatory Research in More-than-Human Worlds*, ed. by Michelle Bastian et al., Routledge Studies in Human Geography, 67 (London; New York: Routledge, 2017).
26 Davies et al., 'Developing a Collaborative Agenda for Humanities and Social Scientific Research on Laboratory Animal Science and Welfare'.

27 Vinciane Despret (trans. Brett Buchanan), *What Would Animals Say If We Asked the Right Questions?* (Minneapolis, MN: University of Minnesota Press, 2016).
28 Donna J. Haraway, *When Species Meet*, 3 (Minneapolis, MN: University of Minnesota Press, 2008).
29 Renelle McGlacken, '(Not) Knowing and (Not) Caring About Animal Research: An Analysis of Writing From the Mass Observation Project', *Science & Technology Studies*, 35.3 (2021), 2–20, DOI: 10.23987/sts.102496.
30 McLeod and Hobson-West, 'Opening up Animal Research and Science-Society Relations?'.
31 Pru Hobson-West, 'The Role of "Public Opinion" in the UK Animal Research Debate', *Journal of Medical Ethics*, 36.1 (2010), 46–49, DOI: 10.1136/jme.2009.030817.
32 Andrew Sayer, *Why Things Matter to People: Social Science, Values and Ethical Life* (Cambridge: Cambridge University Press, 2011), DOI: 10.1017/CBO9780511734779.
33 McGlacken, '(Not) Knowing and (Not) Caring About Animal Research'.
34 Renelle McGlacken and Pru Hobson-West, 'Critiquing Imaginaries of "the Public" in UK Dialogue around Animal Research: Insights from the Mass Observation Project', *Studies in History and Philosophy of Science*, 91 (2022), 280–287, DOI: 10.1016/j.shpsa.2021.12.009.
35 Gail Davies et al., 'Animal Research Nexus: A New Approach to the Connections between Science, Health and Animal Welfare', *Medical Humanities*, 46.4 (2020), 499–511 (p. 499), DOI: 10.1136/medhum-2019-011778.

Afterword

Carrie Friese

There are key books on the subject of laboratory animals that represent, for me at least, key turning points in the intertwined social processes involved in using, regulating, contesting, and understanding animals in science and society.[1] Many of these books are referenced in this edited volume, and range from French's (1975) *Antivivisection and Medical Science in Victorian Society*,[2] Russell and Burch's (1959) *The Principles of Humane Experimental Technique*,[3] Kean's (1998) *Animal Rights*,[4] and Birke, Arluke and Michael's (2007) *The Sacrifice*[5] among others. The work of the Animal Research Nexus Programme, in *Researching Animal Research*, articulates another turning point in my mapping of the social space of laboratory animals, which includes research regarding that social space. Nexus, or connection, analytically instantiates social processes that forego polarised political conflict, and thus opens up new ways to both conduct and research animal research. I want to consider some directions that this conceptualisation of research animals opens up and makes possible for the future.

One of the findings that emerges throughout this book is that a nexus, as a site of connection, is not straightforward – analytically or in practice. Amy Hinterberger (Chapter 4, p. 114) states: 'if we can't connect, we can't care'. And I am inclined to agree with this statement. But the connections explored and enacted in this book also work to render decipherable disconnects (Gorman, Chapter 2), borderlands (Anderson and Hobson-West, Chapter 9), gaps (Message, Chapter 7), and incommensurables (Giraud, Chapter 8). These disconnections create vulnerabilities (e.g., for horseshoe crabs in Chapter 2), and reproduce hierarchies of knowledge (e.g., for

Named Veterinary Surgeons vis-à-vis the veterinary profession in Chapter 9, and for citizen scientists vis-à-vis wildlife researchers in Palmer, Chapter 10). Indeed, we see in Palmer's chapter a cautionary statement regarding the problems that may arise as the long-standing connections between Home Office inspectors and research establishments are being splintered in the name of efficiency within the UK. But incommensurables also articulate, and thus hold out for, a world in which things could be otherwise (see Giraud, Chapter 8). And so, while I am inclined to agree with Hinterberger, I am also inclined to agree with Giraud's argument that sometimes some people care by *not* connecting (see also Mackenzie, Chapter 17). Being able to hold these two possibilities together, side by side and in the context of animal research, has only become possible with the publication of this book. This allows us to begin to move out of and beyond 'the polarisation cycle', as Dennison puts it (Chapter 13, p. 321).

The disconnects that arise through a focus on connection are important because it is easy to turn a nexus into a normative project, assuming that connection is inherently good. While I, for one, would much rather see the practices of a nexus at work than a rigid hierarchy, this volume shows me that a nexus is still, nonetheless, a political project where power relations take shape. *Invisibility* is one modality through which power operates, in varying ways for different actors and across several case studies in this book. Anyone who is opposed to the use of animals in research has historically been excluded from the animal research nexus in Britain (see Myelnikov, Chapter 1; Tyson, Chapter 4; Davies, Gorman, and King, Chapter 11). Horseshoe crabs are invisible and thus vulnerable as (wild animal) 'replacements' for the use of (laboratory) rabbits in toxicity testing (see Gorman, Chapter 2). The values associated with different species are difficult to render visible in making ethical decisions regarding practices like rehoming laboratory animals (see Skidmore, Chapter 3) or including fish in the orbit of sentient species (see Message, Chapter 7). By rendering the invisible visible, the book is able to ask how animal research might be organised differently, and more justly. With Carbone (Chapter 13) we can ask: why cannot bird ringers, patient participants, and even research animals be co-authors of scientific articles?

But where invisibility renders some vulnerable, when linked to a political economic (or cultural economic) analysis, we as readers also begin to see how invisibility benefits other actors (see Peres and Roe, Chapter 12). People working in industry and in science benefit from the invisibility of the horseshoe crab (see Gorman, Chapter 2; Tyson, Chapter 4). Scientists benefit from the creation of 1.45 million mice that were bred but not used in the UK in 2018 alone. This surplus exists *because* scientists want to be able to order mice on demand, with as little as 24 hours' notice (see Peres and Roe, Chapter 12). One area that *Researching Animal Research* opens up is the need for further political economic analyses: who and what is being rendered invisible in the changing configurations of research animals, where outsourcing is creating new sites of invisibility through elongated supply chains rooted in animal life? Carbone uses the term 'alert mode' to signal the worries that these political economic readings give rise to. This term nicely articulates the affective response I had in reading these chapters, and the urgency I felt regarding the need for further research of this kind.

Such an approach would extend the theme of *subjugation* that also cuts across many of the chapters of the book, and similarly expresses the power relations that are necessarily at play. Kirk (Chapter 5) shows how 'the laboratory animal' and 'the animal technician' are both mutually constituted subjects of the twentieth century, inventions of a 'modern' science that was rooted in objectivity and the subjugation of feelings like love. This configuration made animal care a career in science, but it also emplaced any conflict that love and use give rise to onto the animal technician as a person who is called upon to subjugate their emotions (see Greenhough and Roe, Chapter 6; Message, Chapter 7; Dennison, Chapter 13). This is a conflict that veterinarians also experience (see Tremoleda and Kerton, Chapter 8; Dennison, Chapter 13), but that takes on a further regulatory dimension, as shown by Anderson and Hobson-West (Chapter 9). While a culture of care is being developed within laboratory animal facilities to address this as a site of workplace stress (Chapter 6; Chapter 8), we as readers can also become concerned about the conditions of not only the horseshoe crabs but also those workers who remove their blood in Gorman's case study. As some forms of subjugations are rendered visible, there is the need to ask what other sites of subjugation are taking place.

The question of how to hold the *fraught conversations* that such issues necessarily give rise to – where people won't agree with one another but can be open to one another's perspectives – is a key question that arises across this book (see Kirk, Chapter 4; Greenhough & Roe Chapter 6; Greenhough, Chapter 8; Davies, Gorman, and King, Chapter 11; Dennison, Chapter 13; Lear, Chapter 17). This question takes on a specific kind of meaning in the context of Myelnikov's opening analysis of The Animals (Scientific Procedures) Act 1986 (ASPA), where compromise and consensus required the systematic exclusion of certain voices. But the public representation of certain groups – veterinarians in Anderson and Hobson-West (Chapter 9), patients in Davies, Gorman, and King (Chapter 11) – can work to silence experiences and ambivalences in less systematic ways. Meanwhile, the notion of a nexus itself can work to exclude those actors who worry about becoming compromised should they become part of the research animal nexus, which can include both veterinarians (see Myelnikov, Chapter 1) and abolitionists (see Giraud, Chapter 8). In reading this book, it became clear to me that rendering the animal research nexus visible, and making heterogeneous voices legible, requires a move away from the consensus approach, rooted in control, that has long been a hallmark of the British approach to research animals. But what might these new forums for discussion look like?

This book usefully ends by answering precisely this question, with three case studies in doing experimental work as part of the social sciences. The Mouse Exchange, labelling medicines project, and *Vector* project are all experiments in making new kinds of socialities. The authors respectively foreground embodiment (Roe, Peres, and Crudgington, Chapter 14), dissensus (McGlacken and Hobson-West, Chapter 15), and deep play (Crudgington, Scott, Thorpe, and Fleming, Chapter 16), which contrast with more established practices in the public understanding of science. The authors thus move away from the logic of control, and risk letting people who are outside of the research animal nexus speak (see Lear, Chapter 17). In the process, the question shifts from ferreting out unheard and invisible but nonetheless present and existing perspectives, to instead create new conditions through which new things might be sayable and said. After reading these chapters, I felt that we could all, with Mackenzie, ask: 'Can I be honest?'

To conclude, *Researching Animal Research* makes it possible to understand the polarised debates regarding animal research as a structure that shapes but does not determine the research animal nexus. This makes it possible to articulate contradictory and heterodox thoughts and experiences. For example, patient participants in Davies, Gorman, and King (Chapter 11) can be more than their embodied diagnosis. Sociologists can be more than an academic researcher (see Hobson-West, Chapter 13). And in the process, the theme of connection in this book becomes a practice not only of the authors but also of its readers.

Kirk closes his commentary to the first section (Chapter 4, pp. 116–117) by stating: 'if I was asked to identify a single theme that characterises this volume, I would choose connections. How different elements relate, become entangled, and reshape each other to drive historical change in what we refer to as the "animal research nexus"'. This book marks out a fundamental shift in the animal research nexus, wherein the polarisation of vivisection versus anti-vivisection was complicated by the enrolment and invention of a greater number of actors. But *Researching Animal Research* is entangled as well, reshaping that which it has studied and inviting us as readers to also ask how things might be otherwise as part of historical trajectories. The book allows readers to ask: what new worldly imaginations become possible by considering our own research animal nexus, through the lens of the case studies and commentaries that connect this book?

Notes

1 I am a sociologist – a discipline that is concerned with inequalities and that conducts empirical research in order to understand how inequalities operate and are reproduced over time. In my research, I have explored how inequalities operate with regards to animals in ways that intersect with humans. This has included zoo animals and laboratory animals. I do not start with a position regarding the use of animals in these

institutional settings, but rather seek to understand how animals are used in bioscience and biomedicine in ways that reproduce inequalities between humans and animals and between differently positioned humans. For example, not all humans benefit from the knowledge or pharmaceuticals produced with laboratory animals.

2 Richard D. French, *Antivivisection and Medical Science in Victorian Society* (Princeton, NJ: Princeton University Press, 2019).

3 William Moy Stratton Russell and Rex Leonard Burch, *The Principles of Humane Experimental Technique* (London: Methuen, 1959).

4 Hilda Kean, *Animal Rights: Political and Social Change in Britain since 1800* (London: Reaktion Books, 1998).

5 Lynda I. A. Birke et al., *The Sacrifice: How Scientific Experiments Transform Animals and People* (Lafayette, IN: Purdue University Press, 2007).

Select bibliography

This bibliography was put together in early 2023. It includes the collaborative work directly preceding and emerging from the Wellcome Trust-funded Animal Research Nexus Programme [grant no 205393], which ran between 2017 and 2023.

Foundational collaborative work

Davies, Gail, Beth J. Greenhough, Pru Hobson-West, Robert G. W. Kirk, Ken Applebee, Laura C. Bellingan et al., 'Developing a Collaborative Agenda for Humanities and Social Scientific Research on Laboratory Animal Science and Welfare', *PLOS ONE*, 11.7 (2016), 1–12, DOI: 10.1371/journal.pone.0158791.

Peer reviewed articles and editorials

Anderson, Alistair, and Pru Hobson-West, 'Animal Research, Ethical Boundary-Work, and the Geographies of Veterinary Expertise', *Transactions of the Institute of British Geographers* 48 (2023), 491–505, DOI: 10.1111/tran.12594.

Anderson, Alistair, and Pru Hobson-West, '"Refugees from Practice"? Exploring Why Some Vets Move from the Clinic to the Laboratory', *Veterinary Record*, 190.1 (2022), e773, DOI: 10.1002/vetr.773.

Davies, Gail. 'Harm–benefit Analysis: Opportunities for Enhancing Ethical Review in Animal Research', *Laboratory Animals* 47.3 (2018), 57–58, DOI: 10.1038/s41684-018-0002-2.

Davies, Gail, 'Locating the "Culture Wars" in Laboratory Animal Research: National Constitutions and Global Competition', *Studies in History and Philosophy of Science Part A*, 89 (2021), 177–187, DOI: 10.1016/j.shpsa.2021.08.010.

Davies, Gail, Richard Gorman, Beth Greenhough, Pru Hobson-West, Robert G. W. Kirk, Reuben Message, Dmitriy Myelnikov et al.,

'Animal Research Nexus: A New Approach to the Connections between Science, Health and Animal Welfare.' *Medical Humanities*, 46.4 (2020), 499–511, DOI: 10.1136/medhum-2019-011778.

Davies, Gail, Richard Gorman, Renelle McGlacken, and Sara Peres, 'The Social Aspects of Genome Editing: Publics as Stakeholders, Populations and Participants in Animal Research', *Laboratory Animals*, 56.1 (2021), 88–96, DOI: 10.1177/0023677221993157.

García-Sancho, Miguel, and Dmitriy Myelnikov, 'Between Mice and Sheep: Biotechnology, Agricultural Science and Animal Models in Late-Twentieth Century Edinburgh', *Studies in History and Philosophy of Science Part C: Studies in History and Philosophy of Biological and Biomedical Sciences*, 75 June (2019), 24–33, DOI: 10.1016/j.shpsc.2019.01.002.

Gorman, Richard. 'Atlantic Horseshoe Crabs and Endotoxin Testing: Perspectives on Alternatives, Sustainable Methods, and the 3Rs (Replacement, Reduction and Refinement)', *Frontiers in Marine Science*, 7 (2020), DOI: 10.3389/fmars.2020.582132.

Gorman, Richard. 'What Might Decapod Sentience Mean for Policy, Practice, and Public?' *Animal Sentience*, 7.32 (2022), DOI: 10.51291/2377-7478.1720.

Gorman, Richard, and Gail Davies. 'When "Cultures of Care" Meet: Entanglements and Accountabilities at the Intersection of Animal Research and Patient Involvement in the UK', *Social & Cultural Geography*, September (2020), 1–19, DOI: 10.1080/14649365.2020.1814850.

Greenhough, Beth, and Emma Roe. 'Exploring the Role of Animal Technologists in Implementing the 3Rs: An Ethnographic Investigation of the UK University Sector', *Science, Technology, & Human Values*, 43.4 (2017), DOI: 10.1177/0162243917718066.

Greenhough, Beth, and Emma Roe. 'Attuning to Laboratory Animals and Telling Stories: Learning Animal Geography Research Skills from Animal Technologists', *Environment and Planning D: Society and Space*, 37.2 (2019), 367–384, DOI: 10.1177/0263775818807720.

Greenhough, Beth, Gail Davies, and Sophie Bowlby. 'Why "Cultures of Care"?' *Social & Cultural Geography*, 24.1 (2023), 1–10, DOI: 10.1080/14649365.2022.2105938.

Hobson-West, Pru, and Annemarie Jutel. 'Animals, Veterinarians and the Sociology of Diagnosis', *Sociology of Health & Illness* 42.2 (2020), 393–406, DOI: 10.1111/1467–9566.13017.

Hobson-West, Pru, and Ashley Davies. 'Societal Sentience: Constructions of the Public in Animal Research Policy and Practice', *Science, Technology, & Human Values* 4.4 (2018), 671–693, DOI: 10.1177/0162243917736138.

Jenkins, Nicholas, Richard Gorman, Cristina Douglas, Vanessa Ashall, Louise Ritchie, and Anna Jack-Waugh, 'Multi-Species Dementia Studies:

Contours, Contributions and Controversies', *Journal of Aging Studies*, 59 (2021), 100975, DOI: 10.1016/j.jaging.2021.100975.

Kirk, Robert G. W. 'Recovering the Principles of Humane Experimental Technique: The 3Rs and the Human Essence of Animal Research', *Science, Technology, & Human Values*, 43.4 (2018), 622–648, DOI: 10.1177/0162243917726579.

Kirk Robert G. W., and Edmund Ramsden. '"Havens of Mercy": Health, Medical Research, and the Governance of the Movement of Dogs in Twentieth-Century America', *History and Philosophy of the Life Sciences*, 43.4 (2021), 126, DOI: 10.1007/s40656-021-00478-4.

Kirk, Robert G. W., and Dmitriy Myelnikov. 'Governance, Expertise, and the "Culture of Care": The Changing Constitutions of Laboratory Animal Research in Britain, 1876–2000', *Studies in History and Philosophy of Science*, 93 (2022), 107–122, DOI: 10.1016/j.shpsa.2022.03.004.

Lowe, James W. E., Sabina Leonelli, and Gail Davies. 'Training to Translate: Understanding and Informing Translational Animal Research in Pre-Clinical Pharmacology', *TECNOSCIENZA: Italian Journal of Science & Technology Studies*, 10.2 (2020), 5–30.

McGlacken, Renelle. '(Not) Knowing and (Not) Caring About Animal Research: An Analysis of Writing From the Mass Observation Project', *Science & Technology Studies*, 35.3 (2021), 2–20, DOI: 10.23987/sts.102496.

McGlacken, Renelle. 'Negotiating the Necessity of Biomedical Animal Use Through Relations With Vulnerability', BioSocieties (2023), DOI: doi.org/10.1057/s41292-022-00295-3.

McGlacken, Renelle, and Pru Hobson-West. 'Critiquing Imaginaries of "the Public" in UK Dialogue around Animal Research: Insights from the Mass Observation Project', *Studies in History and Philosophy of Science*, 91 (2022), 280–287, DOI: 10.1016/j.shpsa.2021.12.009.

Message, Reuben. '"The Disadvantages of a Defective Education": Identity, Experiment and Persuasion in the Natural History of the Salmon and Parr Controversy, c. 1825–1850', *Science in Context*, 32.3 (2019), 261–284, DOI: 10.1017/S0269889719000255.

Message, Reuben. 'Animal Welfare Chauvinism in Brexit Britain: A Genealogy Care and Control', *BioSocieties* (2022), DOI: 10.1057/s41292-022-00282-8.

Message, Reuben, and Beth Greenhough. '"But It's Just a Fish": Understanding the Challenges of Applying the 3Rs in Laboratory Aquariums in the UK', *Animals*, 9.12 (2019), DOI: 10.3390/ani9121075.

Myelnikov, Dmitriy. 'Cuts and the Cutting Edge: British Science Funding and the Making of Animal Biotechnology in 1980s Edinburgh', *The British Journal for the History of Science*, 50.4 (2017), 701–728, DOI: 10.1017/S0007087417000826.

Myelnikov, Dmitriy. 'Tinkering with Genes and Embryos: The Multiple Invention of Transgenic Mice c. 1980', *History and Technology*, 35.4 (2019), 425–452, DOI: 10.1080/07341512.2019.1694126.

Myelnikov, Dmitriy, and Peres, Sara. 'The Cold Futures of Mouse Genetics: Modes of Strain Cryopreservation Since the 1970s', *Science, Technology, & Human Values* (2022), DOI: 10.1177/01622439221138341.

Ormandy, Elisabeth H., Daniel M. Weary, Katarina Cvek, Mark Fisher, Kathrin Herrmann, Pru Hobson-West, Michael McDonald et al. 'Animal Research, Accountability, Openness and Public Engagement: Report from an International Expert Forum', *Animals*, 9.9 (2019), 622, DOI: 10.3390/ani9090622.

Palmer, Alexandra, and Beth Greenhough. 'Out of the Laboratory, into the Field: Perspectives on Social, Ethical and Regulatory Challenges in UK Wildlife Research', *Philosophical Transactions of the Royal Society B: Biological Sciences*, 376.1831 (2021), DOI: 10.1098/rstb.2020.0226.

Palmer, Alexandra, Beth Greenhough, Pru Hobson-West, Gail Davies, and Reuben Message. 'Can Research Animals Volunteer?' *Society & Animals*. In press.

Palmer, Alexandra, Beth Greenhough, Pru Hobson-West, Reuben Message, James N. Aegerter, Zoe Belshaw, Ngaire Dennison et al. 'Animal Research beyond the Laboratory: Report from a Workshop on Places Other than Licensed Establishments (POLEs) in the UK', *Animals*, 10.10 (2020), 1868, DOI: 10.3390/ani10101868.

Palmer, Alexandra, Reuben Message, and Beth Greenhough. 'Edge Cases in Animal Research Law: Constituting the Regulatory Borderlands of the UK's Animals (Scientific Procedures) Act', *Studies in History and Philosophy of Science Part A*, 90 (2021), 122–130, DOI: 10.1016/j.shpsa.2021.09.012.

Palmer, Alexandra, S. James Reynolds, Julie Lane, Roger Dickey, and Beth Greenhough. 'Getting to Grips with Wildlife Research by Citizen Scientists: What Role for Regulation? *People and Nature*, 3.1 (2021), 4–16, DOI: 10.1002/pan3.10151.

Palmer, Alexandra, Tess Skidmore, and Alistair Anderson. 'When Research Animals Become Pets and Pets Become Research Animals: Care, Death, and Animal Classification', *Social & Cultural Geography* (2022), 1–19, DOI: 10.1080/14649365.2022.2073465.

Peres, Sara, and Emma Roe. 'Laboratory Animal Strain Mobilities: Handling with Care for Animal Sentience and Biosecurity', *History and Philosophy of the Life Sciences* 44.3 (2022), 1–22, DOI: 10.1007/s40656-022-00510-1.

Roe, Emma, and Beth Greenhough. 'A Good Life? A Good Death? Reconciling Care and Harm in Animal Research', *Social & Cultural Geography*, 24.1 (2021), 49–66, DOI: 10.1080/14649365.2021.1901977.

Skidmore, Tess, and Emma Roe. 'A Semi-Structured Questionnaire Survey of Laboratory Animal Rehoming Practice across 41 UK Animal Research Facilities', ed. by I. Anna S. Olsson. *PLOS ONE*, 15.6 (2020), e0234922, DOI: 10.1371/journal.pone.0234922.

Vanderslott, Samantha, Alexandra Palmer, Tonia Thomas, Beth Greenhough, Arabella Stuart, John A. Henry, Marcus English et al. 'Co-Producing Human and Animal Experimental Subjects: Exploring the Views of UK COVID-19 Vaccine Trial Participants on Animal Testing', *Science, Technology, & Human Values* 48.4 (2023), 909–937, DOI: 10.1177/016224392110 57084.

Edited journal special issues

Ankeny, Rachel, and Gail Davies (eds), 'Constituting Animal Research: International Perspectives on the Governance of Laboratory Animal Use and Care', *Studies in History and Philosophy of Science* (2022), www.sciencedirect.com/journal/studies-in-history-and-philosophy-of-science/special-issue/10DN9KM2237.

Davies, Gail, Beth Greenhough, Pru Hobson-West, and Robert G. W. Kirk (eds), 'Science, Culture, and Care in Laboratory Animal Research: Interdisciplinary Perspectives on the History and Future of the 3Rs', *Science, Technology, & Human Values*, 43.4 (2018), 603–621, DOI: 10.1177/016224391875703.

Friese, Carrie, Tarquin Holmes, and Reuben Message (eds), 'National Cultures of Animals, Care and Science', *BioSocieties* (in press).

Greenhough, Beth, Gail Davies, and Sophie Bowlby (eds), 'Special Issue on Cultures of Care', *Social and Cultural Geography*, 24.1, www.tandfonline.com/toc/rscg20/24/1.

Kirk, Robert G. W., Sabina Leonelli, and Dmitriy Myelnikov (eds), 'Circulating Bodies: Human–Animal Movements in Science and Medicine', *History and Philosophy of the Life Sciences* (2022) https://link.springer.com/journal/40656/topicalCollection/AC_6688b1fccc519ccb765ff950e541bde8.

Book chapters and conference proceedings

Ashall, Vanessa, and Pru Hobson-West. 'The Vet in the Lab: Exploring the Position of Animal Professionals in Non-Therapeutic Roles', in *Professionals in Food Chains*, ed. by Svenja Springer and Herwig Grimm (Wageningen: Wageningen Academic Publishers, 2018), pp. 291–295, DOI: 10.3920/978-90-8686-869-8_45.

Davies Gail, and Scalway Helen. 'Diagramming', in *Routledge Handbook of Interdisciplinary Research Methods*, ed. by C. Lury,

R. Fensham, A. Heller-Nicholas, S. Lammes, A. Last, M. Michael, and E. Uprichard (London: Routledge, 2018). www.taylorfrancis.com/books/9781315714523 [accessed 1 February 2023].

Davies, Gail, Richard Gorman, and Bentley Crudgington. 'Which Patient Takes Centre Stage? Placing Patient Voices in Animal Research', in *GeoHumanities and Health*, ed. by Sarah Atkinson and R. Hunt (Cham: Springer, 2020), pp. 141–155, www.ncbi.nlm.nih.gov/books/NBK550914/ [accessed 1 February 2023].

Gorman, Richard. 'Animal Geographies in a Pandemic', in *COVID-19 and Similar Futures*, ed. by Gavin J. Andrews, Valorie A. Crooks, Jamie R. Pearce, and Jane P. Messina, Global Perspectives on Health Geography (Cham: Springer International Publishing, 2021), pp. 207–212, DOI: 10.1007/978-3-030-70179-6_27.

Kirk, Robert G. W. 'The Experimental Animal: In Search of a Moral Ecology of Science?' in *The Routledge Companion to Animal–Human History*, ed. by Hilda Kean and Philip Howell (London: Routledge, 2022), DOI: 10.4324/9780429468933-6.

PhD theses

McGlacken, Renelle. 2021. 'Exploring Everyday Relations with Animal Research – A Sociological Analysis of Writing from the Mass Observation Project', University of Nottingham. https://eprints.nottingham.ac.uk/66576/ [accessed 1 February 2023].

Skidmore, Tess, Alexandra, 'A Life after the Laboratory: Exploring the Policy and Practice of Laboratory Animal Rehoming', University of Southampton, unpublished doctoral thesis (2020), 274 pp. https://eprints.soton.ac.uk/450190/ [accessed 1 February 2023].

Stakeholder and policy reports

Crudgington, Bentley, Beth Greenhough, Alexandra Palmer, and Reuben Message. 'Public Engagement with Fish: Workshop Report', *An Animal Research Nexus Report*, https://animalresearchnexus.org/publications/public-engagement-fish-workshop-report [accessed 1 February 2023].

Crudgington, Bentley, Sara Peres, Paul Hurley, and Emma Roe. 'The Mouse Exchange Toolkit', *An Animal Research Nexus Report* (2021), https://animalresearchnexus.org/publications/mouse-exchange-toolkit [accessed 1 February 2023].

Davies, Gail, Richard Gorman, and Gabrielle King. 'Informing Involvement around Animal Research: Report and Resources from the Animal Research Nexus Project', *An Animal Research Nexus Report* (2021), https://animalresearchnexus.org/publications/informing-involvement-around-animal-research [accessed 1 February 2023].

Gorman, Richard, 'Horseshoe Crabs and the Pharmaceutical Industry: Challenges and Alternatives – Project Report', *An Animal Research Nexus Report* (2020). https://animalresearchnexus.org/publications/horseshoe-crabs-and-pharmaceutical-industry-challenges-and-alternatives-project-report [accessed 1 February 2023].

Gorman, Richard, and Gail Davies. 'Patient and Public Involvement and Engagement (PPIE) with Animal Research', *An Animal Research Nexus Report* (2019), https://animalresearchnexus.org/index.php/publications/patient-and-public-involvement-and-engagement-ppie-animal-research [accessed 1 February 2023].

Greenhough, Beth, and Hibba Mazhary. 'Care-Full Stories: Innovating a New Resource for Teaching a Culture of Care in Animal Research Facilities', *An Animal Research Nexus Report* (2021), https://animalresearchnexus.org/publications/care-full-stories-innovating-new-resource-teaching-culture-care-animal-research [accessed 1 February 2023].

Greenhough, Beth, Mazhary, Hibba and Berdoy, Manuel. 'Care-Full Stories Phase II: Developing a New Resource for Teaching a Culture of Care in Animal Research Facilities', *An Animal Research Nexus Report*, www.geog.ox.ac.uk/research/tl/projects/care-full-stories/ [accessed 1 February 2023].

Hobson-West, Pru, Renelle McGlacken, Julie Brownlie, Nickie Charles, Rebekah Fox, Anne-Marie Kramer, and Kirsty Pattrick. 'Mass Observation: Emotions, Relations and Temporality', *An Animal Research Nexus Report* (2019), https://animalresearchnexus.org/index.php/publications/mass-observation-emotions-relations-and-temporality [accessed 1 February 2023].

Palmer, Alexandra, Beth Greenhough, Reuben Message, and Pru Hobson-West. 'Summary Notes from POLEs Workshop, 30th Sept–1st Oct 2019', *An Animal Research Nexus Report* (2019), https://animalresearchnexus.org/index.php/publications/summary-notes-poles-workshop-30th-sept-1st-oct-2019.

Skidmore, Tess. 'A Life after the Laboratory: Exploring the Policy and Practice of Laboratory Animal Rehoming', *An Animal Research Nexus Report* (2021), https://animalresearchnexus.org/publications/life-after-laboratory-exploring-policy-and-practice-laboratory-animal-rehoming [accessed 1 February 2023].

Welsman, Jo, Richard Gorman, and Gail Davies. 'Lay Members in Biomedical Research Report | Animal Research Nexus', *An Animal Research Nexus Report* (2018), https://animalresearchnexus.org/index.php/publications/lay-members-biomedical-research-report [accessed 1 February 2023].

Public engagement websites

Davies, Gail, and Helen Scalway. 'Micespace' www.micespace.org/ [accessed December 2022].
Hobson-West, Pru, and Renelle McGlacken. 'Labelling Animal Research?' www.labanimallabels.co.uk/ [accessed December 2022].
Message, Reuben, Beth Greenhough, and Bentley Crudgington. 'Psychic Fish' https://psychicfish.co.uk/ [accessed December 2022].
Roe, Emma, Sara Peres, and Bentley Crudgington. 'The Mouse Exchange' https://themouseexchange.org/ [accessed December 2022].

Blog posts

The Animal Research Nexus Programme published regular blogs on their website (https://animalresearchnexus.org/blogs) from 2017 to 2023, allowing us to reflect on our evolving research practices, bring debates about public representation of research animals to a wider readership, and respond to events like the impact of Covid-19 on our and others' work. The following selection features some of these posts (excluding workshop reports, event announcements, publication notices, and material written up elsewhere), alongside other invited blog posts authored by members of the Animal Research Nexus Programme team over this period.

Crudgington, Bentley, and Gail Davies. 'Publics, Pandemics, and the Performance of Ethical Review', *The Polyphony Blog*, 15 May 2020, https://thepolyphony.org/2020/05/15/publics-pandemics-and-the-performance-of-ethical-review/ [accessed 1 February 2023].
Davies, Gail, 'Figuring it Out: Questions of Comparison, Culture, and Care in Animal Use Statistics', *The Animal Research Nexus Blog* (2019), https://animalresearchnexus.org/blogs/figuring-it-out-questions-comparison-culture-and-care-animal-use-statistics [accessed 1 February 2023].
Davies, Gail, 'Architectures of Contagion and Collaboration: How do you Stop an Infection Circulating, Whilst Still Allowing Ideas to Flow?' *The Animal Research Nexus Blog* (2020), https://animalresearchnexus.org/blogs/architectures-contagion-and-collaboration [accessed 1 February 2023].
McGlacken, Renelle. 'The Medical and the Cosmetic: Reflections on "Corrective" Surgery and the Scientific Use of Animals', *The Animal Research Nexus Blog* (2012), https://animalresearchnexus.org/blogs/medical-and-cosmetic-reflections-corrective-surgery-and-scientific-use-animals [accessed 1 February 2023].
McGlacken, Renelle. 'Having Conversations with the Past and the Present: Visiting an Archive', *The Enquire Blog* (22 July 2019), https://blogs.

nottingham.ac.uk/enquire/2019/07/22/winner-of-the-2nd-prize-of-the-enquire-blog-post-competition-2019-renelle-mcglacken-having-conver sations-with-the-past-and-the-present-visiting-an-archive/ [accessed 1 February 2023].

Message, Reuben. 'Becoming Legal Animals: Larval Forms under ASPA', *The Animal Research Nexus Blog* (2019), https://animalresearchnexus.org/blogs/becoming-legal-animals [accessed 1 February 2023].

Message, Reuben. 'Policy-Based Evidence Making', *The Index of Evidence Blog* (2019), www.indexofevidence.org/policybased-evidence [accessed 1 February 2023].

Message, Reuben. 'Poster Critters of Animal Research? Representations of Fish in Public Engagement with Animal Research', *The Animal Research Nexus Blog* (2021), https://animalresearchnexus.org/blogs/poster-crit ters-animal-research [accessed 1 February 2023].

Myelnikov Dmitriy. 'Mouse Jokes', *The Animal Research Nexus Blog* (2019), https://animalresearchnexus.org/blogs/mouse-jokes [accessed 1 February 2023].

Myelnikov Dmitriy. 'The Smoking Beagles', *The Animal Research Nexus Blog* (2021), https://animalresearchnexus.org/blogs/smoking-beagles [accessed 1 February 2023].

Palmer, Alexandra. 'COVID-19 and Wildlife Research Ethics' *The Animal Research Nexus Blog* (2020), https://animalresearchnexus.org/blogs/covid-19-and-wildlife-research-ethics [accessed 1 February 2023].

Palmer, Alexandra. 'The Afterlives of Wildlife Tracking Devices', *Digital Ecologies Blog* (6 July 2021), www.digicologies.com/2021/07/06/alex andra-palmer/ [accessed 1 February 2023].

Palmer, Alexandra, and Sara Peres. 'Present, Not Used (Part 1): A Spectrum of Visibility', *The Animal Research Nexus Blog* (2020), https://ani malresearchnexus.org/blogs/present-not-used-part-1-spectrum-visibility [accessed 1 February 2023].

Palmer, Alexandra, and Sara Peres. 'Present, Not Used (Part 2): Caring for Less Visible Animals', *The Animal Research Nexus Blog* (2021), https://animalresearchnexus.org/blogs/present-not-used-part-2-caring-less-visi ble-animals [accessed 1 February 2023].

Peres, Sara. 'Research in Practice: Developing Ways to Share Data within the Animal Research Nexus Team', *The Animal Research Nexus Blog* (2018), https://animalresearchnexus.org/blogs/research-practice [accessed 1 February 2023].

Peres, Sara. 'Representing Technicians' Work', *The Animal Research Nexus Blog* (2020), https://animalresearchnexus.org/blogs/representing-technicians-work [accessed 1 February 2023].

Book reviews

Reuben Message, 'Model Behavior: Animal Experiments, Complexity and the Genetics of Psychiatric Disorders [Book Review]', *New Genetics and Society*, 38.2 (2019), 243–246, DOI: 10.1080/14636778.2018.1515622.

Miranda Joseph, Abigail H. Neely, Gail Davies, Matthew Sparke, and Susan Craddock, 'Compound Solutions: Pharmaceutical Alternatives for Global Health', The AAG Review of Books, 7.1 (2019), 47–58, DOI: 10.1080/2325548X.2019.1546034.

Data sets

The interview dataset from the Animal Research Nexus Programme has been anonymised and deposited in the UK Data Archive. The archived dataset includes 163 interviews, some of which were with more than one individual. Some individuals preferred not to allow their interviews to be archived. Interviews completed as part of PhD projects were not archived. Access is subject to a ten-year embargo, as agreed with research participants, given the ongoing sensitivities associated with this topic.

Davies, G., B. Greenhough, P. Hobson-West, R. Kirk, and E. Roe, Animal Research in the UK, 2017–2023. UK Data Service. SN: 8942 (2033), DOI: 10.5255/UKDA-SN-8942-1.

Index

Figures and tables shown with numbers in *italics*

Milton Keynes UK
Ingram Content Group UK Ltd.
UKHW021328010324
438760UK00007B/91